Evolutionary Genetics: Concepts and Applications

Evolutionary Genetics: Concepts and Applications

Edited by Lauren Acosta

SYRAWOOD
PUBLISHING HOUSE

New York

Published by Syrawood Publishing House,
750 Third Avenue, 9th Floor,
New York, NY 10017, USA
www.syrawoodpublishinghouse.com

Evolutionary Genetics: Concepts and Applications
Edited by Lauren Acosta

International Standard Book Number: 978-1-68286-795-2 (Hardback)

Cataloging-in-Publication Data

Evolutionary genetics : concepts and applications / edited by Lauren Acosta.
    p. cm.
Includes bibliographical references and index.
ISBN 978-1-68286-795-2
1. Evolutionary genetics. 2. Genetics. 3. Evolution (Biology). I. Acosta, Lauren.
QH390 .E86 2019
572.838--dc23

TABLE OF CONTENTS

PREFACE

This book aims to highlight the current researches and provides a platform to further the scope of innovations in this area. This book is a product of the combined efforts of many researchers and scientists, after going through thorough studies and analysis from different parts of the world. The objective of this book is to provide the readers with the latest information of the field.

The study of the changes in an organism's genome expressed with time and the influence the organism's evolutionary past has on it, is studied under evolutionary genetics. Such changes occur within and between populations. This area of genetic study is under the domain of population genetics. It is vital to the development of modern evolutionary synthesis. Adaptation, population structure, speciation, dominance, epistasis, etc. are fundamental areas in the understanding of evolutionary genetics. Studies in these fields allow an understanding of the levels of genetic variation, demographic inference, evolution of genetic systems and detecting the genes undergoing selection. This book discusses the fundamentals as well as modern approaches of evolutionary genetics. Also included herein is a detailed explanation of the various concepts and applications of evolutionary genetics. It aims to serve as a resource guide to population geneticists, evolutionary geneticists, biologists, researchers and students involved in this area of study.

I would like to express my sincere thanks to the authors for their dedicated efforts in the completion of this book. I acknowledge the efforts of the publisher for providing constant support. Lastly, I would like to thank my family for their support in all academic endeavors.

Editor

The two-fold cost of sex: Experimental evidence from a natural system

Amanda K. Gibson,[1,2,3] Lynda F. Delph,[1] and Curtis M. Lively[1]

[1] *Department of Biology, Indiana University, Bloomington, Indiana 47405*

[2] *Department of Biology, Emory University, Atlanta, Georgia 30322*

[3] *E-mail: amanda.gibson@emory.edu*

Over four decades ago, John Maynard Smith showed that a mutation causing asexual reproduction should rapidly spread in a dioecious sexual population. His reasoning was that the per-capita birth rate of an asexual population would exceed that of a sexual population, because asexual females do not invest in sons. Hence, there is a cost of sexual reproduction that Maynard Smith called the "cost of males." Assuming all else is otherwise equal among sexual and asexual females, the cost is expected to be two-fold in outcrossing populations with separate sexes and equal sex ratios. Maynard Smith's model led to one of the most interesting questions in evolutionary biology: why is there sex? There are, however, no direct estimates of the proposed cost of sex. Here, we measured the increase in frequency of asexual snails in natural, mixed population of sexual and asexual snails in large outdoor mesocosms. We then extended Maynard Smith's model to predict the change in frequency of asexuals for any cost of sex and for any initial frequency of asexuals. Consistent with the "all-else equal" assumption, we found that the increase in frequency of asexual snails closely matched that predicted under a two-fold cost.

KEY WORDS: All-else-equal assumption, asexual reproduction, evolution of sex, experimental evolution, paradox of sex, *Potamopyrgus antipodarum*, sexual reproduction, two-fold cost of males.

Impact Summary

A rare asexual mutant arises in an otherwise sexual population. This asexual female need not mate with a male to produce sons and daughters. Instead, she simply clones herself, producing asexual daughters. Evolutionary theory predicts that this asexual lineage will spread rapidly through the population, driving the sexual lineages rapidly extinct. The reason is that sexual females must spend ~50% of their resources making sons, which cannot themselves make offspring. The growth rate of the sexual population is thus predicted to be half that of the asexual population. This cost is called the "two-fold cost of males." Yet sex abounds in nature. Since the development of this theory, evolutionary biologists have sought advantages for sex that could explain its paradoxical persistence. In this study, we take a step back and ask: do sexuals actually pay a two-fold cost? Though the cost of sex is a critical assumption of the paradox of sex, there are no direct estimates of the cost. To estimate the cost of sex, we conducted an experiment using snails collected from a natural population where sexual and asexual individuals coexist. The snails were reared in large, outdoor mesocosms, and the experiment was replicated in four separate years. We found that the asexual snails increased in frequency in all four years. We then extended a previous model on the two-fold cost so that we could estimate the cost of sex based upon our experimental data. We found that the observed increase in asexual frequency matched that predicted for a two-fold cost of sex. Our results are thus consistent with theoretical predictions. Hence, for sex to be maintained in natural populations, there must be strong selection favoring sexual over asexual reproduction.

Introduction

The cost of males (Maynard Smith 1971, 1978), along with Williams' "cost of meiosis" (1971, 1975), sparked an enduring paradox in evolutionary biology: sexual reproduction is costly but is maintained in most eukaryotic species. In his original model, Maynard Smith (1971, 1978) assumed that sexual females invest

50% of their resources into sons, while asexual females invest 100% of their resources into clonal daughters. He also assumed that sexual and asexual females are equally fecund and that the survivorship of their offspring is equal ("all-else-equal" assumption). Under these conditions, the model predicts that the per-capita birth rate of an asexual population would be twice that of a sexual population (two-fold cost). An asexual mutant should therefore double in frequency when rare and rapidly replace the sexual population. Sex, however, abounds. This inconsistency between theoretical expectation and nature instigated the ongoing hunt for forces that counterbalance the short-term costs of sex.

Though the cost of sex is the foundation of the paradox of sex, the costs are vastly understudied relative to its benefits. Without a cost of sex, there is no need to test hypotheses for the maintenance of sex. For example, intrinsic differences between sexual and asexual females can violate the all-else-equal assumption (e.g., costs of elevated ploidy in asexuals), reducing the cost of sex predicted by theory (Lehtonen et al. 2012; Meirmans et al. 2012). Alternately, intrinsic differences may augment the cost of sex (e.g., energetic costs associated with mating), in which case a stronger selective advantage for sex would be required to explain coexistence.

There are no direct experimental measures of the cost of sex. Several empirical studies have addressed the critical all-else-equal assumption of Maynard Smith's model, with mixed support. Asexual and sexual females have similar fecundity and/or offspring survival in only five of the ten cases reviewed in Meirmans et al. (2012) (e.g., snails—Jokela et al. (1997b); Crummett and Wayne (2009); rotifers—Stelzer (2011)). In the other cases, the transition to asexuality is accompanied by reduced fecundity or survival, violating the all-else-equal assumption. A few additional studies have shown that the frequency of asexual individuals increases over time in mixed populations, suggesting that sex is indeed costly to some extent (Browne and Halanych 1989; Jokela et al. 1997b; Stelzer 2011). But a crucial question remains: exactly how costly is sex?

Here, we directly estimated the cost of sex in a natural system. To do so, we established seminatural mesocosms using *Potamopyrgus antipodarum* snails collected directly from a natural population in which asexual and sexual females coexist. In the experimental populations, the frequency of asexual snails increased significantly from parent to offspring generation. We then expanded upon Maynard Smith's original model by relaxing his assumption that asexuals are rare. To estimate the net cost of sex in *P. antipodarum*, we fit this simple model to our experimental data. We found that the change in the frequency of asexuals matched that predicted under a two-fold cost of sex. As such, the net cost of sex in this system is consistent with Maynard Smith's critical "all-else-equal" assumption.

Methods
NATURAL HISTORY

Obligately sexual lineages of *P. antipodarum* coexist with obligately asexual lineages. Sexual males and females are diploid. Asexual lineages arise by mutation from local sexual genotypes (Neiman et al. 2005) and are primarily triploid females (higher ploidies have been found) (Neiman et al. 2011). Prior studies supported the all-else-equal assumption for fecundity in *P. antipodarum*: sexual and asexual females are similar in size at reproductive maturity, brood at similar rates, and have an equal number of eggs per brood (Jokela et al. 1997a; Jokela et al. 1997b; Paczesniak 2012). We conducted the same comparisons for snails in our experimental populations and in the field population from which they were derived (S.I. I). Additional intrinsic fitness differences may exist. Specifically, we did not know whether sexual and asexual females are equally likely to survive to reproduction, or if they produce an equivalent number of viable offspring. We tested this assumption in the present study.

EXPERIMENTAL TEST
Establishing seminatural mesocosms

We established outdoor seminatural mesocosms to experimentally measure the change in frequency of asexual snails. In each of four years, we obtained data from six mesocosms initiated with 800 juvenile snails. Experimental mesocosms were populated each year with field-collected snails, giving 24 total replicates. By using field collections, we maintained the relative frequencies and genetic diversity of clonal and sexual lineages of the natural population. This is important, as the asexual population of snails consists of many genetically distinct clones (Dybdahl and Lively 1995) and their frequencies change rapidly (Jokela et al. 2009; Paczesniak et al. 2014). Our experimental results therefore reflect the natural variation present in the field.

In January 2012–2015, juvenile *P. antipodarum* were collected by passing a net through *Isoetes kirkii* vegetation (~1 meter depth) at sites along the southwestern coast of Lake Alexandrina (Fig. S1A). In 2012, the sampled sites were Swamp and Camp. In 2013 and 2014, the sampled sites were 1st Fence, Swamp, 2nd Fence and West Point. In 2015, we substituted Halfway for West Point. These sites have been well-studied since 1994 (Jokela et al. 1997b), so we knew that large numbers of snails could be found there and that both reproductive modes would be represented. Ecological conditions are similar at these nearby sites. Hosts from these sites are also undifferentiated at neutral loci, consistent with substantial gene flow (Fox et al. 1996; Paczesniak et al. 2014).

We transferred all field samples to the University of Canterbury's Edward Percival Field Station in Kaikoura, NZ and sieved them with a 1.7 mm sieve to obtain juveniles. Body size reflects age, and the 1.7 mm sieve effectively differentiates reproductively mature males and females (>2.5 mm in length) from juveniles. We used juveniles to establish mesocosms, because we aimed to minimize selection by parasites. Juveniles have low rates of infection with sterilizing trematodes (Levri and Lively 1996), the dominant parasites of *P. antipodarum* (Hechinger 2012). Therefore, using juveniles minimized the proportion of our starting populations that was castrated. Importantly, the few infected snails in the mesocosms were unable to transmit the infection. Trematodes require additional host species (e.g., waterfowl) to complete their life cycle (Hechinger 2012).

In 2013–2015, we counted out and combined 200 snails from each of the four sites to give 800 snails per mesocosm. In 2012, we similarly combined snails from the two sampled sites. Each mesocosm was thus representative of the whole sampled region of the lake. We transferred experimental replicates to 1000 L Dolav box pallets, filled with ~800 L of water (Fig. S1B). These were located outside the Edward Percival field station in NZ, so experimental populations experienced natural seasonal variation in temperature, weather, and photoperiod. We covered the mesocosms with shade cloth and fed the snails with spirulina for ~2 weeks. The mesocosms were then left unattended from mid-February until early January of the next year. We added no additional food during this time. Snails obtained food from algae growing in the tanks and from other environmental inputs. Under natural temperature conditions, a year is sufficient time for juveniles to mature and reproduce, but insufficient time for their offspring to reproduce. Therefore, only two generations were present in the mesocosms at the end of the experimental year. At this point, we emptied the mesocosms and sieved the experimental populations at 1.4 mm to separate parent and offspring snails into discrete generations. This small size cut-off enabled us to effectively separate parents from offspring, which were very young and thus very small. Occasionally, offspring failed to pass through the sieve and remained in the parental collection. These offspring were easily identified by size and shell morphology.

Data collection

After separating parent and offspring snails at the end of an experimental year, we collected a random sample of 150 parents from each mesocosm. These parents were immediately dissected under a microscope to determine shell length, sex, brooding status, brood size, and infection status. Mean infection frequency with sterilizing trematodes was $9.82 \pm 0.79\%$ SEM. These individuals were infected as juveniles, prior to field collection. No infections were acquired in the mesocosms because the trematodes' definitive hosts were absent. Because we aimed to measure intrinsic differences in birth rates between asexual and sexual females, we excluded infected (i.e., castrated) individuals from our anal-

yses. This removed an obvious force that can alter the relative fitness of sexual and asexual females. The heads of all dissected females were individually frozen, shipped to Indiana University (IN, USA), and stored at –80°C. Males were assumed to be sexual diploids (Neiman et al. 2011).

We retained a random sample of the offspring (>200) from each mesocosm. This sample was maintained at the Edward Percival Field Station for ~5 weeks, with regular water changes and *ad libitum* spirulina feedings. Offspring were then transported alive to Indiana University and promptly frozen at –80°C for storage.

Flow cytometry was conducted as in Gibson et al. (2016). Triploid asexual females can be differentiated from diploid sexual females and males because their ~50% larger genome size is detected as elevated fluorescence of nuclei. We analyzed 3000 nuclei per sample for parents and 2000 for offspring. For the parental generation, we ran flow cytometry on females only. Parents were sufficiently developed to differentiate males from females, and male snails are exclusively diploid and sexual at Lake Alexandrina (Neiman et al. 2011). For each mesocosm, we analyzed 62.13 ± 4.16 SEM females randomly subsampled from those dissected and frozen. For the offspring generation, we ran flow cytometry on both males and females, because the offspring were too young to sex. For each mesocosm, we analyzed 70.38 ± 1.62 snails randomly subsampled from the frozen offspring. Samples were excluded if there were fewer than 1000 nuclei obtained for a parental snail or fewer than 400 for an offspring snail, if there was no clear peak in fluorescence, or if the peak fell between the gates that designate regions consistent with diploid versus triploid nuclei. We excluded $6.01 \pm 1.49\%$ of parents and $5.38 \pm 0.913\%$ of offspring.

Statistical analysis

We first determined the number of triploid females versus diploid males and females in our subsamples of the parental and offspring generations. For the offspring generation, we obtained these numbers directly from the flow cytometry results because we ran flow cytometry on male and female offspring. For the parental generation, we ran flow cytometry on a subset of females only. We obtained the number of triploid and diploid female parents directly from the flow cytometry results. We used the ratio of male to female snails in the entire mesocosm sample to calculate the number of males consistent with a subsample of females of this size. We then calculated the number of diploid individuals (females + males) (Table S1). Infected individuals were excluded from these calculations to remove differential selection due to parasites.

To determine if the frequency of asexual individuals increased from the parent to offspring generation, we fit a logistic model with the number of triploid (female) and diploid (male and female) individuals in a replicate generation as the binomial response variable (logit link function). Generation (parent, offspring), year (2012–2015), and their interaction were categorical predictor variables. We initially fit this model as a generalized estimating equation (function geeglm in geepack) (R Core Team 2013) to account for autocorrelation of parent and offspring snails derived from the same experimental population (Liang and Zeger 1986; Zeger and Liang 1986; Ziegler and Vens 2010). There was no autocorrelation of generations from the same mesocosm. Therefore, we fit a simpler generalized linear model (function glm in R). We tested the significance of each effect using a likelihood ratio test of models with and without the effect of interest. We calculated the fold-increase and 95% confidence intervals using the odds ratios for generation from a logistic model that excluded the interaction term, the profile likelihood confidence intervals for the odds ratios, and the mean proportion of asexual individuals in the parental generation. In the Supporting Information (II), we report the results of the logistic model fitted with the quasi-binomial distribution to account for overdispersion. The results are qualitatively identical to those with the binomial distribution. We therefore report the binomial results in the main text for ease of interpretation.

MODEL FIT
Candidate models
In the Results ("Extended Model"), we used basic population genetic theory to predict the increase in the frequency of asexuals as a function of their initial frequency and the cost of sex. We applied equation (2) of this model (see Results) to our mesocosm data to ask if the observed proportion of asexual offspring (q_{t+1}) was consistent with a two-fold cost of sex ($c = 2$) given the initial (parental) proportion of snails that were asexual in experimental mesocosms (q_t). We formulated four candidate models: (1) no cost of sex ($c = 1$), (2) a twofold cost ($c = 2$), (3) the maximum likelihood estimate (MLE) of the cost, and (4) the MLE of costs that vary with year (Table 1). We proposed model 4 because the composition of clones and/or the accuracy of the experiment may have differed among years. To model yearly variation, we specified the cost of sex (c) as a function of the baseline cost of sex (c_0) in 2012 and a deviation term (d) indexed by year.

Likelihood function
Our four candidate models specified different probabilities of observing the number of triploid offspring in the total number of offspring analyzed, given the proportion of individuals that were triploid in the parental generation. Offspring numbers were obtained directly from the flow cytometry subsample. The proportions of parents that were triploid were obtained from the flow cytometry subsample and the estimated number of males, as described above (Table S1). We assumed a beta-binomial distribution (R package emdbook, Bolker 2008) for our likelihood function (S.I. III). We used the mle2 function (package bbmle, R) to find maximum likelihood estimates of the parameters. We then obtained the likelihood of each model given our experimental data (24 paired estimates of proportion asexual in parental and offspring generations, q_t and q_{t+1} respectively).

Model comparison
We compared models using Akaike's information criterion (Akaike 1973), corrected for small sample sizes (Sugiura 1978; Hurvich and Tsai 1991) (AIC$_c$) and ΔAIC, the difference in AIC$_c$ of the focal model and the best model (lowest AIC$_c$). (Burnham and Anderson 1998). Roughly, models with ΔAIC values below two have substantial support, models with ΔAIC from 4 to 7 have considerably less support, and models with ΔAIC above 10 have no support (Burnham and Anderson 1998). Because these cut-offs are only rules of thumb, Burnham and Anderson (1998) (pp. 128–129) advise calculating the ΔAIC value that delineates a 95% confidence set of models. We followed their recommended procedure by bootstrapping our data set 10,000 times with replacement. For each bootstrapped dataset, we fit our four candidate models and calculated ΔAIC for model 2 (best model) by subtracting the minimum AIC value from the AIC value for model 2 in each bootstrap replicate. We identified the value of ΔAIC$_2$ that was greater than or equal to 95% of the ΔAIC$_2$ values obtained in the bootstrapping analysis. The confidence set of models is defined as those having ΔAIC less than or equal to this limit in the actual data analysis.

We also calculated Akaike weights, w, which can be interpreted as the probability that model i is the best model among the set of R candidate models (Akaike 1978; Burnham and Anderson 1998). Values near 0 indicate that a model is very unlikely to be the best model in the set of candidate models. Lastly, we bootstrapped our data set 10,000 times with replacement and re-ran model fitting to estimate 95% confidence intervals for parameter estimates.

ALL-ELSE-EQUAL ASSUMPTION
Our estimates of the net cost of sex (c) allowed us to test Maynard Smith's original assumption that sexual and asexual females produce the same number of surviving offspring. We calculated the ratio of surviving asexual offspring to surviving sexual offspring (r) using our estimates of c and the primary sex ratio (s) (see "Extended model" in the Results section). We do not know the primary sex ratio of sexual *P. antipodarum* at Lake Alexandrina. Our a priori prediction is a sex ratio of 50% female ($s = 0.5$). The population of *P. antipodarum* is large (Paczesniak et al. 2014), so we predict a Fisherian sex ratio (Hamilton 1967). In addition, related prosobranch snails have chromosomal sex determination with females heterogametic (Barŝiene et al. 2000; Yusa

Table 1. Results of model inference and selection.

Model			logL	Parameters[b]	AIC_c	ΔAIC_c	w
1	No cost	$q_{t+1} = q_t$	-100.78	1 (θ)	203.74	29.01	0.00
	$c = 1$						
2	2-fold	$q_{t+1} = \dfrac{2q_t}{1 + q_t}$	-86.27	1 (θ)	174.73	0.00	0.72
	$c = 2$						
3	Estimate	$q_{t+1} = \dfrac{cq_t}{1 + q_t(c - 1)}$	-86.07	2 (θ, c)	176.71	1.98	0.27
	$c = MLE$						
4		$q_{t+1} = \dfrac{cq_t}{1 + q_t(c - 1)}$	-85.16	4 (θ, c_0, d_2, d_3, d_4)	183.65	8.93	0.01
	By year[a]	2012: $c = c_0$					
	$c = MLE*year$	2013: $c = c_0 + d_2$					
		2014: $c = c_0 + d_3$					
		2015: $c = c_0 + d_4$					

We proposed four candidate models for our experimental data. These four models assume different values of the cost of sex: (1) no cost ($c=1$); (2) a two-fold cost ($c=2$); (3) the maximum likelihood estimate (MLE) of the cost; and (4) the MLE of costs that vary with year. Each model is represented in the form of equation (2). We ranked models according to ΔAIC_c and evaluated the weight of evidence for each model using w, the Akaike weight.

[a] For model 4, the cost was indexed by experimental year. Maximum likelihood estimates of d_j that significantly deviate from 0 indicate that the cost of sex in experimental year j differed from that estimated in 2012 (S.I. V).

[b] Total number of estimated parameters. To fit models to experimental data, we assumed a beta-binomial distribution for the likelihood functions and thus estimated an additional overdispersion parameter θ.

2007). To estimate a range for the primary sex ratio, we calculated the tertiary sex ratio in our experimental and field populations (i.e., the proportion of females in the adult sexual subpopulations) (Tables S2 and S3). Further details are in the Supporting Information (IV).

Results

EXPERIMENTAL TEST

We experimentally measured the change in asexual frequency in a single generation of the freshwater snail *P. antipodarum*. Our goal was to directly estimate the net cost of sex in a natural system. To do this, we added juvenile snails sampled from Lake Alexandrina (South Island, New Zealand, Fig. S1A) to six 800-liter mesocosms (Fig. S1B). The use of field collections ensured the relevance of our results to natural populations. The juveniles matured and reproduced over the course of one year. We then separated parents and offspring by size into discrete generations (t and $t + 1$, respectively) to estimate the proportion of asexual individuals in the parents (q_t) and the offspring (q_{t+1}) generations (Fig. 1A). We replicated the experiment for four years, for a total of 24 independent replicates.

The frequency of asexuals increased 1.60-fold (95% CI [1.48, 1.73]) from an initial frequency of 29% in the parental generation (Fig. 1B; logistic model, generation: likelihood ratio D = 123.40, df = 1, p < 0.001). There was no variation in the direction of change between years (interaction: D = 2.37, df = 3, p = 0.499),

but the overall frequency of asexuals was highest in 2013 and 2014 (odds ratio vs 2012: 2013, 1.70 [1.39, 2.07]; 2014, 1.43 [1.17, 1.75]; 2015, 1.15 [0.94, 1.41]; year: D = 31.90, df = 3, p < 0.001).

The increase in asexual frequency was substantial (1.60-fold), but clearly less than the two-fold increase predicted for a two-fold cost under Maynard Smith's original model. However, asexual snails were not rare at the beginning of the experiment (Fig. 1B), as assumed by Maynard Smith's model. For a two-fold cost of sex, how much should asexuals increase in frequency when they are not initially rare? In the next section, we answer this question by using basic population genetic theory to predict the increase in frequency of asexuals given any initial frequency and any cost of sex. We then fit this model to our experimental data to estimate the cost of sex for *P. antipodarum*.

EXTENDED MODEL

We constructed a simple model that relaxes several assumptions of Maynard Smith's model. First, we relaxed the assumption that asexuals are rare. Secondly, we allowed the primary sex ratio of the sexual population to deviate from 0.5. Finally, we relaxed the "all-else-equal" assumption that asexual and asexual females produce on average the same number of surviving offspring.

Basic population genetic theory shows that the change in frequency of an allele over one generation is a function of its initial frequency and the strength of selection (Gillespie 1998).

A Experimental mesocosms

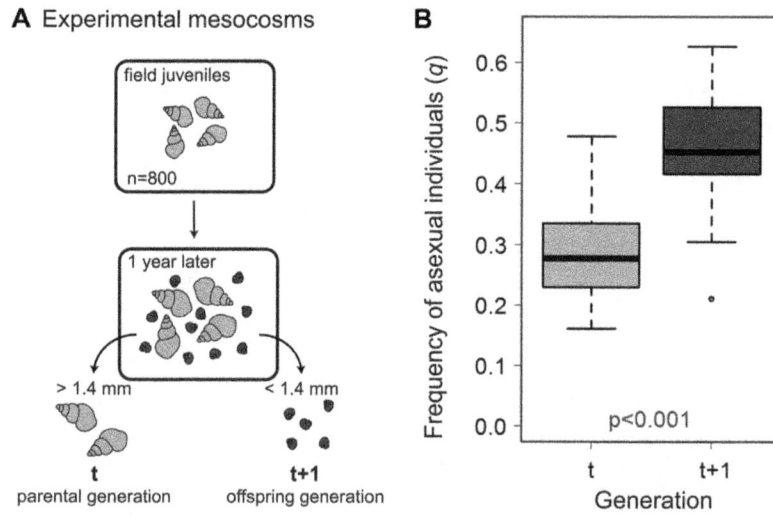

Figure 1. Increase in asexual frequency in experimental mesocosms. (A) Mesocosms were initiated with 800 field-collected juveniles (gray), which matured to adulthood and produced offspring (black) over the course of one year. Parents (originally juveniles) and offspring were separated by size and split into discrete generations (*t* and *t+1*, respectively). We then estimated the frequency of asexual individuals in parent (q_t) and offspring (q_{t+1}) generations. (B) The frequency of asexuals increased from the parent (*t*) to offspring (*t+1*) generation. Box plot shows median (black bar), upper, and lower quartiles (limits of box), minimum and maximum (whiskers, excluding outliers), and outliers (dots). The measure of significance is derived from the logistic model reported in the text. Each generation is represented by 24 mesocosms. The numbers of triploid females represented by each mesocosm are: 28.33 ± 1.50 SEM for parents and 23.67 ± 3.60 for offspring for the six mesocosms in 2012; 21.00 ± 1.97 for parents and 37.00 ± 2.29 for offspring in 2013 mesocosms; 16.67 ± 2.75 for parents and 34.33 ± 2.03 for offspring in 2014 mesocosms; and 16.67 ± 1.52 for parents and 27.83 ± 2.82 for offspring in 2015 mesocosms.

We applied this theory to predict the change in asexual frequency as a function of the parental frequency and the net cost of sex. We first write the frequency of asexuals in the next generation (q_{t+1}) as:

$$q_{t+1} = q_t \frac{W_{asex}}{\bar{W}}. \tag{1}$$

where q_t is the initial frequency of asexuals, W_{asex} is the per-capita birth rate for asexual females, and $\bar{W}$ is the mean per-capita birth rate for the mixed population of sexual and asexual individuals: $\bar{W} = q_t W_{asex} + (1 - q_t)W_{sex}$. Here W_{sex} is the per-capita birth rate for the sexual population, which includes males and females.

Let the per-capita birth rate of the sexual population be a fraction ($1/c$) of the asexual birth rate, where c represents the net cost of sex (Fig. 2A), such that $W_{sex} = W_{asex}/c$. An estimated value of two for c would be consistent with a two-fold cost of sex, while a value of one would mean that sexual females incur no net cost. By substituting, equation (1) becomes:

$$q_{t+1} = \frac{cq_t}{1 + q_t(c - 1)}. \tag{2}$$

Dividing both sides of equation (2) by q_t, we can calculate the fold-increase in the frequency of asexuals as:

$$\frac{q_{t+1}}{q_t} = \frac{c}{1 + q_t(c - 1)}. \tag{3}$$

Given a two-fold cost, equation (3) illustrates that the proportional increase in asexual frequency declines from two to one as the frequency of asexuals (q_t) moves from rarity to fixation (Fig. 2B). For a cost of sex equal to 2 ($c = 2$), a rare asexual mutant will double in frequency, as shown by Maynard Smith. The predicted increase is far less when asexuals are common (e.g., 1.54-fold for $q_t = 30\%$) (Fig. 2B).

Equation (2) can be rearranged to directly estimate the cost of sex from any starting frequency of asexuals:

$$c = \frac{q_{t+1}(1 - q_t)}{q_t(1 - q_{t+1})}. \tag{4}$$

In this model, the parameter c represents the net cost of sex, which includes the cost of males weighted by any fecundity-survival asymmetries in sexual versus asexual females. It is important to deconstruct c into its component parts, because the net cost of sex is a function of the cost of males plus many additional factors that may generate asymmetries in the fitness of sexual and asexual females (e.g., costs of mating, costs or benefits associated with ploidy differences). We represent these potential asymmetries using the parameter r, the ratio of the mean number of surviving offspring produced by asexual females divided by the mean number produced by sexual females. A value of one for r is consistent with the all-else-equal assumption: sexual and asexual females produce the same number of surviving offspring. Let the variable

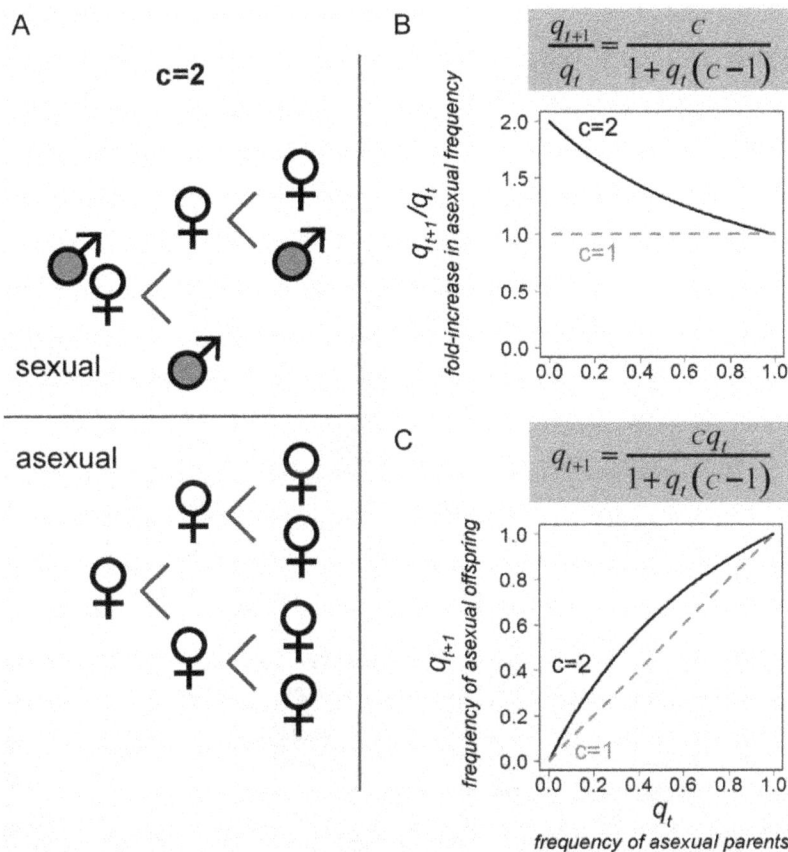

Figure 2. Theoretical predictions for the cost of sex. (A) Under a two-fold cost of sex ($c = 2$), asexual females can produce twice as many childbearing offspring (females) as sexual females. The net cost c is the product of the female fecundity-survival ratio r and the cost of males. Here, sexual and asexual females produce an equivalent number ($n = 2$) of surviving offspring (fecundity-survival ratio $r = 1$), consistent with the all-else-equal assumption. Sexual females make 50% daughters ($s = 0.5$), so the cost of males is two ($1/s = 2$). The total cost of sex is then two ($c = r * 1/s$). (B) Equation (3) shows the fold-increase in asexual reproduction: under a two-fold cost ($c = 2$, black solid line), doubling is observed only at very low starting frequencies of asexual individuals. The proportional increase in asexual frequency declines from two to one as the initial frequency of asexuals (q_t) increases from rarity to fixation. Equation (2)'s corresponding prediction for the frequency of asexual individuals in the offspring generation (q_{t+1}) is shown in (C). We use equation (2) when fitting models to experimental data. When there is no net cost to sexual reproduction ($c = 1$, gray dashed line), asexuals have no intrinsic birth rate advantage and will not change in frequency.

s be the proportion of resources allocated by sexual mothers to daughters. Assuming that sons and daughters are equally costly (Fisher 1930), s represents the proportion of offspring that are daughters in broods of sexual females (i.e., the primary sex ratio). The cost of males is then $1/s$. The total cost of sex is simply the product of the cost of males and the female fecundity-survival ratio (Fig. 2A): $c = r/s$.

MODEL FIT

From our experiment in seminatural mesocosms, we observed that the proportional increase in asexual frequency was less than two-fold. We also observed that asexuals were initially common, not rare. From the basic theory outlined above, we know that, for a two-fold cost of sex, the proportional increase in asexual frequency is predicted to be less than two-fold when asexuals

are common. Here, we combine our theory and data to ask: is a two-fold cost the best approximation to our experimental data? Specifically, we used equation (2) (Fig. 2C) to ask if the observed frequency of asexual offspring was consistent with a two-fold cost of sex given the initial (parental) frequency of asexual snails in experimental mesocosms.

We formulated four candidate models: (1) no cost of sex ($c = 1$), (2) a two-fold cost ($c = 2$), (3) the maximum likelihood estimate (MLE) of the cost, and (4) the MLE of costs that vary with year. A two-fold cost of sex (model 2) and the maximum likelihood estimate of the cost (model 3) were the best approximations to our experimental measures of q_t and q_{t+1} (Table 1). These two models had low ΔAIC and a high weight of evidence in their favor. The likelihood of model 3 was maximized at a cost of sex that slightly exceeds two ($c = 2.14$, 95% CI [1.81, 2.55]), but

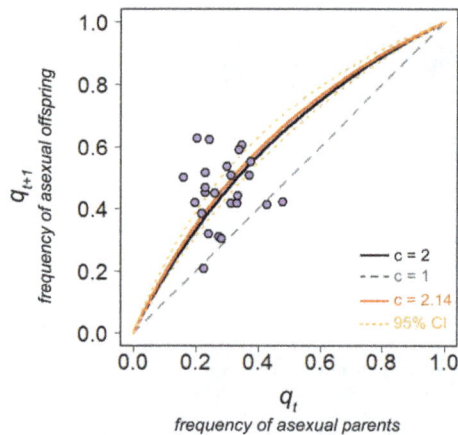

Figure 3. Experimental data are consistent with model predictions of a two-fold cost of sex. We fit our simple model (Fig. 2C; eq. (2)) to experimental data (Fig. 1B) on the frequency of asexuals q in generations t and $t+1$ in 24 seminatural mesocosms (purple points). We used standard maximum likelihood techniques and Akaike's information criterion to compete different estimates of the cost of sex c in *P. antipodarum*. The predicted frequency of asexual offspring (q_{t+1}) for a given frequency of asexual parents (q_t) is shown for three values of the cost of sex: no cost ($c = 1$, gray dashed line), a twofold cost ($c = 2$, black solid line), and the maximum likelihood estimate ($c = 2.14$, solid orange line). The 95% confidence intervals of the maximum likelihood estimate include two ([1.81, 2.55], dotted orange lines). Each point represents one mesocosm. For each mesocosm, the average number of triploid parents was 20.67 ± 1.57 SEM and the average number of triploid offspring was 30.71 ± 1.95.

not significantly so (Fig. 3). We concluded that, given the initial frequency of asexuals in our experimental mesocosms (q_t), the observed frequency of asexuals in the offspring generation (q_{t+1}) was consistent with a two-fold cost of sex.

The analysis firmly rejected model 1's assumption of equivalent per capita birth rates of sexual and asexual populations (i.e., no cost of sex). There was also little support for temporal variation in the cost of sex: the ΔAIC for model 4 was relatively large, exceeding the upper limit for our 95% confidence set of models (ΔAIC = 3.94), and the weight of evidence was 0 (Table 1). Parameter estimates also gave little support for yearly variation in cost (S.I. V; Fig. S2).

ALL-ELSE-EQUAL ASSUMPTION

By applying theory to our empirical data, we estimated the net cost of sex c to be equal to, or slightly greater than, 2. The net cost of sex c is a product of the cost of males ($1/s$; s being the primary sex ratio of broods of sexual females) and the ratio of surviving offspring produced by asexual versus sexual females (r, the female fecundity-survival ratio). Under the all-else-equal assumption of Maynard Smith's original model, asexual and sexual female make

an equivalent number of surviving offspring ($r = 1$). There are many reasons to expect r to deviate from 1, such as the energetic costs associated with outcrossing. We were able to test the all-else-equal assumption for *P. antipodarum* using our estimates of the net cost of sex c and the sex ratio s (S.I. IV).

For our *a priori* prediction of s equal to 0.5, our estimate of the fecundity-survival ratio r is 1 for model 2 ($c = 2$) and 1.07 (95% CI [0.91, 1.28]) for model 3, consistent with the all-else-equal assumption. We also calculated r assuming that the primary sex ratio is equal to the tertiary sex ratio calculated from parents in our experimental mesocosms: $s = 0.61$ (S.I. IV). Our estimate of r is then 1.22 for model 2 and 1.31 [1.10, 1.56] for model 3. Estimates of r above 1 are consistent with our observation that, in the mesocosms, the average brood of an asexual female contained 21% more embryos than that of a sexual female (S.I. I). We conclude that asexual females produce at least as many (i.e., $r = 1$), if not more (e.g., $r = 1.31$), surviving offspring than sexual females. Clearly, a net reduction in fitness does not accompany the transition to asexuality and elevated ploidy in *P. antipodarum*. In fact, this analysis and our life-history comparisons (S.I. I) suggest that sexual females may pay additional fitness costs beyond just the cost of males.

Discussion

Here, we provide a direct estimate of the net cost of sexual reproduction in a mixed population of freshwater snails. First, we conducted an experiment in semi-natural mesocosms (Fig. 1A) to show that asexual snails increase substantially in frequency (1.6-fold) from parent to offspring generation (Fig. 1B). However, this increase in asexual frequency is less than the two-fold increase predicted for the two-fold cost of males under Maynard Smith's original model. We resolve this apparent inconsistency between theory and data by using a standard population genetic approach. The results show that a two-fold cost of sex manifests as a two-fold increase in asexual frequency only when asexuals are very rare. When asexuals are common, as in our experiment, a two-fold cost manifests as a smaller increase in asexual frequency (Fig. 2B). We then applied this model to our experimental data. We found that, given the initial frequency of asexual snails in our mesocosms, the observed frequency of asexual offspring is consistent with that predicted under a two-fold cost of sex (Fig. 3, Table 1). We conclude that asexual *P. antipodarum* produce at least as many viable offspring as sexual females, resulting in at least a two-fold fitness cost for sexual reproduction. Our estimate of the net cost of sex in *P. antipodarum* is thus consistent with Maynard Smith's two-fold cost of males.

Given the net two-fold cost of sex observed here, asexual lineages should rapidly outcompete sexual lineages. Sexual individuals, however, comprised 71.2 ± 1.6% SEM of our field-collected juveniles from 2012–2015. Previous studies directly demonstrate the long-term persistence of sexual lineages in *P. antipodarum* (Jokela et al. 2009; Gibson et al. 2016 together span 20 years at Lake Alexandrina). Why do sexual and asexual *P. antipodarum* coexist in nature? One possibility (known as the Red Queen hypothesis) is that coevolving parasites select against common clonal genotypes, thereby giving an advantage to sexual reproduction (Jaenike 1978; Hamilton 1980; Bell 1982; Hamilton et al. 1990). Consistent with this idea, a long-term field study (Jokela et al. 2009) and a controlled laboratory experiment (Koskella and Lively 2009) both showed that common *P. antipodarum* clones declined in frequency over time after they became disproportionately infected by the sterilizing trematode *Microphallus*. Thus parasite-mediated frequency-dependent selection may maintain sexual snail lineages in the face of competition with multiple asexual clones.

What little research there is suggests that the cost of sex varies substantially among systems (Meirmans et al. 2012). Estimating the cost of sex is thus a critical starting point for evaluating the paradox of sex in a natural system. The cost of sex is nonetheless overlooked: hypotheses for the maintenance of sex are often tested in systems for which the cost of sex is unclear. The present study provides a simple framework for estimating the net cost of sex in other species. Our approach has two key requirements. First, it must be possible to separate individuals of different generations (parent vs offspring). Second, the experimental environment must limit selection by extrinsic factors that are known to differentially impact sexual vs. asexual fitness. For example, we made an effort to eliminate selection by coevolving trematode parasites in our seminatural mesocosms. Though we cannot exclude the possibility that our estimate of the net cost of sex in part reflects selection by extrinsic factors, many aspects of the mesocosm environment (e.g., reduced predation, competition) should have reduced differential selection, allowing an estimate of the intrinsic cost of sex.

The long-term maintenance of sex is one of the core anomalies in evolutionary biology, and the two-fold cost of sex is the foundational assumption of the paradox. Here, we have provided a straightforward approach to measuring the net cost of sex. Our results provide a quantitative validation of the two-fold cost of males in a natural system. This large and immediate fitness disadvantage justifies the search for a large and sustained short-term advantage to cross-fertilization.

ACKNOWLEDGMENTS

We thank Spencer R. Hall for invaluable assistance with the statistical approach, Samantha Klosak, Peyton Joachim, and Julie Xu for help with flow cytometry, and Daniela Vergara for technical assistance. We greatly appreciate assistance from the University of Canterbury's Edward Percival Field Station in Kaikoura, New Zealand and thank, in particular, Ngaire Perrin. Lastly, we acknowledge the Indiana University-Bloomington Flow Cytometry Core Facility and its manager Christiane Hassel for facilitating the flow cytometry work. This work was funded by a US National Science Foundation grant to CML and Jukka Jokela (DEB-0640639), an award from Indiana University to LD, and awards to AKG from the American Society of Naturalists (Student Research Award, Ruth Patrick Student Poster Award), the Society for the Study of Evolution (Rosemary Grant Student Research Award), Indiana University (Provost's Travel Award for Women in Science), the National Science Foundation (DDIG-1401281; GRFP), and the US National Institutes of Health (IU's Common Themes in Reproductive Diversity Traineeship).

AUTHOR CONTRIBUTIONS

A.K.G. designed and conducted the experiment, collected, and analyzed the data, and wrote the manuscript. L.F.D. designed and conducted the experiment and collected the data. C.M.L. designed and conducted the experiment, developed the theoretical model, and wrote the manuscript.

LITERATURE CITED

Akaike, H. 1973. Information theory as an extension of the maximum likelihood principle. Pp. 267–281 *in* B. Petrov and F. Csaki, eds. Second international symposium on information theory. Akademiai Kiado, Budapest.

———. 1978. A Bayesian analysis of the minimum AIC procedure. Ann. Inst. Stat. Math. 30:9–14.

Barŝiene, J., G. Ribi, and D. Barŝyte. 2000. Comparative karyological analysis of five species of *Viviparus* (Gastropoda: Prosobranchia. J. Moll. Stud. 66:259–271.

Bell, G. 1982. The masterpiece of nature: The evolution and genetics of sexuality. California Univ. Press, Berkeley, CA.

Bolker, B. 2008. Ecological models and data in R. Princeton Univ. Press, Princeton.

Browne, R. A., and K. M. Halanych. 1989. Competition between sexual and parthenogenetic *Artemia*: a re-evaluation (Branchiopoda, Anostraca). Crustaceana 57:57–71.

Burnham, K. P., and D. R. Anderson. 1998. Model selection and inference: A practical information-theoretic approach. Springer, New York.

Crummett, L., and M. Wayne. 2009. Comparing fecundity in parthenogenetic versus sexual populations of the freshwater snail *Campeloma linum*: is there a two-fold cost of sex. Invert. Biol. 128:1–8.

Dybdahl, M. F., and C. M. Lively. 1995. Diverse endemic and polyphyletic clones in mixed populations of the freshwater snail, *Potamopyrgus antipodarum*. J. Evol. Biol. 8:385–398.

Fisher, R. A. 1930. The genetical theory of natural selection. Oxford Univ. Press, Oxford, U. K.

Fox, J., M. F. Dybdahl, J. Jokela, and C. M. Lively. 1996. Genetic structure of coexisting sexual and clonal subpopulations in a freshwater snail (*Potamopyrgus antipodarum*). Evolution 50:1541–1548.

Gibson, A. K., J. Y. Xu, and C. M. Lively. 2016. Within-population covariation between sexual reproduction and susceptibility to local parasites. Evolution. 70:2049–2060.

Gillespie, J. H. 1998. Population genetics: A concise guide. John Hopkins Univ. Press, Baltimore.

Hamilton, W. 1967. Extraordinary sex ratios. Science 156:477–488.

———. 1980. Sex versus non-sex versus parasite. Oikos 35:282–290.

Hamilton, W. D., R. Axelrod, and R. Tanese. 1990. Sexual reproduction as an adaptation to resist parasites (a review). Proc. Natl. Acad. Sci. 87:3566–3573.

Hechinger, R. F. 2012. Faunal survey and identification key for the trematodes (Platyhelminthes: Digenea) infecting *Potamopyrgus antipodarum* (Gastropoda: Hydrobiidae) as first intermediate host. Zootaxa 3418:1–27.

Hurvich, C., and C.-L. Tsai. 1991. Bias of the corrected AIC criterion for underfitted regression and time series models. Biometrika 76:297–307.

Jaenike, J. 1978. An hypothesis to account for the maintenance of sex within populations. Evol. Theory 3:191–194.

Jokela, J., M. F. Dybdahl, and C. M. Lively. 2009. The maintenance of sex, clonal dynamics, and host-parasite coevolution in a mixed population of sexual and asexual snails. Am. Nat. 174:S43–S53.

Jokela, J., C. Lively, J. Fox, and M. Dybdahl. 1997a. Flat reaction norms and "frozen" phenotypic variation in clonal snails (*Potamopyrgus antipodarum*). Evolution 51:1120–1129.

Jokela, J., C. M. Lively, M. F. Dybdahl, and J. Fox. 1997b. Evidence for a cost of sex in the freshwater snail *Potamopyrgus antipodarum*. Ecology 78:452–460.

Koskella, B., and C. M. Lively. 2009. Evidence for negative frequency-dependent selection during experimental coevolution of a freshwater snail and a sterilizing tremtaode. Evolution 63:2213–2221.

Lehtonen, J., M. D. Jennions, and H. Kokko. 2012. The many costs of sex. Trends Ecol. Evol. 27:172–178.

Levri, E. P., and C. M. Lively. 1996. The effects of size, reproductive condition, and parasitism on foraging behaviour in a freshwater snail, *Potamopyrgus antipodarum*. Anim. Behav. 51:891–901.

Liang, K.-Y., and S. Zeger. 1986. Longitudinal data analysis using generalized linear models. Biometrika 73:13–22.

Maynard Smith, J. 1971. The origin and maintenance of sex. Pp. 163–175 *in* G. C. Williams, ed. Group selection. Aldine Atherton, Chicago.

———. 1978. The evolution of sex. Cambridge Univ. Press, Cambridge, U.K.

Meirmans, S., P. G. Meirmans, and L. R. Kirkendall. 2012. The costs of sex: facing real-world complexities. Quart. Rev. Biol. 87:19–40.

Neiman, M., J. Jokela, and C. M. Lively. 2005. Variation in asexual lineage age in *Potamopyrgus antipodarum*, a New Zealand snail. Evolution 59:1945–1952.

Neiman, M., D. Paczesniak, D. M. Soper, A. T. Baldwin, and G. Hehman. 2011. Wide variation in ploidy level and genome size in a New Zealand freshwater snail with coexisting sexual and asexual lineages. Evolution 65:3202–3216.

Paczesniak, D. 2012. Ecological and ecological dynamics in natural populations of co-existing sexual and asexual lineages. ETH Zurich, Zurich.

Paczesniak, D., S. Adolfsson, K. Liljeroos, K. Klappert, C. M. Lively, and J. Jokela. 2014. Faster clonal turnover in high-infection habitats provides evidence for parasite-mediated selection. J. Evol. Biol. 27:417–428.

R Core Team. 2013. R: a language and environment for statistical computing. R Foundation for Statistical Computing Vienna, Austria. Available at http://www.R-project.org.

Stelzer, C.-P. 2011. The cost of sex and competition between cyclical and obligate parthenogenetic rotifers. Am. Nat. 177:E43–E53.

Sugiura, N. 1978. Further analysis of the data by Akaike's information criterion and the finite corrections. Comm. Stat. Theory Meth. A7:13–26.

Williams, G. C. 1971. Introduction. Pp. 1–15 *in* G. C. Williams, ed. Group selection. Aldine Atherton, Chicago.

———. 1975. Sex and EVOLUTION. Princeton Univ. Press, Princeton, NJ.

Yusa, Y. 2007. Causes of variation in sex ratio and modes of sex determination in the Mollusca—an overview. Am. Malacol. Bull. 23:89–98.

Zeger, S., and K.-Y. Liang. 1986. Longitudinal data analysis for discrete and continuous outcomes. Biometrics 42:121–130.

Ziegler, A., and M. Vens. 2010. Generalized estimating equations: notes on the choice of the working correlation matrix. Methods Inf. Med. 49:421–425.

Integrating viability and fecundity selection to illuminate the adaptive nature of genetic clines

Susana M. Wadgymar,[1] S. Caroline Daws,[2] and Jill T. Anderson[1,3]

[1]Department of Genetics and Odum School of Ecology, University of Georgia, Athens, Georgia 30602

[2]Department of Ecology, Evolution and Behavior, University of Minnesota, St. Paul, Minnesota 55108

[3]E-mail: jta24@uga.edu

Genetically based trait variation across environmental gradients can reflect adaptation to local environments. However, natural populations that appear well-adapted often exhibit directional, not stabilizing, selection on ecologically relevant traits. Temporal variation in the direction of selection could lead to stabilizing selection across multiple episodes of selection, which might be overlooked in short-term studies that evaluate relationships of traits and fitness under only one set of conditions. Furthermore, nonrandom mortality prior to trait expression can bias inferences about trait evolution if viability selection opposes fecundity selection. Here, we leveraged fitness and trait data to test whether phenotypic clines are genetically based and adaptive, whether temporal variation in climate imposes stabilizing selection, and whether viability selection acts on adult phenotypes. We monitored transplants of the subalpine perennial forb, *Boechera stricta* (Brassicaceae), in common gardens at two elevations over 2–3 years that differed in drought intensity. We quantified viability, and fecundity fitness components for four heritable traits: specific leaf area, integrated water-use efficiency, height at first flower, and flowering phenology. Our results indicate that genetic clines are maintained by selection, but their expression is context dependent, as they do not emerge in all environments. Moreover, selection varied spatially and temporally. Stabilizing selection was most pronounced when we integrated data across years. Finally, viability selection prior to trait expression targeted adult phenotypes (age and size at flowering). Indeed, viability selection for delayed flowering opposed fecundity selection for accelerated flowering; this result demonstrates that neglecting to account for viability selection could lead to inaccurate conclusions that populations are maladapted. Our results suggest that reconciling clinal trait variation with selection requires data collected across multiple spatial scales, time frames, and life-history stages.

KEY WORDS: Elevational gradient, flowering phenology, invisible fraction, stabilizing selection, specific leaf area, water-use efficiency.

Impact Summary

Natural selection has produced extraordinary diversity in adaptations to natural environments. Many species are distributed broadly across climatic gradients, such as elevation or latitude. Natural populations of these species often exhibit continuous clines in morphology and physiology in response to environmental variation. Clines provide supurb opportunities to study natural selection and adaptation. We performed a field experiment replicated across space and over time to evaluate the environmental context of clinal trait variation and natural selection in the perennial montane plant *Boechera stricta*. We transplanted $N = 4510$ individuals derived from $N = 24$ maternal families into natural communities in two experimental gardens. Over 2–3 growing seasons, we measured survival, flowering success, fruit production, and four ecologically relevant traits. By incorporating the broad range of trait values found in populations distributed across a steep environmental gradient, our experiment allowed us to test whether selection favored trait values associated with local populations. When we modeled data from all years in one analysis, we found evidence for stabilizing selection on all traits; that is, selection favored intermediate trait values, consistent with average values for local plants. In contrast, analyses for each individual year of data typically identified only weak, directional selection. Lastly, our experiment revealed that selection during juvenile stages can act on adult traits that have yet to be expressed, demonstrating the need to account for selection acting throughout a species's life cycle to characterize adaptation. Our results suggest that accurate depictions of natural selection depend on the suite of traits being examined, the duration of the experimental study, and the number of study sites under observation.

Introduction

Adaptive evolution generates phenotypic innovation, enables species to colonize new environments, and promotes speciation (Kocher 2004; Prentis et al. 2008; Gilbert et al. 2015). Genetically based clines in functional traits can signify adaptation to continuous environmental variation along gradients, yet phenotypic divergence along gradients can also arise from plasticity and neutral processes (Vasemägi 2006; Montesinos-Navarro et al. 2011; Kooyers et al. 2014). Testing whether genetic clines are adaptive requires evaluating selection in the natural environments in which species evolve; however, accurately estimating selection gradients can be challenging when patterns of selection vary temporally or across life history.

Environmental conditions differ across the landscape and change over time, exposing natural populations to spatial and temporal variation in selection (Grant and Grant 2002; Morrissey and Hadfield 2012; Siepielski et al. 2013; Ågren et al. 2017). Spatially variable selection can drive adaptive population divergence (Hall and Willis 2006), and temporal shifts can maintain variation within populations (Calsbeek et al. 2010). In natural populations, the presence of extensive genetic variation in ecologically relevant traits (Mousseau and Roff 1987; Paaby and Rockman 2014), in concert with pervasive directional selection, should facilitate adaptation (Kingsolver et al. 2001; Thurman and Barrett 2016) and lead to rapid evolutionary change. Paradoxically, many populations also exhibit evolutionary stasis, such that trait values do not change appreciably with time (Estes and Arnold 2007; Haller and Hendry 2014). Fluctuating selection can, theoretically, lead to evolutionary stasis if the direction of selection reverses sign frequently (Siepielski et al. 2009; Bell 2010). These changes in the sign of directional selection over time could manifest in stabilizing selection across multiple episodes of selection as populations repeatedly traverse the summit of a fitness peak (Phillips and Arnold 1989; McGlothlin 2010). However, natural populations that appear highly adapted to contemporary conditions often show evidence for directional, not stabilizing, selection in studies that evaluate only one episode of selection (Kingsolver and

Diamond 2011; Kingsolver et al. 2012). Evolutionary stasis will arise if the position of the peak within phenotypic space is stable (Arnold et al. 2008). Therefore, temporal variation in selection could explain the stability of genetic clines despite frequent observations of short-term directional selection.

Evidence for temporally varying selection in wild populations is limited (Morrissey and Hadfield 2012). Subtle fluctuations in linear selection gradients may be the only indicators of nonlinear selection, as stabilizing selection can be difficult to detect directly when sample sizes are low or populations inhabit a broad or stable fitness peak (Haller and Hendry 2014). By incorporating a wider range of phenotypes than typically found in a single population, multiyear common garden experiments provide powerful tests of the role of temporal variation in generating stabilizing selection over time, particularly in perennial species for which trait evolution is influenced by multiple episodes of selection. Furthermore, replicate gardens across environmental gradients can reveal whether selection maintains clinal trait variation.

Detecting selection is particularly challenging when a nonrandom portion of individuals die prior to trait expression (Hadfield 2008). Nevertheless, studies that neglect viability selection could generate biased estimates of selection and inaccurate predictions about trait evolution (Hadfield 2008; Mojica and Kelly 2010). For example, despite fecundity selection favoring large flowers, low survival of large-flowered *Mimulus guttatus* genotypes drives selection for smaller flowers via lifetime fitness (Mojica and Kelly 2010). Mortality events generally prevent experiments from retaining sufficient statistical power to estimate viability selection. Multivariate genotypic selection analyses (Rausher 1992) are well-suited for evaluating selection because they account for missing data generated from mortality (i.e. the "invisible fraction") by modeling family-level trait and fitness data (Hadfield 2008).

We conducted a multiyear field study to disentangle the contributions of adaptation, plasticity, and neutral evolutionary processes to clinal trait variation and to characterize the nature of selection on functional traits in the context of spatiotemporal environmental variation. Elevation gradients present superb opportunities for investigating adaptation and spatial variation in selection. Climatic conditions change appreciably over short spatial scales in mountainous systems (Körner 2007), which can result in consistent patterns of adaptation to local climate (Byars et al. 2007; Kim and Donohue 2013). Indeed, populations of the subalpine perennial forb *Boechera stricta* (Brassicaceae) experience spatially restricted gene-flow and have adapted locally to variation in selective regimes that occur over relatively short spatial scales (Anderson et al. 2015).

We hypothesize that (1) genetically based clines in flowering phenology, morphology, and ecophysiology are consistent with clinal trait variation in natural populations (Anderson and Gezon

2015) and are maintained by selection. By assessing spatial and temporal patterns of natural selection, we tested whether (2) natural selection fluctuates around local phenotypic optima. Finally, we estimated selection at two life-history stages to evaluate whether (3) viability selection operating prior to trait expression influences the evolutionary trajectories of adult phenotypes. Despite numerous estimates of selection in nature (Kingsolver et al. 2012), studies rarely examine spatial and temporal dynamics simultaneously (Siepielski et al. 2013), especially across multiple life-history stages.

Methods

SYSTEM

We conducted fieldwork in subalpine meadows around the Rocky Mountain Biological Laboratory (RMBL, Gunnison County, Colorado). *Boechera stricta* is a short-lived, self-fertilizing perennial forb native to the Rocky Mountains, where it occurs across a broad elevational gradient (700–3900 m) (Al-Shehbaz and Windham 2010). We estimate the generation time of this species to be 2–3 years (Anderson et al. 2012) and individuals experience multiple episodes of selection across their lifetimes. In Gunnison county and other regions in Colorado, robust populations of *B. stricta* occur from ~2700 m to ~3600 m in elevation, with sparse populations in elevations as low as 2500 m, but we have not located populations at lower elevations (Anderson, pers. obs.). High elevation sites have later spring snowmelt, cooler temperatures, greater water availability, and shorter growing seasons than low elevation locales (Dunne et al. 2003; Anderson and Gezon 2015).

Our study captured striking differences in climate from 2012–2014 (Harte et al. 2015). These years followed a sequence of decreasing abiotic stress, with snowfall levels of 640, 788, and 1177 cm from 2012 to 2014, and snowmelt occurring earlier than the historical average by 44 days, 23 days, and 13 days in those same years. Snowfall averaged 1088.3 ± 273.0 cm annually from 1974–2015 (mean $\pm$SD; b. barr, Gothic Long-Term Weather Data; http://www.gothicwx.org/long-term-snow.html) (Harte et al. 2015).

COMMON GARDENS

To test unresolved hypotheses about natural selection and genetic clines, we transplanted maternal families collected across a broad elevational gradient into two common gardens: a hot and dry low-elevation site (2891 m, 38°57.086"N, 106°59.4645"W) and a cool and wet high-elevation site (3133 m, 39°02.346"N, 107°03.818"W). As expected based on global and regional patterns (Körner 2007), weather stations near our field sites show that mean annual temperature declines by 0.0023–0.0046°C and total annual precipitation increases by 0.042 cm for every meter gain in elevation (Anderson and Gezon 2015). Snow melted 11 days ear-lier in the low versus high elevation garden in 2013 and 2014, and Decagon 5TM sensors installed in 2014 showed that soil temperature was $4.52°C \pm 0.04$ greater in the lower versus higher garden from July 2–August 28, 2014 (Anderson and Gezon 2015).

We examined selection on four traits associated with climatic adaptation: specific leaf area (SLA), stable Carbon isotopes ($\delta^{13}C$), height at first flower, and flowering phenology (e.g., Campbell et al. 2010; Ward et al. 2012; Pratt and Mooney 2013; Read et al. 2014; Ågren et al. 2017). Natural populations of *B. stricta* exhibit significant phenotypic clines in these ecologically relevant traits across elevational gradients (Anderson and Gezon 2015). Foliar stable carbon isotopes ($\delta^{13}C$) provide robust data on water-use efficiency (WUE) integrated over the lifespan of the leaf (Farquhar et al. 1989). In regions such as the Colorado Rockies where aridity and growing season length decrease with elevation, we expect selection to favor drought tolerance at low elevations and rapid development at high elevations, in which case WUE ($\delta^{13}C$) and flowering time would decline with source elevation. However, similar clines in $\delta^{13}C$ could emerge from declining partial pressure of atmospheric CO_2 with elevation (Körner 2007).

The power to test adaptive population divergence is greater in studies that include few families from a large number of populations, rather than studies that include more families from fewer populations (Goudet and Buchi 2006; Blanquart et al. 2013). Therefore, we selected one maternal family from each of 24 local populations, which ranged in elevation from 2869–3682 m (Table S1), enabling tests of selection using accessions that experienced substantially different environmental conditions during their evolutionary histories (Anderson and Gezon 2015; Anderson et al. 2015). For the 24 source populations, we extracted climatic data from WorldClim using a bilinear interpolation with the highest spatial resolution available (30 arc-second). In accordance with data from sensors installed within our field sites, temperature, and aridity decline with elevation ($R^2 = 0.985$ and 0.96, respectively, $p < 0.0001$), while precipitation increases with elevation ($R^2 = 0.96$, $p < 0.0001$). Therefore, source elevation can serve as a reliable proxy for climate of origin in this region.

Since *B. stricta* self-pollinates and has limited within population genetic variation (Song et al. 2006), our sampling design maximized genetic diversity and is effective for quantifying genetic clines and selection. This experiment has revealed local adaptation to elevation and genetic isolation by distance (Anderson et al. 2015) as well as plasticity in foliar morphology and genetically based clines in flowering phenology in the 2013 growing season (Anderson and Gezon 2015). Here, we integrate trait and fitness data from a larger database covering 2012–2014 to assess temporal variation in genetic clines and selection on phenology, morphology, and ecophysiology.

To reduce maternal effects and generate families via self-fertilization, we grew field-collected seeds for one generation in the greenhouse and allowed them to self-pollinate. We then germinated seeds from these full sibling maternal lines and reared seedlings in the greenhouse for three months prior to outplanting. In September 2011, we transplanted $N = 2293$ juvenile plants into the low elevation garden (hereafter: 2011 cohort; ~95.5 full siblings per family). In September 2012, we transplanted a second cohort into both gardens (hereafter: 2012 cohort): $N = 1096$ individuals into the lower garden and $N = 1121$ individuals into the higher garden (~46 full siblings per family per garden). We used seeds of the same generation and the same 24 families in both cohorts and gardens, and planted two full siblings per family into blocks of 48 individuals. Extensive replication of siblings within each family allowed us to generate precise and accurate estimates of heritabilities, family-level trait means, and ultimately selection gradients. We disturbed the natural community minimally during transplanting and did not manipulate the biotic or abiotic environment. This experiment captured the full lifespan of most experimental individuals, as only 21.9% (2011 cohort) and 49% (2012 cohort) of individuals remained alive in the final census (September 2014).

FITNESS COMPONENTS AND PHENOTYPES

In May–August 2012–2014, we visited each garden 3–4 times/week and recorded survival, plant size, numbers of flowers and fruits, and the length of the longest fruit. We collected all mature fruits. For plants that flowered between censuses, we estimated the day of first flowering from data on the number of fruits and fruit elongation rates (Table S2). Analysis of the raw flowering time data (not shown) generated similar results. We calculated flowering time as the number of elapsed days between snowmelt and first flowering. Fitness components included flowering success (viability) and fecundity. At the end of each season, we calculated the sum of the lengths of all mature fruits for each plant, excluding any aborted fruits lacking seeds. We used total mature fruit length as our metric of fecundity, as it correlated with seed number (Pearson correlation coefficient: $r = 0.71$, $p < 0.0001$, $N = 72$) and weight ($r = 0.74$, $p < 0.0001$, $N = 72$).

We quantified integrated water-use efficiency via natural abundance stable Carbon isotopes ($\delta^{13}C$) using samples from ~3 siblings per family collected each year from both gardens (Table S3). We pulverized leaf tissue using a GenoGrinder (SPEX SamplePrep), weighed 3.000–3.200 mg of ground foliar tissue on an ultramicrobalance (UMX2, Mettler-Toledo) and loaded tissue into tin capsules. The Cornell University Stable Isotope Laboratory combusted samples on an isotope ratio mass spectrometer connected to an elemental analyzer. We report $\delta^{13}C$ relative to Vienna Pee Dee Belemnite. Traits in this study were moderately correlated ($|r| < 0.3$, Table S3).

ANALYSES
Genetic variation
In *B. stricta* and other selfing species, responses to selection depend on total genetic variance rather than additive genetic variance (Roughgarden 1979). We used restricted maximum likelihood to estimate broad-sense heritability (H^2) for each garden and cohort as genetic variance (V_G) divided by phenotypic variance (V_P = family variance + block variance + error variance) in models that included random effects for family and block and repeated effects for plant identity across seasons (Proc Mixed, SAS ver. 9.4). To estimate the heritability of flowering success, we used a binary distribution with a logit link (Proc Glimmix).

Genetically based clines and plasticity
To assess genetic clines and plasticity, we modeled all traits simultaneously in a repeated measures multivariate regression with a Kenward-Roger degree of freedom approximation. We analyzed the two cohorts separately because we have three years of data from one garden (2011 cohort) and two years of data from both gardens (2012 cohort). We first calculated family level averages (LSMEANS) by modeling each trait in each season as a function of fixed effects for family (2011 cohort) or family × garden (2012 cohort) with a random effect for block (Proc Mixed). We then analyzed multivariate family-level LSMEANS as a function of source elevation, season, and source elevation × season (full results in Tables S4 and S5). For the 2012 cohort, we also included the main effect of garden, as well as 2- and 3-way interactions among garden, source elevation, and season. Analyses of individual-level data generated quantitatively similar results (Table S6). Multivariate repeated measures models use direct (Kronecker) product structures to specify the covariance structure of the R matrix [type = UN@AR(1)], which fits multiple response variables (unstructured covariance matrix, UN) measured on the same plants across years [autoregressive covariance matrix, AR(1)] (Galecki 1994). We standardized trait values to a mean of 0 and standard deviation of 1 but present unstandardized data in figures.

We directly estimated genetic clines by incorporating source elevation of maternal families into our analyses. Interactions between garden or season and source elevation indicate that genetic clines vary across space or through time. We used preplanned estimate statements to calculate the slope and test the significance of genetic clines for each trait, controlling for multiple tests with the Benjamini–Hochberg procedure (1995); we report corrected P-values in the results. Significant main effects of season and garden reveal temporal and spatial plasticity, respectively. Table S6 contains complementary analyses correcting for neutral population genetic structure using 13 microsatellite loci genotyped previously (Anderson et al. 2015).

Genotypic selection analyses

To test whether clines are adaptive, we used genotypic selection analyses (Rausher 1992). As traits and fitness were highly heritable (Table S3), we can link family-level trait data (averaged over individuals that expressed traits) with family-level fitness (averaged over all individuals, including those that died or failed to reproduce). These analyses test if selection operates prior to trait expression by incorporating the invisible fraction into fitness components (Hadfield 2008). We analyzed selection via three fitness components: probability of flowering among plants alive at the beginning of each season (viability), the length of fruit (a proxy for seed set) among plants that successfully flowered (fecundity), and fruit length among all individuals (cumulative). Genotypic selection analysis and artificial selection experiments are currently the only mechanisms to evaluate viability selection when individuals die before trait expression (Hadfield 2008). Our estimated selection gradients are likely conservative because the most maladapted individuals probably die earlier, such that their (maladapted) trait values would not contribute to family mean phenotypes.

To investigate whether selection fluctuates around local phenotypic optima, we conducted genotypic selection analyses using Aster models (Geyer et al. 2007). Aster assesses lifetime selection by modeling the dependence of fitness components on those expressed earlier in life history (Shaw et al. 2008). This approach can integrate episodes of selection throughout the life cycle, allowing us to compare the contributions of selection acting through viability and fecundity to cumulative patterns of selection (Table S7).

To the best of our knowledge, Aster analyses have not yet been used to conduct genotypic selection analyses. Although able to integrate multiple episodes of selection, Aster models discard data for families with incomplete trait data for any year, resulting in five dropped families from the 2011 cohort and two from the 2012 cohort. Data from maternal families that fail to reproduce in a given year are inherently valuable for quantifying viability selection. As such, we have replicated all analyses in a repeated measures framework in SAS, which only removes families from the specific years with missing trait data. SAS models resulted in more conservative estimates of selection as compared to Aster in some cases. We treat the few discrepancies between results generated from both platforms as tentative (see Supplemental Materials for details on Aster and SAS methodology).

We conducted cumulative, viability, and fecundity selection analyses separately for each cohort, evaluating direct selection by analyzing fitness as a function of all four traits simultaneously. We included source elevation as a covariate in in our models to account for unmeasured traits that may also vary across elevational gradients, and we evaluated quadratic effects of source elevation through loglikelihood tests (for a similar approach, see Colautti

and Barrett 2013). To test for spatial variation in selection, we included fixed effects for garden and trait × garden interactions in analyses of the 2012 cohort. Significant interactions indicate that selection varies spatially, in which case we extracted selection gradients from overall Aster models, but estimated the significance of selection gradients in separate models for each garden. We then analyzed each growing season separately in SAS models (Proc Glimmix and Proc Mixed, Table S11). We standardized all traits to a mean of 0 and standard deviation of 1 and we relativized fecundity and cumulative fitness to a mean of 1. We converted logistic regression coefficients to selection gradients following Janzen and Stern (1998). We estimated standardized linear selection gradients and statistical significance in models that contained linear effects only; we extracted quadratic selection gradients and significance from second-order polynomial models including linear and quadratic effects of traits (and other relevant fixed effects) (Lande and Arnold 1983). We doubled all quadratic regression coefficients and standard errors to estimate quadratic selection gradients (Stinchcombe et al. 2008).

Results

Phenotypes and fitness components were highly heritable in this study, with H^2 ranging from 0.083–0.43 for the 18 estimates of heritability (average: 0.25; all $P < 0.0001$ after correction for multiple testing; Table S3). We summarize predictions and results in Table 1.

GENETICALLY BASED CLINES ARE CONSISTENT WITH CLINAL TRAIT VARIATION IN NATURAL POPULATIONS

In both gardens, we detected significant genetically based clines in traits, which were consistent with the direction and magnitude of trait variation across elevation in natural populations (Anderson and Gezon 2015) and were generally robust to correction for neutral population differentiation (Tables S4–S6, Figs. 1 and S1). Furthermore, genetic clines were concordant with spatial plasticity in these traits (Table S4, Fig. S2). Nevertheless, genetic clines did not emerge in all years. In accordance with our predictions (Table 1), integrated WUE (as measured via stable Carbon isotopes, $\delta^{13}C$) declined significantly with source elevation in 2012, the year of extreme early snowmelt ($t_{51.6} = -3.65, P = 0.0019$; Fig. 1A; Table S5). This genetic cline persisted when drought was less severe in 2013 ($t_{51.6} = -2.62, P = 0.017$), but disappeared in the relatively benign year of 2014 ($t_{51.6} = -1.23, P = 0.24$; Fig. 1 inset). Context dependency became even more apparent after correcting for neutral population structure, as the genetic cline was only significant for the driest year of 2012 (Table S6). The 2012 cohort showed no evidence for genetic clines in $\delta^{13}C$ (Table S5B, Fig. S2).

Table 1. Predictions and results for genetic clines, plasticity, and selection.

Trait	Predicted: genetic cline	Predicted: plasticity	Results: genetic clines	Results: plasticity	Results: selection and clines	Results: viability vs fecundity
Water-use efficiency (δ^{13}C)	(−)	Greater trait values in dry environments (low vs high elevation)	Concordant and context dependent	Concordant	Concordant	No
Specific leaf area (cm^2/g)	(+)[*]	Lower trait values in dry environments (low vs high elevation)	Concordant and context dependent	Concordant	Concordant	No
Flowering time (relative to snowmelt date)	(−)	Earlier flowering under short seasons (high vs low elevation)	Concordant and context dependent	Concordant	Concordant	Yes
Height at flowering	(−)	Shorter height at flowering in high elevations	Concordant and context dependent	Concordant	Concordant	No

We base predictions are on phenotypic variation across natural *Boechera stricta* populations (Anderson and Gezon 2015) and on climatic adaptation from other systems (e.g., Campbell et al. 2010; Leonardi et al. 2012; Ward et al. 2012; Pratt and Mooney 2013; Read et al. 2014). "Predicted: Genetic cline" indicates the predicted relationship between trait values and source elevation in common gardens, with (−) predicting a negative slope and (+) predicting a positive slope. "Predicted: Plasticity" provides expectations for trait variation with environment. In the two corresponding results columns, we indicate whether genetic clines and phenotypic plasticity were concordant or discordant with predictions. "Results: Selection and clines" conveys whether selection is generally concordant with genetic clines, with fitness optima integrated across time and life history corresponding to local mean trait values. Lastly, "Results: Viability vs fecundity" illustrates whether viability and fecundity selection favor contrasting trait values.

[*]Note: This prediction is based on natural *B. stricta* populations (Anderson and Gezon 2015).

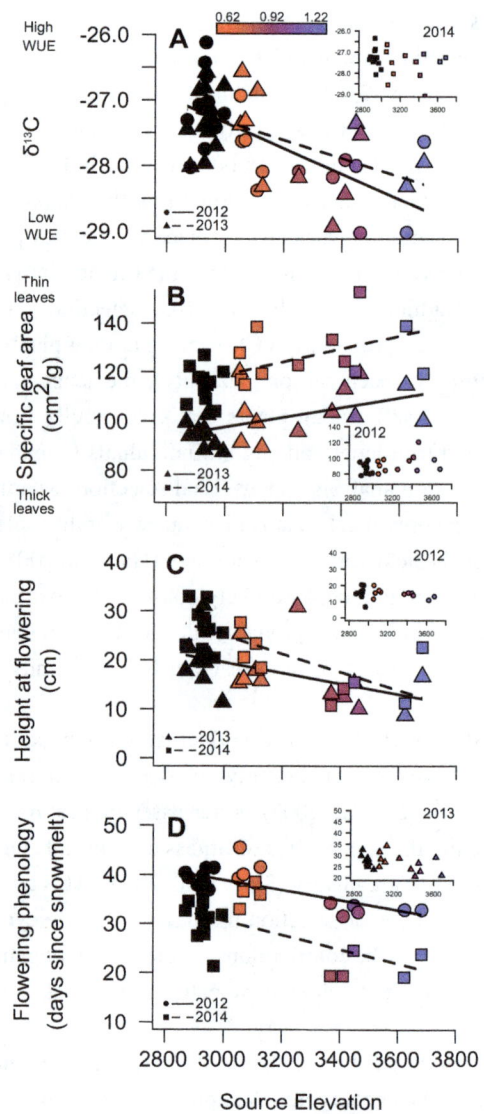

Figure 1. Genetically based clines in four functional traits measured over three growing seasons for the 2011 cohort planted into the low elevation garden: (A) Carbon isotope discrimination (integrated water-use efficiency, WUE), (B) specific leaf area, (C) height at flowering, and (D) flowering phenology. Data points represent family-level averages for these heritable traits. Lines depict the shape of the association between source elevation and trait values (only shown for significant relationships after correction for multiple tests). Inset panels reflect seasons in which we found no significant genetic clines. Figure S1 presents genetically based clines for the 2012 cohort. The color coding represents calculated Aridity Index based on climatic data extracted from WorldClim, ranging from arid low elevation populations (red) to less arid high elevation populations (blue).

Genetic clines in specific leaf area (SLA) were concordant with trait variation in natural *B. stricta* populations (Anderson et al. 2015): SLA increased significantly with source elevation in two years for the 2011 cohort (2013: $t_{61.2} = 2.74$, $P = 0.014$; 2014: $t_{61.2} = 3.69$, $P = 0.0019$, Table S5; Fig. 1B), and we saw a similar positive relationship in the lower garden of the 2012 cohort in 2014 ($t_{82.6} = 2.63$, $P = 0.026$, Fig. S1 and Table S5). In the higher garden, there was a marginal trend in the same direction in 2013 ($t_{84.5} = 2.25$, $P = 0.053$). Clines in SLA were nonsignificant after correcting for neutral population structure (Table S6), yet we found evidence that selection favors local trait values in both cohorts (see below).

Growing seasons are limited in duration in high elevation sites; therefore, we predicted that plants in high elevation locales would flower rapidly at small stature to complete their reproductive cycles prior to the onset of inclement conditions. Genetic clines were consistent with expectations and generally stable across space and time. For the 2011 cohort, we saw the expected decline in flowering time with source elevation in two years (2012: $t_{52.6} = -3.22$, $P = 0.0052$; 2013: $t_{52.5} = -1.7$, $P = 0.11$; 2014: $t_{53.9} = -4.34$, $P = 0.0004$, Fig. 1C, Table S5) and the expected reduction in height at flowering with source elevation in two of the three years (2012: $t_{57.1} = -0.68$, $P = 0.50$; 2013: $t_{57} = -2.86$, $P = 0.012$; 2014: $t_{58.1} = -4.79$, $P = 0.0001$, Fig. 1D). In three of four cases, the clines in size at reproduction and flowering time are also apparent in the 2012 cohort (Table S5, Fig. S1). Reproductive phenology clines generally remain significant after correction for neutral population structure (Table S6).

Natural selection

(1) Selection fluctuates around local phenotypic optima and (2) viability selection influences adult trait evolution

We evaluated direct selection via three fitness components: the probability of flowering (viability selection), fruit length (a proxy for seedset) among plants that flowered (fecundity selection), and fruit length among all plants (cumulative selection). Temporal fluctuations in environmental conditions alter not only the distribution of phenotypes in our study (Fig. S2 and Table S4), but also the magnitude and direction of selection. Indeed, multivariate genotypic selection analyses uncovered significant linear and quadratic direct selection. Local families typically expressed trait values close to trait optima, suggesting that genetic clines are adaptive (Fig. 2). Furthermore, nonlinear selection was primarily apparent in models that integrated data across years rather than models investigating selection separately for each growing season (Tables S8–S12).

2011 cohort

Specific leaf area was under stabilizing selection across life history, with similar intermediate trait optima favored by viability and fecundity selection (viability: $\chi^2(1) = 32.7$, $P < 0.0001$; fecundity: $\chi^2(1) = 55.2$, $P < 0.0001$; Fig. 2A). The synergy between viability and fecundity selection resulted in cumulative selection

favoring SLA values similar to the phenotypes of local families ($\chi^2(1) = 7.51$, $P = 0.006$; Fig. 2B).

Patterns of selection on the other vegetative trait ($\delta^{13}C$) were not as consistent. Aster models detected stabilizing viability selection favoring intermediate $\delta^{13}C$ values ($\chi^2(1) = 7.21$ $P = 0.007$, Fig. S3A), but SAS failed to find evidence for significant viability selection on this trait (Table S8). Additionally, Aster models uncovered weak nonlinear significant selection on $\delta^{13}C$ via fecundity ($\chi^2(1) = 9.4$, $P = 0.002$, Fig. S3A), which again was not evident in SAS analysis (Table S8). As SAS was unable to detect selection on $\delta^{13}C$, we treat these Aster results as tentative.

Mulitivariate models revealed stabilizing viability selection on both adult traits, with optimal trait values corresponding with phenotypes of local families (flowering phenology: $\chi^2(1) = 8.07$, $P = 0.005$, Fig. 2C; height at flowering: $\chi^2(1) = 8.27$, $P = 0.004$; Fig. 2E, Table S8). In contrast, nonlinear fecundity selection showed curvilinearity in directional selection for earlier flowering ($\chi^2(1) = 17.4$, $P < 0.0001$; Fig. 2C). Directional fecundity selection favored taller plants at flowering ($\chi^2(1) = 32.6$, $P < 0.0001$; Fig. 2E). Finally, consistent with other studies, directional cumulative selection favored earlier flowering ($\chi^2(1) = 9.08$, $P = 0.003$, Figs. 2D, 3C) and greater height at flowering ($\chi^2(1) = 8.1$, $P = 0.005$, Fig. 2F, 3D).

2012 cohort

As with the 2011 cohort, stabilizing selection emerged principally in models that integrated across episodes of selection. Viability and fecundity selection on specific leaf area (SLA) varied spatially (Table S10). In the high elevation garden, stabilizing selection operated on SLA consistently across life history, with viability, fecundity, and cumulative selection favoring slightly different intermediate trait optima (viability: $\chi^2(2) = 8.78$, $P = 0.012$; fecundity: $\chi^2(1) = 189.8$, $P < 0.0001$; cumulative: $\chi^2(1) = 32.4$, $P < 0.0001$; Fig. 2G, H, Table S9). In contrast, in the lower garden, selection favored reduced SLA across all fitness components, although weak nonlinear selection via viability and fecundity indicated slight curvilinearity in patterns of directional selection (viability: $\chi^2(1) = 8.78$, $P = 0.012$; fecundity: $\chi^2(1) = 116.7$, $p < 0.0001$; linearcumulative: $\chi^2(2) = 26.4$, $P < 0.0001$; Table S9, Fig. S3 C, D).

Viability models detected spatially variable nonlinear selection on water-use efficiency (WUE) as measured through $\delta^{13}C$ ($\chi^2(1) = 22.9$, $P < 0.0001$, Table S10), with positive quadratic selection in the lower elevation garden favoring greater WUE values than expressed in local families ($\chi^2(1) = 7.3$, $P = 0.012$, Fig. S3E) and stabilizing selection for WUE values similar to those of local families in the higher elevation garden ($\chi^2(1) = 3.94$, $P = 0.047$, Fig. S3I). This pattern was counterbalanced by fecundity selection for reduced WUE in the high elevation garden ($\chi^2(1) = 76.4$, $P < 0.0001$, Fig. S3I), but there was no evidence for

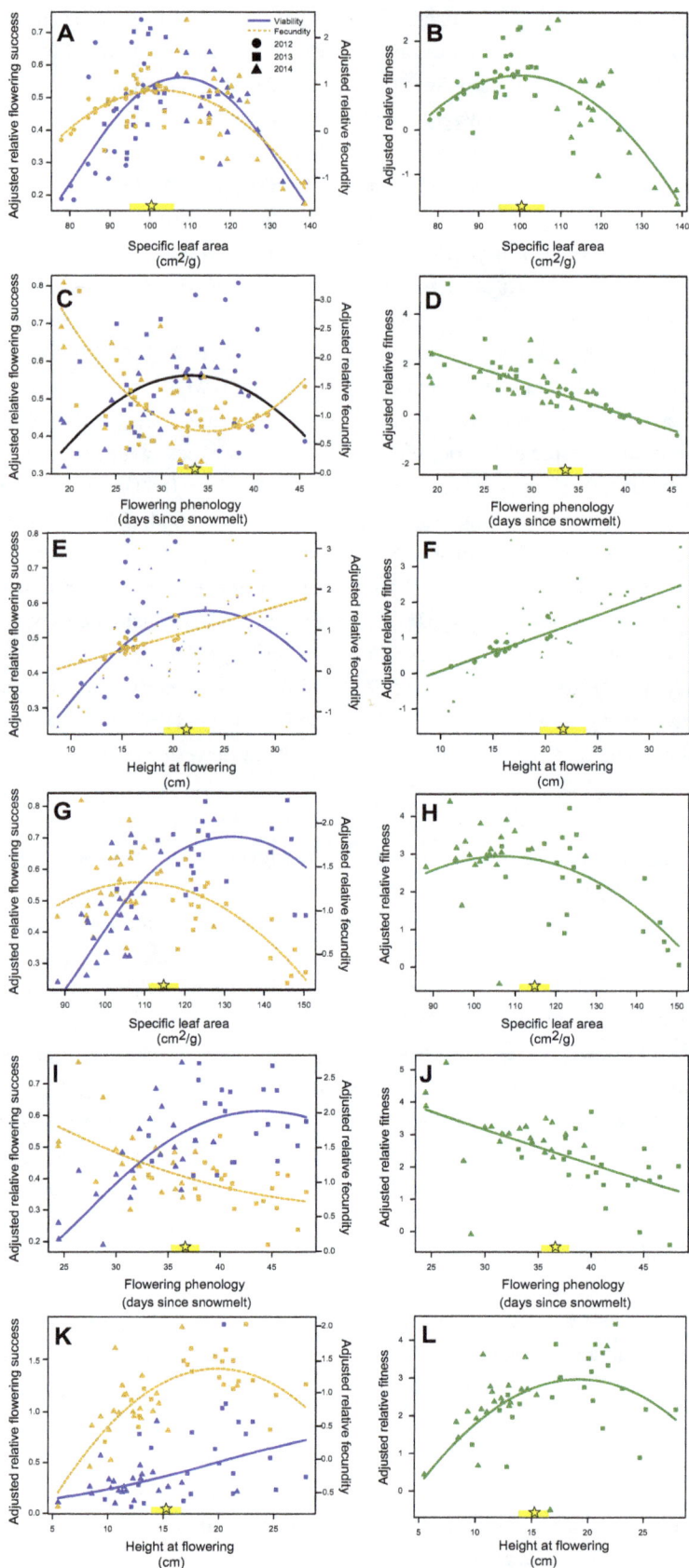

Figure 2. Patterns of viability and fecundity selection (left column) and the resulting cumulative selection (right column) on (A, B) SLA in the 2011 cohort, (C, D) flowering phenology in the 2011 cohort, (E, F) height at flowering in the 2011 cohort, (G, H) SLA in the 2012

fecundity selection in the lower garden. Models detected cumulative stabilizing selection on WUE favoring local trait values in the higher elevation garden ($\chi^2(1) = 16.2$, $P < 0.0001$, Fig. S3J), but no evidence for cumulative selection on WUE in the lower garden.

As with the 2011 cohort, multivariate models revealed viability selection acting on adult traits. Viability selection favored delayed flowering time in the higher elevation garden ($\chi^2(1) = 7.91$, $P = 0.005$) while other fitness components displayed the more customary pattern of selection for earlier flowering with moderate curvilinearity (fecundity: $\chi^2(1) = 36.5$, $P = 0.0001$; cumulative: $\chi^2(1) = 7.31$, $P = 0.007$; Fig. 2I, J). In the lower elevation garden, selection was congruent across fitness components, favoring earlier flowering phenology at all life stages from viability (slight curvilinearity; $\chi^2(1) = 7.58$, $P = 0.006$) to fecundity (slight curvilinearity $\chi^2(1) = 45.2$, $P < 0.0001$) and cumulative fitness (linear relationship: $\chi^2(2) = 15.1$, $P < 0.001$, Fig. 3C, Fig. S3G, H). However, the viability results should be treated as tentative as SAS reported viability selection for delayed flowering even in the lower garden ($\beta = 0.44 \pm 0.16$, $t_{58} = -2.7$, $P = 0.009$; Table S9).

Height at flowering was also subject to spatially variable linear selection ($\chi^2(1) = 52.6$, $P < 0.0001$), as viability selection favored taller plants at flowering in the higher elevation garden (Fig. 2K), but did not operate on this trait in the lower garden (Fig. S3K, Table S9). Stabilizing fecundity selection for greater size at flowering differed in magnitude across the two sites ($\chi^2(1) = 63.8$, $P < 0.0001$), being stronger in the higher elevation garden (Fig. 2K) than the lower garden (Fig. S3K). Cumulative selection also varied spatially ($\chi^2(1) = 32.1$, $P < 0.0001$), with strong stabilizing selection in the high garden (Fig. 2L), but only weak linear selection in the lower garden (Fig. S3L).

For both cohorts, separate analyses of each year revealed weaker patterns of natural selection with less evidence for stabilizing selection than models of datasets combined across years (Fig. 2, Fig. S4, Table S12). Furthermore, in the few cases where we detected stabilizing selection in separate analyses of each growing season, local trait values lie farther away from the optima than in integrated analyses, as seen for cumulative selection on SLA (Fig. 2, Fig. S4).

Discussion

Natural selection drives adaptive evolution; however, we have yet to reconcile the stability and extent of clinal trait variation with patterns of selection. In our study, *Boechera stricta* exhibited significant genetically based clines in ecophysiological, morphological, and phenological traits that are likely maintained by selection. In all cases, the direction of genetic clines and spatial plasticity was concordant with a priori expectations based on trait variation in natural populations (Anderson and Gezon 2015), and selection often favored trait values expressed by local families. Genetically based clines likely evolved in response to long-standing spatial variation in selection across steep elevational gradients. Here, we discuss three major findings that advance our ability to characterize adaptation in heterogeneous environments: Genetically based clines are context dependent, temporally fluctuating selection can favor intermediate phenotypic optima, and viability selection prior to trait expression can influence the evolution of adult phenotypes.

CONTEXT DEPENDENCY OF GENETICALLY BASED CLINES

Changing environmental conditions can impose novel selection, altering the shape of genetic clines in ecologically relevant traits (e.g., Brakefield and de Jong 2011). In our study, genetic clines occurred in the same direction relative to source elevation in both gardens and cohorts, but some clines only emerged in years with limited snowpack, early snowmelt, and dry growing seasons (integrated water-use efficiency, $\delta^{13}C$). For other traits (specific leaf area and height at flowering), genetic clines only became apparent in more benign years. Laboratory experiments may not manipulate the agents of selection that shape or maintain clinal trait variation, and the magnitude of selection can differ between lab and field environments (Kellermann et al. 2015). Thus, clinal variation may not manifest under benign controlled conditions or short-term field studies. This context-dependency in genetic clines and trait expression indicates that multiyear and multisite investigations in natural settings are critical for characterizing natural selection and adaptation.

Adaptive context of genetically based clines

Life-history information and data from common gardens can be useful for discerning the adaptive context of genetic clines.

cohort from the high-elevation garden, (I, J) flowering phenology in the 2012 cohort from the high-elevation garden (2012 cohort), and (K, L) height at flowering in the 2012 cohort from high-elevation garden. These analyses include data from all years of the study. Y-axes display adjusted fitness values, which were statistically corrected for other variables included in the models by adding residuals from full models to predicted fitness values estimated from regression coefficients for the trait of interest. We depict fitness curves using linear and (undoubled) quadratic regression coefficients from multivariate models, but we doubled quadratic regression coefficients to calculate nonlinear selection gradients in Tables S8 and S10 (Stinchcombe et al. 2008). Mean trait values for local genotypes are displayed with a yellow star and are bracketed in yellow by 2 x SE. Local values were extracted from models of genetically based clines (Figs. 1 and S1) for families with a source elevation of 2891 m (the elevation of the lower garden) or 3133 m (the elevation of the higher garden).

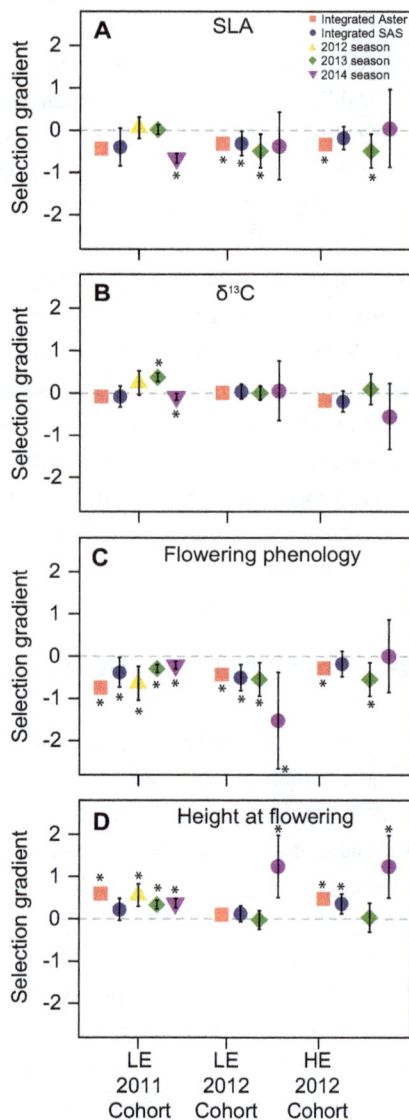

Figure 3. Direct linear selection gradients from cumulative selection analyses in the 2011 and 2012 cohorts in the low elevation (LE) and high elevation (HE) gardens for (A) SLA, (B) carbon isotope discrimination (integrated water-use efficiency, WUE), (C) flowering phenology, and (D) height at flowering. We show the direction and magnitude of selection from analyses integrated across years (Aster and SAS models) as well as from separate analyses for each season of study (SAS models). Estimating standard errors from Aster models is not straightforward. Thus, stars for integrated Aster models indicate significant selection as revealed by log likelihood ratio tests (Table S7–S12).

Congruent with expectations from studies of other speices along elevational and latitudinal gradients (Monty and Mahy 2008; Kawakami et al. 2011; Woods et al. 2012, but see Stinchcombe et al. 2004), we found consistent genetic clines in phenological traits, indicating that families from low elevation populations consistently flower later and at larger sizes than high elevation families. Tradeoffs between age and size at flowering likely constrain the joint evolution of these traits (e.g., Anderson et al. 2011).

In concert, these genetic clines suggest evolutionary responses to short growing seasons in populations inhabiting cooler poleward latitudes and upslope elevations.

The warmer and drier conditions at low elevation sites can favor traits that improve drought-resistance. In *Boechera stricta*, steep (in 2012) to nonexistant (in 2014) genetic clines in $\delta^{13}C$ likely reflect differential water-use efficiency (WUE) across the elevational range in this region, as is true in other systems in which aridity declines with elevation (e.g., Lajtha and Getz 1993; Van de Water et al. 2002; Reed and Loik 2016). Had $\delta^{13}C$ clines been driven primarily by partial pressure of CO_2, we would have expected the clines to be stable across space and time. Our data suggest that low-elevation genotypes of *B. stricta* exhibit high WUE, as is the case in the model plant *A. thaliana* (Wolfe and Tonsor 2014). Furthermore, stabilizing selection favored relatively low water-use efficiency in the high elevation garden. Specific leaf area (SLA) increased with source elevation in our common gardens (this study) and across elevation in natural populations (Anderson and Gezon 2015). Although genetic clines in SLA were not robust to correction for neutral population genetic structure, selection favored local trait values in both gardens. This discrepancy indicates that the adaptive nature of clinal trait variation may only become evident after analysis of selection at relevant life-history stages. Our data also suggest that high elevation *B. stricta* families invest in rapid resource acquisition because of short growing seasons, while low elevation families have evolved drought tolerance (this study) and resistance to insect herbivory (Anderson et al. 2015) because of extended dry periods and abundant herbivores. These results highlight the potential for complex suites of conditions to contribute to trait clines and selection regimes across elevational gradients.

FLUCTUATING SELECTION FAVORS INTERMEDIATE TRAIT VALUES

In fluctuating environments, the evolutionary trajectory of a population will depend on the cumulative effects of variable selection regimes (Bell 2010). Our study was replicated for just three years, and yet we still detected temporally fluctuating selection that culminated in strong patterns of stabilizing selection on both juvenile and adult traits. In contrast, analyses considering each year of data separately indicated that patterns of selection were generally weak and variable. Studies that neglect or underestimate fluctuating selection can falsely conclude that traits are not targeted by selection or erroneously inflate expected evolutionary responses to selection. Our data demonstrate that examining natural selection over the course of a single growing season may uncover patterns and generate predictions that are at odds with those derived from multiyear studies. By conducting our study over multiple growing seasons, we were able to quantify how temporal environmental variation shifts phenotypic distributions and components of fitness, and how these shifts alter the selective landscape.

The literature provides mixed evidence for temporal variability in directional selection. Selection on beak size and shape oscillated over three decades of observations in Darwin's finches (Grant and Grant 2002), but comparatively long-term datasets are exceedingly rare. In a review of 5519 estimates of selection from 89 temporally replicated studies, only 25% of linear selection gradients fluctuated in sign over time (Siepielski et al. 2009). Furthermore, temporal fluctuations were primarily found for traits that experienced relatively weak selection (Kingsolver et al. 2012), with sampling error inflating heterogeneity in most estimates of selection (Morrissey and Hadfield 2012). While current empirical evidence suggests that temporal variation in selection may not be ubiquitous, it undoubtedly maintains phenotypic and genetic variation in some systems. Considering data from only one season in a study of selection can obscure the insights gained from a more integrated approach (McGlothlin 2010).

VIABILITY SELECTION AND THE INVISIBLE FRACTION

Mortality events can present considerable challenges when characterizing natural selection (Hadfield 2008; Nakagawa and Freckleton 2008). This mortality-generated missing data, known as the "invisible fraction," can be nonrandom with respect to ontogeny, habitat of origin, experimental treatments, and size classes (Bennington and McGraw 1995; Sinervo and McAdam 2008; Engen et al. 2012). However, most studies of natural selection do not account for viability selection, either because the family structure of focal individuals is unknown or because early survival and trait data are unavailable (but see Mojica and Kelly 2010; Tarwater and Beissinger 2013).

By monitoring survival to reproduction, we determined that viability selection targeted flowering onset and size at flowering. Furthermore, viability selection can both enforce and oppose fecundity selection. For instance, for the 2012 cohort in the high elevation garden, viability selection favored delayed flowering, and larger plant size while fecundity selection favored earlier flowering and an intermediate plant size. The onset of flowering often coincides with a shift from resource investment in vegetative growth toward reproduction. The potential fitness tradeoff between flowering early for reproductive assurance and flowering at a larger size for increased resource acquisition has been studied in relation to fruit or seed production (Bolmgren and Cowan 2008). Our results suggest that viability selection can exacerbate or alleviate this tradeoff, depending on the environmental context.

The evolution of reproductive phenology appears to represent a compromise between the opposing forces of viability and fecundity selection (Tarwater and Beissinger 2013). We found fecundity selection for early flowering, which is consistent with

a large-scale meta-analysis of phenotypic selection on flowering phenology across systems (Munguía-Rosas et al. 2011). Nevertheless, local trait values are much closer to the optimum from viability than fecundity selection for the 2011 cohort (Fig. 2C) and lie near the intersection of viability and fecundity selection gradients in the high elevation environment (Fig. 2I). Therefore, neglecting to account for viability selection could lead to inaccurate conclusions that populations are maladapted with trait values that fall far from fitness optima. When the action of selection via different fitness components is not additive or reinforcing, reconciling patterns of natural selection with evidence of adaptation will require evaluating selection at multiple life-history stages (Mojica and Kelly 2010; Tarwater and Beissinger 2013).

Conclusions

Our results illustrate that accounting for variation in trait expression and fitness across space, time, and ontogeny can improve characterizations of genetically based clines and natural selection. In our system, adaptive genetically based clines were stable across cohorts and gardens, but they were not apparent in all years. This context-dependency indicates that temporal variation in conditions can obscure genetic clines, which may limit the ability of short-term field studies, or experiments conducted under laboratory conditions, to detect important clines in ecologically relevant traits. Furthermore, stabilizing selection was most prominent after integrating fitness and trait data across years, suggesting that data derived from single-year studies may not capture cumulative patterns of selection. Finally, viability selection analyses revealed that failing to account for the "invisible fraction" could lead to an incomplete view of selection acting on both juvenile and adult traits.

AUTHOR CONTRIBUTIONS
J.A. designed research, J.A. and S.C.D. conducted fieldwork, J.A. and S.W. analyzed data, S.W. and J.A. wrote paper and S.C.D. edited.

ACKNOWLEDGMENTS
We thank N. Lowell, C. Way, A. Battiata, and B. Chowdhury for help with data collection, and K. Springer and T. Mitchell-Olds for rearing plants in the Duke greenhouses. We thank A. Weis, M. Johnson, T. Pendergast, E. Austen, R. Mactavish, N. Workman, H. R. Dawson, and three anonymous reviewers for comments on a previous draft and T. Mitchell-Olds for statistical advice. This project was funded by the Universities of Georgia and South Carolina and the National Science Foundation (DEB #1553408).

LITERATURE CITED
Ågren, J., C. G. Oakley, S. Lundemo, and D. W. Schemske. 2017. Adaptive divergence in flowering time among natural populations of *Arabidopsis*

thaliana: estimates of selection and QTL mapping. Evolution 71:550–564

Al-Shehbaz, I. A., and M. D. Windham. 2010. *Boechera*. Pp. 348–412 *in* FoN. A. E., Committee, ed. Floral of North America North of Mexico. Oxford Univ. Press, New York and Oxford.

Anderson, J., and Z. Gezon. 2015. Plasticity in functional traits in the context of climate change: a case study of the subalpine forb *Boechera stricta* (Brassicaceae). Global Change Biol. 21:1689–1703.

Anderson, J., D. Inouye, A. McKinney, R. Colautti, and T. Mitchell-Olds. 2012. Phenotypic plasticity and adaptive evolution contribute to advancing flowering phenology in response to climate change. Proc. R Soc. B 279:3843–3852.

Anderson, J., N. Perera, B. Chowdhury, and T. Mitchell-Olds. 2015. Microgeographic patterns of genetic divergence and adaptation across environmental gradients in *Boechera stricta* (Brassicaceae). Am. Nat. 186:S60–S73.

Anderson, J. T., C. R. Lee, and T. Mitchell-Olds. 2011. Life-history QTLs and natural selection on flowering time in *Boechera stricta*, a perennial relative of *Arabidopsis*. Evolution 65:771–787.

Arnold, S. J., R. Burger, P. A. Hohenlohe, B. C. Ajie, and A. G. Jones. 2008. Understanding the evolution and stability of the G-matrix. Evolution 62:2451–2461.

Bell, G. 2010. Fluctuating selection: the perpetual renewal of adaptation in variable environments. Philos. Trans. R. Soc. Lond. Ser. B Biol. Sci. 365:87–97.

Benjamini, Y., and Y. Hochberg. 1995. Controlling the false discovery rate: a practical and powerful approach to multiple testing. J. R Stat. Soc. 57:289–300.

Bennington, C. C., and J. B. McGraw. 1995. Phenotypic selection in an artificial population of impatiens pallida: the importance of the invisible fraction. Evolution 49:317–324.

Blanquart, F., O. Kaltz, S. L. Nuismer, and S. Gandon. 2013. A practical guide to measuring local adaptation. Ecol. Lett. 16:1195–1205.

Bolmgren, K., and P. Cowan. 2008. Time—size tradeoffs: a phylogenetic comparative study of flowering time, plant height and seed mass in a north-temperate flora. Oikos 117:424–429.

Brakefield, P. M., and P. W. de Jong. 2011. A steep cline in ladybird melanism has decayed over 25 years: a genetic response to climate change? Heredity 107:574–578.

Byars, S., W. Papst, and A. A. Hoffmann. 2007. Local adaptation and cogradient selection in the alpine plant, *Poa hiemata*, along a narrow altitudinal gradient. Evolution 61:2925–2941.

Calsbeek, R., L. Bonvini, and R. M. Cox. 2010. Geographic variation, frequency-dependent selection, and the maintenance of a female-limited polymorphism. Evolution 64:116–125.

Campbell, D. R., C. A. Wu, and S. Travers. 2010. Photosynthetic and growth response of reciprocal hybrids to variation in water and Nitrogen availability. Am. J. Bot. 97:925–933.

Colautti, R. I. and S. C. H. Barrett. 2013. Rapid adaptation to climate facilitates range expansion of an invasive plant. Science 342:364.

Dunne, J. A., J. Harte, and K. J. Taylor. 2003. Subalpine meadow flowering phenology responses to climate change: integrating experimental and gradient methods. Ecol. Monogr. 73:69–86.

Engen, S., B. E. SÆTher, T. Kvalnes, and H. Jensen. 2012. Estimating fluctuating selection in age-structured populations. J. Evol. Biol. 25:1487–1499.

Estes, S., and S. J. Arnold. 2007. Resolving the paradox of stasis: models with stabilizing selection explain evolutionary divergence on all timescales. Am. Nat. 169:227–244.

Farquhar, G. D., J. R. Ehleringer, and K. T. Hubick. 1989. Carbon isotope discrimination and photosynthesis. Annu. Rev. Plant Physiol. Plant Mol. Biol. 40:503–537.

Galecki, A. T. 1994. General class of covariance structures for two or more repeated factors in longitudinal data analysis. Comm. Stat. 23:3105–3119.

Geyer, C. J., S. Wagenius, and R. G. Shaw. 2007. Aster models for life history analysis. Biometrika 94:415–426.

Gilbert, S. F., T. C. G. Bosch, and C. Ledon-Rettig. 2015. Eco-evo-devo: developmental symbiosis and developmental plasticity as evolutionary agents. Nat. Rev. Genet. 16:611–622.

Goudet, J. and L. Buchi. 2006. The effects of dominance, regular inbreeding and sampling design on QST, an estimator of population differentiation for quantitative traits. Genetics 172:1337–1347.

Grant, P. R. and R. Grant. 2002. Unpredictable evolution in a 30-year study of Darwin's Finches. Science 296:707–711.

Hadfield, J. D. 2008. Estimating evolutionary parameters when viability selection is operating. Proc. R. Soc. Lond., Ser. B: Biol. Sci. 275:723–734.

Hall, M. C., and J. H. Willis. 2006. Divergent selection on flowering time contributes to local adaptation in *Mimulus guttatus* populations. Evolution 60:2466–2477.

Haller, B. C., and A. P. Hendry. 2014. Solving the paradox of stasis: squashed stabilizing selection and the limits of detection. Evolution 68:483–500.

Harte, J., S. R. Saleska, and C. Levy. 2015. Convergent ecosystem responses to 23-year ambient and manipulated warming link advancing snowmelt and shrub encroachment to transient and long-term climate—soil carbon feedback. Global Change Biol. 21:2349–2356.

Janzen, F. J., and H. S. Stern. 1998. Logistic Regression for empirical studies of multivariate selection. Evolution 52:1564–1571.

Kawakami, T., T. J. Morgan, J. B. Nippert, T. W. Ocheltree, R. Keith, P. Dhakal, and M.C. Ungerer. 2011. Natural selection drives clinal life history patterns in the perennial sunflower species, *Helianthus maximiliani*. Mol. Ecol. 20:2318–2328.

Kellermann, V., A. A. Hoffmann, T. N. Kristensen, N. N. Moghadam, and V. Loeschcke. 2015. Experimental evolution under fluctuating thermal conditions does not reproduce patterns of adaptive clinal differentiation in *Drosophila melanogaster*. Am. Nat. 186:582–593.

Kim, E., and K. Donohue. 2013. Local adaptation and plasticity of *Erysimum capitatum* to altitude: its implications for responses to climate change. J. Ecol. 101:796–805.

Kingsolver, J., and S. E. Diamond. 2011. Phenotypic selection in natural populations: what limits directional selection? Am. Nat. 177:346–357.

Kingsolver, J., S. E. Diamond, A. M. Siepielski, and S. M. Carlson. 2012. Synthetic analyses of phenotypic selection in natural populations: lessons, limitations and future directions. Evol. Ecol. 26:1101–1118.

Kingsolver, J., H. Hoekstra, J. Hoekstra, D. Berrigan, S. Vignieri, C. Hill, et al. 2001. The strength of phenotypic selection in natural populations. Am. Nat. 157:245–261.

Kocher, T. D. 2004. Adaptive evolution and explosive speciation: the cichlid fish model. Nat. Rev. Genet. 5:288–298.

Kooyers, N. J., L. R. Gage, A. Al-Lozi, and K. M. Olsen. 2014. Aridity shapes cyanogenesis cline evolution in white clover (*Trifolium repens* L.). Mol. Ecol. 23:1053–1070.

Körner, C. 2007. The use of altitude in ecological research. Trends Ecol. Evol. 22:569–574.

Lajtha, K., and J. Getz. 1993. Photosynthesis and water-use efficiency in pinyon-juniper communities along an elevation gradient in northern New Mexico. Oecologia 94:95–101.

Lande, R., and S. J. Arnold. 1983. The measurement of selection on correlated characters. Evolution 37:1210–1226.

Leonardi, S., T. Gentilesca, R. Guerrieri, F. Ripullone, F. Magnani, M. Mendcuccini, et al. 2012. Assessing the effects of nitrogen deposition and climate on carbon isotope discrimination and intrinsic water-use efficiency of angiosperm and conifer trees under rising CO_2 conditions. Global Change Biol. 18:2925–2944.

McGlothlin, J. W. 2010. Combining selective episodes to estimate lifeitme nonlinear selection. Evolution 64:1377–1385.

Mojica, J. P., and J. K. Kelly. 2010. Viability selection prior to trait expression is an essential component of natural selection. Proc. R. Soc. Lond. Ser. B Biol. Sci. 277:2945–2950.

Montesinos-Navarro, A., J. Wig, F. X. Pico, and S. J. Tonsor. 2011. *Arabidopsis thaliana* populations show clinal variation in a climatic gradient associated with altitude. New Phytol. 189:282–294.

Monty, A., and G. Mahy. 2008. Clinal differentiation during invasion: *Senecio inaequidens* (Asteraceae) along altitudinal gradients in Europe. Oecologia 159:305–315.

Morrissey, M. B., and J. D. Hadfield. 2012. Directional selection in temporally replicated studies is remarkably consistent. Evolution 66:435–442.

Mousseau, T. A., and D. A. Roff. 1987. Natural selection and the heritability of fitness components. Heredity 59:181–197.

Munguía-Rosas, M., J. Ollerton, V. Parra-Tabla, and J. De-Nova. 2011. Meta-analysis of phenotypic selection on flowering phenology suggests that early flowering plants are favoured. Ecol. Lett. 14:511–521.

Nakagawa, S., and R. P. Freckleton. 2008. Missing inaction: the dangers of ignoring missing data. Trends Ecol. Evol. 23:592–596.

Paaby, A. B., and M. V. Rockman. 2014. Cryptic genetic variation: evolution's hidden substrate. Nat. Rev. Genet. 15:247–258.

Phillips, P. C., and S. J. Arnold. 1989. Visualizing multivariate selection. Evolution 43:1209–1222.

Pratt, J. D., and K. A. Mooney. 2013. Clinal adaptation and adaptive plasticity in *Artemisia californica*: implications for the response of a foundation species to predicted climate change. Global Change Biol. 19:2454–2466.

Prentis, P. J., J. R. U. Wilson, E. E. Dormontt, D. M. Richardson, and A. J. Lowe 2008. Adaptive evolution in invasive species. Trends Plant Sci. 13:288–294.

Rausher, M. D. 1992. The measurement of selection on quantitative traits: biases due to environmental covariances between traits and fitness. Evolution 46:616–626.

Read, Q. D., L. C. Moorhead, N. G. Swenson, J. K. Bailey, and N. J. Sanders. 2014. Convergent effects of elevation on functional leaf traits within and among species. Funct. Ecol. 28:37–45.

Reed, C. C. and M. E. Loik. 2016. Water relations and photosynthesis along an elevation gradient for *Artemisia tridentata* during an historic drought. Oecologia 181:65–76.

Roughgarden, J. 1979. Theory of population genetics and evolutionary ecology: an introduction. Macmillan, New York.

Shaw, R., C. Geyer, S. Wagenius, H. Hangelbroek, and J. Etterson. 2008. Unifying life history analyses for inference of fitness and population growth. Am. Nat. 172:E35–E47.

Siepielski, A. M., J. D. DiBattista, and S. M. Carlson 2009. It's about time: the temporal dynamics of phenotypic selection in the wild. Ecol. Lett. 12:1261–1276.

Siepielski, A. M., K. M. Gotanda, M. B. Morrissey, S. E. Diamond, J. D. DiBattista, and S. Carlson. 2013. The spatial patterns of directional phenotypic selection. Ecol. Lett. 16:1382–1392.

Sinervo, B. and A. G. McAdam. 2008. Maturational costs of reproduction due to clutch size and ontogenetic conflict as revealed in the invisible fraction. Proc. R Soc. Lond. B Biol. Sci. 275:629–638.

Song, B. H., M. Clauss, A. Pepper, and T. Mitchell-Olds. 2006. Geographic patterns of microsatellite variation in *Boechera stricta*, a close relative of *Arabidopsis*. Mol. Ecol. 15:357–369.

Stinchcombe, J. R., A. F. Agrawal, P. A. Hohenlohe, S. J. Arnold, and M. W. Blows. 2008. Estimating nonlinear selection gradients using quadratic regression coefficients: double or nothing? Evolution 62:2435–2440.

Stinchcombe, J. R., C. Weinig, M. C. Ungerer, K. M. Olsen, C. Mays, S. S. Halldorsdottir, et al. 2004. A latitudinal cline in flowering time in *Arabidopsis thaliana* modulated by the flowering time gene *FRIGIDA*. Proc. Natl. Acad. Sci. USA 101:4712–4717.

Tarwater, C. E., and S. R. Beissinger. 2013. Opposing selection and environmental variation modify optimal timing of breeding. Proc. Natl. Acad. Sci. 110:15365–15370.

Thurman, T. J., and R. D. H. Barrett. 2016. The genetic consequences of selection in natural populations. Mol. Ecol. 25:1429–1448.

Van de Water, P. K., S. W. Leavitt, and J. L. Betancourt. 2002. Leaf $\delta^{13}C$ variability with elevation, slope aspect, and precipitation in the southwest United States. Oecologia 132:332–343.

Vasemägi, A. 2006. The adaptive hypothesis of clinal variation revisited: single-locus clines as a result of spatially restricted gene flow. Genetics 173:2411–2414.

Ward, J., D. S. Roy, I. Chatterjee, C. R. Bone, C. Springer, and J. K. Kelly. 2012. Identification of a major QTL that alters flowering time at elevated $[CO_2]$ in *Arabidopsiss thaliana*. PLoS ONE 7:e49028.

Wolfe, M. D., and S. J. Tonsor. 2014. Adaptation to spring heat and drought in northeastern Spanish *Arabidopsis thaliana*. New Phytol. 201:323–334.

Woods, E. C., A. P. Hastings, N. E. Turley, S. B. Heard, and A. A. Agrawal. 2012. Adaptive geographical clines in the growth and defense of a native plant. Ecol. Monogr. 82:149–168.

Linking speciation to extinction: Diversification raises contemporary extinction risk in amphibians

Dan A. Greenberg[1,2,3,4] and Arne Ø. Mooers[1,3]

[1]Department of Biological Sciences, Simon Fraser University, Burnaby, British Columbia V5A 1S6, Canada
[2]Earth-to-Ocean Research Group, Simon Fraser University, Burnaby, British Columbia V5A 1S6, Canada
[3]Crawford Lab for Evolutionary Studies, Simon Fraser University, Burnaby, British Columbia V5A 1S6, Canada

[4]E-mail: dan.greenberg01@gmail.com

Many of the traits associated with elevated rates of speciation, including niche specialization and having small and isolated populations, are similarly linked with an elevated risk of extinction. This suggests that rapidly speciating lineages may also be more extinction prone. Empirical tests of a speciation-extinction correlation are rare because assessing paleontological extinction rates is difficult. However, the modern biodiversity crisis allows us to observe patterns of extinction in real time, and if this hypothesis is true then we would expect young clades that have recently diversified to have high contemporary extinction risk. Here, we examine evolutionary patterns of modern extinction risk across over 300 genera within one of the most threatened vertebrate classes, the Amphibia. Consistent with predictions, rapidly diversifying amphibian clades also had a greater share of threatened species. Curiously, this pattern is not reflected in other tetrapod classes and may reflect a greater propensity to speciate through peripheral isolation in amphibians, which is partly supported by a negative correlation between diversification rate and mean geographic range size. This clustered threat in rapidly diversifying amphibian genera means that protecting a small number of species can achieve large gains in preserving amphibian phylogenetic diversity. Nonindependence between speciation and extinction rates has many consequences for patterns of biodiversity and how we may choose to conserve it.

KEY WORDS: Amphibia, diversification, extinction risk, IUCN, peripatry, speciation rate, species longevity.

Impact Summary

The rates of speciation and extinction dictate the frequency at which new species arise and are lost over evolutionary time. Characteristics of species that may promote speciation include being highly specialized to particular environments, existing in isolated populations, or having a low population abundance. These same traits are also associated with extinction: specialized species are vulnerable to environmental change, species that exist in isolated pockets lack population connectivity, and small populations can blink out rapidly. This suggests that lineages speciating readily due to these traits may also readily lose species. Assessing whether speciation and extinction rates are correlated is difficult, as measuring extinction based on fossils can be biased for many groups. However, we are currently in the midst of observing numerous extinctions in real time, and observing variation in the species currently at risk of extinction may serve as a proxy measure for extinction rate across groups. In this study, we show in amphibians that lineages that have high ongoing diversification also have a greater share of species threatened with extinction compared to slowly diversifying groups. This supports the idea that speciation and extinction may go hand-in-hand. Comparing this pattern in amphibians to other clades reveals a surprising discrepancy: only plants have been found to show a similar pattern. One mechanism that may produce this link between speciation and extinction could be the mode of speciation—new species arising from isolated populations may be highly specialized, range-restricted, and vulnerable to extinction. In the grand scheme for amphibian conservation, evolutionarily distinct species are less at risk of extinction—and therefore preserving the amphibian tree of life can be achieved with modest conservation goals. If speciation and extinction rise (and fall) in tandem, this might suggest that lineages may fall along a continuum of producing few, long-lived species, or many short-lived species. Linking speciation rates and extinction rates to each other, and to particular modes of speciation, would be an important advance in our understanding of how life on earth diversifies.

Introduction

The evolutionary rates of speciation and extinction, their difference being diversification rate, shape current patterns of diversity across the tree of life. Standing diversity varies considerably across clades, consistent with lineage-specific aspects of biology influencing speciation rates, extinction rates, or both (Jablonski 2008). Some biological characteristics that may increase speciation rates include poor dispersal capability (Claramunt et al. 2012), specialization and narrow niche breadths (Rolland and Salamin 2016), large body size (Liow et al. 2008; Monroe and Bokma 2009), or persistence at low population size (Stanley 1990). In turn, these characteristics are also predicted to increase risk of extinction: poor dispersers have limited abilities to (re)colonize or move to suitable environments (Smith and Green 2005; Sandel et al. 2011), specialists are vulnerable to environmental change (McKinney 1997; Colles et al. 2009), large-bodied species typically have slow life histories (Cardillo et al. 2005; Reynolds et al. 2005), and small populations are subject to demographic stochasticity or extinction from local catastrophies (Lande et al. 2003; Mace et al. 2008). If similar traits drive both speciation and extinction rates, then these rates may be positively correlated across lineages.

Support for a positive speciation-extinction correlation has remained elusive, in part due to the difficulty of estimating either rate. There is some evidence for a positive speciation-extinction relationship from the paleontological record in certain groups (Stanley 1990), but for many clades the fossil record is poor. Under certain assumptions, it is possible to estimate speciation and extinction rates separately from phylogenies of extant lineages (Nee et al. 1994), but resultant extinction rates tend to be sorely underestimated (Rabosky 2010). However, we are currently in an era of unprecedented extinction and this unfortunate state of affairs may allow us to directly compare rates of extinction across clades as biodiversity losses accelerate. For certain taxa, clades that seem to have speciated both rapidly and recently have in turn a greater share of currently rare and threatened species (Schwartz and Simberloff 2001; Lozano and Schwartz

2005; Davies et al. 2011), consistent with the expectation under a general speciation-extinction relationship and suggesting that modern patterns of extinction may serve as a viable surrogate.

Contemporary rates of extinctions are estimated to be magnitudes greater than paleontological rates due to human activities (Pimm et al. 1995; Ceballos et al. 2015). Importantly, although certain drivers of extinction are different in the modern context (Harnik et al. 2012a; Condamine et al. 2013), the same traits associated with modern extinctions have also been linked with species' lifespan and mass extinctions in the fossil record (McKinney 1997). For instance, geographic range size dominates patterns of modern extinction risk across terrestrial vertebrates (Cardillo et al. 2005; Sodhi et al. 2008; Lee and Jetz 2011; Böhm et al. 2016), and similarly is one of the best predictors of species longevity in the fossil record (Kiessling and Aberhan 2007; Harnik et al. 2012b; Orzechowski et al. 2015; Smits 2015). Specialization has been linked to both modern extinction risk and to species durations in terms of both dietary breadth (Boyles and Storm 2007; Olden et al. 2008; Smits 2015) and habitat/environment breadth (Heim and Peters 2011; Harnik et al. 2012b; Ducatez et al. 2014). Both abundance and body size affect modern extinction risks across taxa (Cardillo et al. 2005; Reynolds et al. 2005; Mace et al. 2008). Fossil evidence also suggests that abundance can dictate the longevity of species (Kiessling and Aberhan 2007), and that large-bodied species often have higher background and mass extinction rates (Liow et al. 2008; Sallan and Galimberti 2015; but see Smits 2015). If these traits drive both ancient and modern extinctions, and tend to be conserved within lineages over time, then we may expect that extant clades with high contemporary extinction risk should also have high extinction rates over their history. Temporal changes in threats may shift the traits underlying extinction risk (Bromham et al. 2012; Lyons et al. 2016), but many of these traits appear general enough to create consistent long-term differences in extinction risk (Harnik et al. 2012b; Finnegan et al. 2015; Orzechowski et al. 2015; Smits 2015). Though this concept has yet to be thoroughly tested, emerging evidence suggests that lineages suffering high contemporary extinction risk similarly had high rates of extinction in the fossil record (McKinney 1997; Condamine et al. 2013; Finnegan et al. 2015). Examining modern extinctions may therefore offer an accelerated view of the same patterns that structure paleontological extinction rates across clades.

Net diversification rates are easier to estimate than independent speciation or extinction rates, but diversification is biased towards speciation rates for more recent groups such as genera because extinction must lag speciation (Nee et al. 1994). Therefore diversification rates in extant lineages are often reflective of speciation rates, as is typically inferred through analyses of molecular phylogenies of extant taxa (Rabosky 2010).

If contemporary patterns of extinction reflect paleontological rates, and if diversification rates tend to reflect speciation, then, under the hypothesis of covarying speciation and extinction rates, modern rates of extinction should be positively correlated with diversification rates across young clades. Alternatively, if extinction and speciation are independent then one would expect no correlation between modern rates of extinction and clade diversification rates. Here, we test this hypothesis using patterns of diversification and extinction across 329 genera of Amphibia, a vertebrate group with one of the highest rates of modern extinction (Hoffmann et al. 2010).

Methods

TAXONOMIC AND PHYLOGENETIC DATA

We identified amphibian genera that had both phylogenetic and threat status data available that would allow separate estimates of diversification rate and contemporary extinction risk ($N = 329$ genera). We delineated genera based on the taxonomy from the Amphibian Species of the World database v6.0 (Frost 2016) and included all monophyletic clades that (i) had at least one species assessed for threat status by the International Union for the Conservation of Nature (IUCN) Red List (IUCN 2016), (ii) that had both crown and stem group ages, and (iii) that had more than two representatives on the phylogeny for non-mono/ditypic genera (to mitigate against underestimating crown ages). For each genus we compiled data on extant species richness, and both crown and stem group age. Extant species richness (n) was assessed based on species counts in the Amphibian Species of the World database. Crown and stem group ages (in millions of years) were estimated from one of the most extensive published, time-calibrated phylogenies for amphibians (Pyron 2014). Net diversification rates can be estimated either by crown or stem ages (Magallon and Sanderson 2001). Both estimators have their drawbacks: crown ages exclude monotypic genera, and stem ages are shared between pairs of lineages. We therefore considered both stem and crown diversification-rates using the method-of-moments estimator (Magallon and Sanderson 2001).

EXTINCTION RISK

To characterize the contemporary extinction rate for each clade, we assessed the proportion of species in each genus that are currently threatened with extinction. Each amphibian species that has been assessed by the IUCN Red List ($n = 6460$; IUCN 2016) was classified based on their threat category as either "threatened" (IUCN threat categories: Vulnerable (VU), Endangered (EN), Critically Endangered (CR), Extinct in the Wild (EW), or Extinct (EX)) or "nonthreatened" (species listed as Least Concern (LC) and Near-Threatened (NT)). For each genus our

measure of extinction rate was the proportion of "threatened" species.

RANGE SIZE PATTERNS

Geographic range size is typically the dominant driver of extinction risk for terrestrial vertebrates (Cardillo et al. 2005; Sodhi et al. 2008; Lee and Jetz 2011), and evolutionary processes can shape patterns of geographic distributions considerably (Barraclough and Vogler 2000). Species range-restriction has also been associated with heightened rates of speciation in some taxa (Jablonski and Roy 2003; Price and Wagner 2004), including certain groups of amphibians (Eastman and Storfer 2011; Wollenberg et al. 2011). To investigate whether relationships between extinction risk and diversification might be mediated through species' range size patterns we examined associations between genera diversification rate and the mean logarithmic extent of occurrence across species. Range size, in km^2, was estimated for 6311 species based on extent of occurrence polygons from the IUCN (IUCN 2016).

ANALYSIS

To determine the role of evolutionary diversification on contemporary patterns of extinction across genera we used phylogenetic generalized linear models, which can control for phylogenetic autocorrelation in extinction risk across genera. Extinction risk (proportion of threatened species per genus) was fit with a binomial error distribution. Models were run using uninformative priors for 2×10^6 generations with a 2×10^5 burn-in, and a sampling interval of 1000. We compared models examining the relationship between proportion of species threatened per genus and species richness, crown and stem age, and diversification rate based on stem or crown ages. Species richness and lineage ages were log_e transformed, and crown diversification rate was square root transformed, to improve their distributions. To describe the relationship between mean species' range size (log_e transformed) and diversification rate across genera, we used the same modeling approach with a Gaussian error distribution. The significance of richness, age, and diversification were evaluated based on the 95% credibility intervals (CI) of the coefficient estimates. We calculated the mean correlation coefficient (r) between predicted and observed genus extinction risk to evaluate the fit for each model. Analyses were performed using the package "MCMCglmm" (Hadfield 2010) in R v. 3.3.3.

Results

Extinction risk was distributed unevenly across the amphibian genera, with rapidly diversifying clades having a greater share of threatened species (Fig. 1); this holds true for diversification rates estimated from both stem ages ($\beta = 7.55$, 95% CI = 1.32,

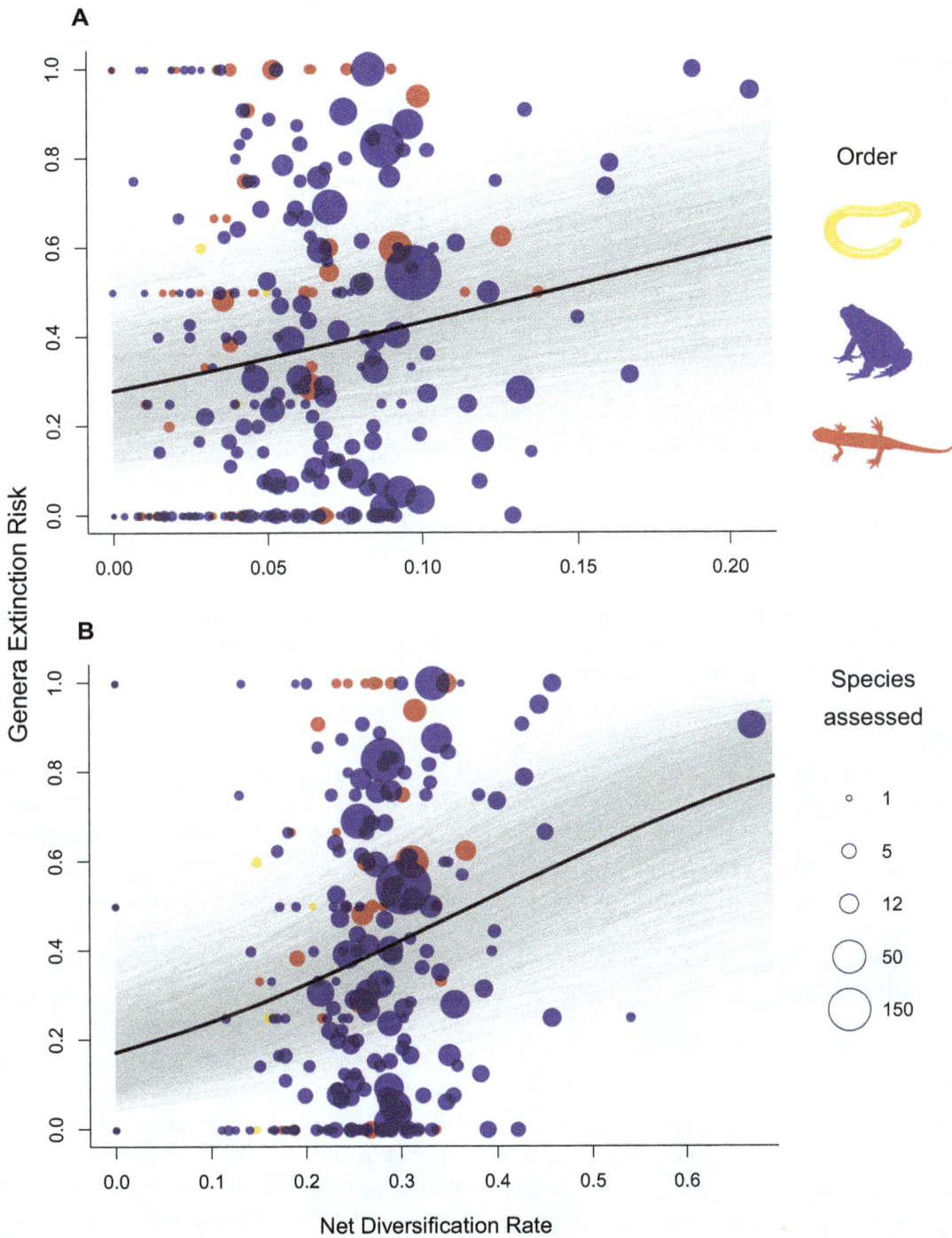

Figure 1. Plot of the proportion of globally threatened species and diversification rate across amphibian genera, showing a positive relationship between extinction risk (proportion species threatened) and net diversification rate calculated using (A) stem age ($n = 329$) and (B) crown group age (square root transformed, $n = 247$). Gray lines indicate the fitted relationships (1800 samples) drawn from the posterior distribution of the models.

14.66, pMCMC $= 0.02$, Fig. 1A) and crown ages ($\beta = 4.16$, 95% CI $= 1.70$, 6.53, pMCMC < 0.001, Fig. 1B; these two diversification estimates were moderately correlated, $r = 0.69$). Diversification rate (for both stem and crown group age) was the best evolutionary descriptor of the distribution of threat across

these clades, as neither species richness, stem age, nor crown age had a significant influence on extinction risk (Table 1).

Considering only the subset of genera that have both crown and stem diversification rates (and so have at least two species) we found that the relationship between contemporary extinction

Table 1. Summary of generalized linear models relating $\log_e$ (genera species richness), $\log_e$ (age), and net diversification rate (square root transformed for crown diversification rate) to patterns of extinction risk (proportion of threatened species) for all genera with stem ages (top, including monotypic genera, $n = 329$) and all genera with crown ages (bottom, $n = 247$).

Variable	β (95% CI)	pMCMC	Pagel's λ (95% CI)	r
Species richness	0.108 (−0.08, 0.26)	0.201	0.43 (0.33, 0.51)	0.047
Stem age	−0.345 (−0.83, 0.16)	0.170	0.44 (0.35, 0.51)	0.081
Stem diversification rate	6.735 (0.91, 12.61)	0.018	0.43 (0.35, 0.51)	0.087
Species richness	0.189 (−0.04, 0.39)	0.094	0.42 (0.33, 0.50)	0.097
Crown age	−0.339 (−0.76, 0.06)	0.120	0.43 (0.33, 0.51)	0.083
Crown diversification rate	4.162 (1.70, 6.53)	< 0.001	0.43 (0.33, 0.51)	0.178

Coefficients represent the posterior mean and correspond to a logit link, and r represents the correlation between observed and model predicted genus extinction risk.

risk and stem diversification was even stronger in this subset (β = 11.42, 95% CI = 3.96, 21.02; pMCMC = 0.01), suggesting that monotypic genera may contribute to uncertainty in the pattern. Although explanatory power was generally modest (Table 1), the models are robust: the proportion of threatened species significantly increases with crown diversification rate when removing when removing both the most rapidly diversifying, and highly threatened, clade *Telmatobius* (β = 3.78, 95% CI = 1.52, 6.40, pMCMC = 0.001), and also when removing the 10% highest diversifying clades ($n = 224$; β = 3.41, 95% CI = 0.49, 6.27, pMCMC = 0.026).

Across these 329 genera there was a strong phylogenetic signal in average species' range size (Pagel's λ = 0.73, 95% CI = 0.54, 0.82), and in addition to having a greater share of threatened species, rapidly diversifying genera also contained species with smaller mean geographic ranges (β = −11.00, 95% CI = −3.08, −18.24, pMCMC = 0.004, Fig. 2A; β = −3.47, 95% CI = −0.79, −6.18, pMCMC = 0.01, Fig. 2B).

Discussion

The positive relationship between the proportion of currently threatened species and their evolutionary diversification across amphibian genera is consistent with theory linking speciation and extinction rates across clades. Importantly, diversification rate had a much stronger influence than lineage age or species richness, suggesting that the process of speciation itself could be driving this relationship.

The causal mechanisms expected to simultaneously drive speciation and extinction rates are general across biodiversity (Stanley 1990), suggesting that this pattern should be widespread. Although evidence of a positive correlation between these rates has been found in fossil data among different groups (Stanley 1979; Jablonski 1986; Gilinsky 1994; Liow et al. 2008), there appears to be little support for a link between diversification and modern extinction risk across other vertebrates. Neither birds

(Jetz et al. 2014), nor mammals (Verde Arregoitia et al. 2013), nor squamate reptiles (Tonini et al. 2016), exhibit any association between evolutionary distinctiveness (a species-level measure of diversification; Jetz et al. 2012) and threat status. The only other group where a direct link between diversification and extinction risk has been demonstrated is within angiosperms from the Cape of South Africa (Davies et al. 2011). In this highly endemic region, the youngest, rapidly diversifying clades also have a greater share of threatened species. This pattern of heightened extinction risk in diversifying plant clades may be a general phenomenon, as species rarity rises in tandem with clade richness in vascular plants across both taxonomic levels and geographic realms (Schwartz and Simberloff 2001; Lozano and Schwartz 2005). This raises a key question: what do amphibians have more in common with plants than with their tetrapod counterparts?

A pattern of positively correlated speciation and extinction may ultimately be driven by mode of speciation. Amphibians often have specialized breeding habitat requirements and are generally poor dispersers (Smith and Green 2005; Wells 2007), which may produce many small, geographically isolated populations that in turn encourage speciation. This form of peripatric speciation may predominate for amphibians, as has been suggested for plant speciations in South Africa and observed in the heightened rates of species rarity in speciose plant families (Schwartz and Simberloff 2001; Lozano and Schwartz 2005; Davies et al. 2011). Under this hypothesis, rapidly speciating clades would produce a preponderance of range-restricted species that are in turn highly threatened by anthropogenic drivers (Sodhi et al. 2008). Indeed, we found that genera diversification rate was negatively correlated with average species' range size, consistent with peripatry being a potential mechanism driving an association between speciation and extinction. Alternatively, it may not be that peripatric speciation dominates in amphibians, but rather that some other biological trait both drives diversification and tends to limit range size, for example small body size or narrow niche breadths (Wollenberg et al. 2011; Slatyer et al. 2013). We might also expect

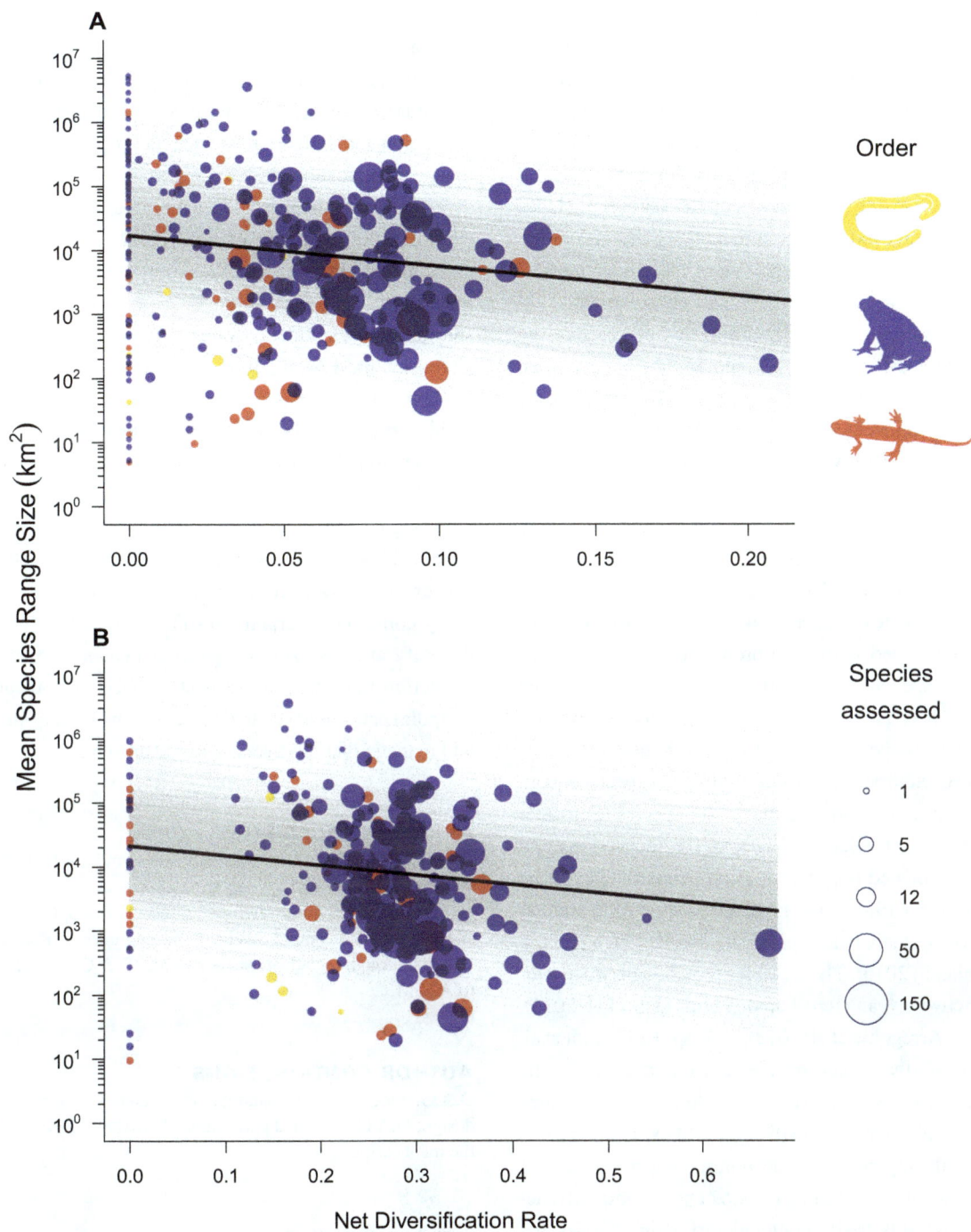

Figure 2. Plot of mean species' geographic range size (km²) and net diversification rate across amphibian genera, calculated using (A) stem age and (B) crown group age (square root transformed). Gray lines indicate the fitted relationships (1800 samples) from the posterior distribution of the models.

that species' geography, and its heritability, could play an important role driving both speciation and extinction across clades if certain physical environments or biomes concurrently drive both processes. Understanding how the form and tempos of speciation relates to species' characteristics will be critical to unraveling these evolutionary patterns of extinction in the amphibians.

Another compelling question concerns how these patterns of impending extinction might shape the future amphibian tree of life. We can estimate the expected loss of phylogenetic diversity based on current patterns of extinction risk: if all currently threatened species were lost across the 329 genera in our dataset, then we would lose 21.55% of genus-level phylogenetic diversity.

However, an even distribution of threat across these same genera would result in significantly less phylogenetic diversity loss at 20.05% (95% CI = 19.0%, 21.1%; see electronic supplementary material). This runs counter to the typical expectation for the loss of evolutionary history when speciation and extinction are positively correlated (Heard and Mooers 2000; but see Parhar and Mooers 2011). Interestingly, our result is due to a subset of clades facing complete lineage extinction, in that all species are threatened. Saving just one species, irrespective of identity, in each of these genera (n = 20) would prevent the loss of an estimated 1.4 billion years of evolutionary history. From this perspective the most effective method to preserve amphibian biodiversity in an age of contemporary mass extinction may entail shifting some focus from species to lineages, even if this means allowing some extinction of phylogenetically redundant species in rapidly diversifying lineages.

A link between speciation and extinction rates has many consequences for shaping past, present, and future patterns of biodiversity. It may suggest that lineages fall along a slow-to-fast continuum for species turnover, where rapidly speciating lineages produce short-lived, extinction-prone species due to shared traits driving both speciation and extinction processes in tandem (Stanley 1990). There is some limited evidence for this including patterns of higher species turnover in large-bodied mammals (Liow et al. 2008; Monroe and Bolkma 2009), that speciose plant clades may both produce and lose many rare species (Schwartz and Simberloff 2001; Davies et al. 2011), the reduced species longevity and heightened origination of range-restricted marine gastropods (Jablonski 1986), and the elevated speciation and extinction rates of specialist taxa generally (Colles et al. 2009; Rolland and Salamin 2016). The lack of association between evolutionary distinctiveness and threat among birds (Jetz et al. 2014), mammals (Verde Arregoitia et al. 2013), and reptiles (Tonini et al. 2016), may indicate that either these patterns do not arise at the taxonomic scale of species or that high clade turnover obscures the relationship between net diversification and extinction risk in these groups. Analyzing this same question at the species-level for amphibians might help resolve this paradox and, importantly, account for other processes driving contemporary extinction risk that may have contributed to the fairly low explanatory power of diversification at the genus level. For instance, a species-level analysis would allow us to assess the role of geography in patterns of diversification and extinction in amphibians (see, e.g., Pyron and Wiens 2013). However, this crucial step is currently precluded by the lack of a fully sampled amphibian phylogeny necessary for such an analysis. To account for turnover, independently estimating speciation and extinction rates, perhaps through combining both fossil and molecular phylogenetic data in well-sampled clades, will be key to assess whether speciation and extinction rates are con-

currently driven by biological characteristics across a diverse set of taxa.

Ecological limits may also be crucial to a positive speciation-extinction correlation. Clades near their carrying capacity, where speciation and extinction balance out, may be expected to exhibit the positive relationship we report here, while clades in their diversity "growth phase" may be able to escape this trade-off (Rabosky 2009; Etienne et al. 2012). This growth phase may be associated with novel ecological opportunities or adaptations that may allow some high turnover clades to temporarily decouple speciation and extinction rates and undergo adaptive radiations (Rabosky and Lovette 2008). Understanding the conditions that maintain, or break down, any relationship between speciation and extinction rates will be key to our understanding of the long-term temporal dynamics of biodiversity.

Here, we demonstrate that net diversification is associated with a greater contemporary extinction risk across amphibian genera. This pattern is consistent with the theory that speciation and extinction rates may be driven by the same suites of traits, or by common geography, resulting in clades that both rapidly diversify and lose species. Nonindependence of speciation and extinction rates would add a new piece to both understanding temporal patterns of biodiversity and how we may aim to prioritize and manage that biodiversity in the present.

ACKNOWLEDGEMENTS
We thank T. J. Davies, various audiences, and the FAB*Lab for discussion. R. A. Pyron and several anonymous reviewers provided valuable comments on previous versions of the manuscript. This research was supported by the Natural Sciences and Engineering Research Council of Canada through a graduate scholarship to D.A.G. and a Discovery Grant to A.Ø.M.

AUTHOR CONTRIBUTIONS
D.A.G. conceived of the study; A.Ø.M. and D.A.G. contributed to study design; D.A.G. collected and analyzed the data; DAG and A.Ø.M. wrote the manuscript.

LITERATURE CITED
Barraclough, T. G., and A. P. Vogler. 2000. Detecting the geographical pattern of speciation from species-level phylogenies. Am. Nat. 155:419–434.

Böhm, M., R. Williams, H. R. Bramhall, K. M. McMillan, A. D. Davidson, A. Garcia, L. M. Bland, J. Bielby, and B. Collen. 2016. Correlates of extinction risk in squamate reptiles: the relative importance of biology, geography, threat and range size. Glob. Ecol. Biogeogr. 25:391–405.

Boyles, J. G., and J. J. Storm. 2007. The perils of picky eating: dietary breadth is related to extinction risk in insectivorous bats. PLoS One 2:e672.

Bromham, L., R. Lanfear, P. Cassey, G. Gibb, and M. Cardillo. 2012. Reconstructing past species assemblages reveals the changing patterns and drivers of extinction through time. Proc. R. Soc. Lond. B 279:4024–4032.

Cardillo, M., G. M. Mace, K. E. Jones, J. Bielby, O. R. P. Bininda-Emonds, W. Sechrest, *et al.* 2005. Multiple causes of high extinction risk in large mammal species. Science 309:1239–1241.

Ceballos, G., P. R. Ehrlich, A. D. Barnosky, A. García, R. M. Pringle, and T. M. Palmer. 2015. Accelerated modern human–induced species losses: entering the sixth mass extinction. Sci. Adv. 1:e1400253.

Claramunt, S., E. P. Derryberry, J. V. Remsen, and R. T. Brumfield. 2012. High dispersal ability inhibits speciation in a continental radiation of passerine birds. Proc. R. Soc. Lond. B 279:1567–1574.

Colles, A., L. H. Liow, and A. Prinzing. 2009. Are specialists at risk under environmental change? Neoecological, paleoecological and phylogenetic approaches. Ecol. Lett. 12:849–863.

Condamine, F. L., J. Rolland, and H. Morlon 2013. Macroevolutionary perspectives to environmental change. Ecol. Lett. 16:72–85.

Davies, T. J., G. F. Smith, D. U. Bellstedt, J. S. Boatwright, B. Bytebier, R. M. Cowling, F. Forest, L. J. Harmon, A. M. Muasya, B. D. Schrire, *et al* 2011. Extinction risk and diversification are linked in a plant biodiversity hotspot. PLoS Biol. 9:e1000620.

Ducatez, S., R. Tingley, and R. Shine. 2014. Using species co-occurrence patterns to quantify relative habitat breadth in terrestrial vertebrates. Ecosphere 5:art152.

Eastman, J. M., and A. Storfer 2011. Correlations of life-history and distributional-range variation with salamander diversification rates: evidence for species selection. Syst. Biol. 60:503–518.

Etienne, R. S., B. Haegeman, T. Stadler, T. Aze, P. N. Pearson, A. Purvis, and A. B. Philimore. 2012. Diversity-dependence brings molecular phylogenies closer to agreement with the fossil record. Proc. R. Soc. Lond. B 279:1300–1309.

Finnegan, S., S. C. Anderson, P. G. Harnik, C. Simpson, D. P. Tittensor, J. E. Byrnes, Z. V. Finkel, D. R. Lindberg, L. H. Liow, R. Lockwood, *et al.* 2015. Paleontological baselines for evaluating extinction risk in the modern oceans. Science 348:567–570.

Frost, D. R. 2016. Amphibian Species of the World: An Online Reference. Version 6.0. Available at: research.amnh.org/vz/herpetology/amphibia.

Gilinsky, N. L. 1994. Volatility and the Phanerozoic decline of background extinction intensity. Paleobiology 20:445–458.

Hadfield, J. 2010. MCMC methods for multi-response generalized linear mixed models: the MCMCglmm R package. J. Stat. Softw. 33:1–22.

Harnik, P. G., H. K. Lotze, S. C. Anderson, Z. V. Finkel, S. Finnegan, D. R. Lindberg, *et al.* 2012a. Extinctions in ancient and modern seas. Trends Ecol. Evol. 27:608–617.

Harnik, P. G., C. Simpson, and J. L. Payne. 2012b. Long-term differences in extinction risk among the seven forms of rarity. Proc. R. Soc. Lond. B 279:4969–4976.

Heard, S. B., and A. Ø. Mooers. 2000. Phylogenetically patterned speciation rates and extinction risks change the loss of evolutionary history during extinctions. Proc. R. Soc. Lond. B 267:613–620.

Heim, N. A., and S. E. Peters. 2011. Regional environmental breadth predicts geographic range and longevity in fossil marine genera. PLoS One 6:e18946.

Hoffmann, M., C. Hilton-Taylor, A. Angulo, M. Böhm, T. M. Brooks, S. H. M. Butchart, K. E. Carpenter, J. Chanson, B. Collen, N. A. Cox, *et al.* 2010. The impact of conservation on the status of the world's vertebrates. Science 330:1503–1509.

IUCN. 2016. The IUCN Red List of Threatened Species. Version 2016-1. Available at: http://www.iucnredlist.org

Jablonski, D. 1986. Larval ecology and macroevolution in marine invertebrates. Bull. Mar. Sci. 39:565–587.

———. 2008. Species selection: theory and data. Annu. Rev. Ecol. Evol. Syst. 39:501–524.

Jablonski, D., and K. Roy. 2003. Geographical range and speciation in fossil and living molluscs. Proc. R. Soc. Lond. B 270:401–406.

Jetz, W., G. H. Thomas, J. B. Joy, K. Hartmann, and A. Ø. Mooers. 2012. The global diversity of birds in space and time. Nature 491:444–448.

Jetz, W., G. H. Thomas, J. B. Joy, D. W. Redding, K. Hartmann, and A. Ø. Mooers. 2014. Global distribution and conservation of evolutionary distinctness in birds. Curr. Biol. 24:919–930.

Kiessling, W., and M. Aberhan. 2007. Geographical distribution and extinction risk: lessons from Triassic-Jurassic marine benthic organisms. J. Biogeogr. 34:1473–1489.

Lande, R., S. Engen, and B.-E. Sæther. 2003. Stochastic Population Dynamics in Ecology and Conservation. Oxford Univ. Press, Oxford, UK.

Lee, T. M., and W. Jetz. 2011. Unravelling the structure of species extinction risk for predictive conservation science. Proc. R. Soc. Lond. B 278:1329–1338.

Liow, L. H., M. Fortelius, E. Bingham, K. Lintulaakso, H. Mannila, L. Flynn, and N. C. Stenseth. 2008. Higher origination and extinction rates in larger mammals. Proc. Natl. Acad. Sci. USA 105:6097–6102.

Lozano, F. D., and M. W. Schwartz. 2005. Patterns of rarity and taxonomic group size in plants. Biol. Conserv. 126:146–154.

Lyons, S. K., J. H. Miller, D. Fraser, F. A. Smith, A. Boyer, E. Lindsey, and A. M. Mychajliw. 2016. The changing role of mammal life histories in Late Quaternary extinction vulnerability on continents and islands. Biol. Lett. 12:20160342.

Mace, G. M., N. J. Collar, K. J. Gaston, C. Hilton-Taylor, H. R. Akçakaya, N. Leader-Williams, E. J. Milner-Gulland, and S. N. Stuart. 2008. Quantification of extinction risk: IUCN's system for classifying threatened species. Conserv. Biol. 22:1424–1442.

Magallon, S., and M. J. Sanderson. 2001. Absolute diversification rates in Angiosperm clades. Evolution 55:1762–1780.

McKinney, M. L. 1997. Extinction vulnerability and selectivity: combining ecological and paleontological views. Annu. Rev. Ecol. Syst. 28:495–516.

Monroe, M. J., and F. Bokma. 2009. Do speciation rates drive rates of body size evolution in mammals? Am. Nat. 174:912–918.

Nee, S., E. C. Holmes, R. M. May, and P. H. Harvey. 1994. Extinction rates can be estimated from molecular phylogenies. Philos. Trans. R. Soc. Lond. B. Biol. Sci. 344:77–82.

Olden, J. D., N. L. Poff, and K. R. Bestgen. 2008. Trait synergisms and the rarity, extirpation, and extinction risk of desert fishes. Ecology 89:847–856.

Orzechowski, E. A., R. Lockwood, J. E. K. Byrnes, S. C. Anderson, S. Finnegan, Z. V. Finkel, P. G. Harnik, D. R. Lindberg, L. H. Liow, H. K. Lotze, *et al.* 2015. Marine extinction risk shaped by trait-environment interactions over 500 million years. Glob. Chang. Biol. 21:3595–3607.

Parhar, R. K., and A. Ø. Mooers. 2011. Phylogenetically clustered extinction risks do not substantially prune the Tree of Life. PLoS One 6:e23528.

Pimm, S. L., G. J. Russell, J. L. Gittleman, and T. M. Brooks. 1995. The future of biodiversity. Science 269:347–350.

Price, J. P., and W. L. Wagner. 2004. Speciation in Hawaiian angiosperm lineages: cause, consequence, and mode. Evolution 58:2185–2200.

Pyron, R. A. 2014. Biogeographic analysis reveals ancient continental vicariance and recent oceanic dispersal in amphibians. Syst. Biol. 63:779–797.

Pyron, R. A., and J. J. Wiens. 2013. Large-scale phylogenetic analyses reveal the causes of high tropical amphibian diversity. Proc. R. Soc. Lond. 280:20131622.

Rabosky, D. L. 2009. Ecological limits and diversification rate: alternative paradigms to explain the variation in species richness among clades and regions. Ecol. Lett. 12:735–743.

————. 2010. Extinction rates should not be estimated from molecular phylogenies. Evolution 64:1816–1824.

Rabosky, D. L., and I. J. Lovette. 2008. Explosive evolutionary radiations: decreasing speciation or increasing extinction through time? Evolution 62:1866–1875.

Reynolds, J. D., N. K. Dulvy, N. B. Goodwin, and J. A. Hutchings. 2005. Biology of extinction risk in marine fishes. Proc. R. Soc. Lond. B 272:2337–2344.

Rolland, J., and N. Salamin. 2016. Niche width impacts vertebrate diversification. Glob. Ecol. Biogeogr. 25:1252–1263.

Sallan, L., and A. K. Galimberti. 2015. Body-size reduction in vertebrates following the end-Devonian mass extinction. Science 350:812–815.

Sandel, B., L. Arge, B. Dalsgaard, R. G. Davies, K. J. Gaston, W. J. Sutherland, and J.-C. Svenning. 2011. The influence of Late Quaternary climate-change velocity on species endemism. Science 334:660–664.

Schwartz, M. W., and D. Simberloff. 2001. Taxon size predicts rates of rarity in vascular plants. Ecol. Lett. 4:464–469.

Slatyer, R. A., M. Hirst, and J. P. Sexton. 2013. Niche breadth predicts geographical range size: a general ecological pattern. Ecol. Lett. 16:1104–1114.

Smith, A. M., and D. M. Green. 2005. Dispersal and the metapopulation paradigm in amphibian ecology and conservation: are all amphibian populations metapopulations? Ecography 28:110–128.

Smits, P. D. 2015. Expected time-invariant effects of biological traits on mammal species duration. Proc. Natl. Acad. Sci. USA 112:13015–13020.

Sodhi, N. S., D. Bickford, A. C. Diesmos, T. M. Lee, L. P. Koh, B. W. Brook, C. H. Sekercioglu, and C. J. A. Bradshaw. 2008. Measuring the meltdown: Drivers of global amphibian extinction and decline. PLoS One 3:e1636.

Stanley, S. M. 1979. Macroevolution, pattern and process. W. H. Freeman, San Francisco, USA.

————. 1990. The general correlation between rate of speciation and rate of extinction: fortuitous causal linkages. Pp. 103–172 in R. M., Ross and W. D., Allmon, eds. Causes of Evolution: A Paleontological Perspective. Chicago Univ. Press, Chicago, USA.

Tonini, J. F. R., K. H. Beard, R. B. Ferreira, W. Jetz, and R. A. Pyron. 2016. Fully-sampled phylogenies of squamates reveal evolutionary patterns in threat status. Biol. Conserv. 204:23–31.

Verde Arregoitia, L. D., S. P. Blomberg, and D. O. Fisher. 2013. Phylogenetic correlates of extinction risk in mammals: species in older lineages are not at greater risk. Proc. R. Soc. Lond. B 280:20131092.

Wells, K. D. 2007. The ecology and behavior of amphibians. Chicago Univ. Press, Chicago, USA.

Wollenberg, K. C., D. R. Vieites, F. Glaw, and M. Vences. 2011. Speciation in little: the role of range and body size in the diversification of *Malagasy mantellid* frogs. BMC Evol. Biol. 11:217.

4

No evidence for maintenance of a sympatric *Heliconius* species barrier by chromosomal inversions

John W. Davey,[1,2,3] (iD) Sarah L. Barker,[1] Pasi M. Rastas,[1] Ana Pinharanda,[1,2] Simon H. Martin,[1] Richard Durbin,[4] W. Owen McMillan,[2] Richard M. Merrill,[1,2] and Chris D. Jiggins[1,2,5]

[1]Department of Zoology, University of Cambridge, Downing Street, Cambridge CB2 3EJ, United Kingdom

[2]Smithsonian Tropical Research Institute, Gamboa, Panama

[3]E-mail: johnomics@gmail.com

[4]Wellcome Trust Sanger Institute, Cambridge CB10 1SA, United Kingdom

[5]E-mail: c.jiggins@zoo.cam.ac.uk

Mechanisms that suppress recombination are known to help maintain species barriers by preventing the breakup of coadapted gene combinations. The sympatric butterfly species *Heliconius melpomene* and *Heliconius cydno* are separated by many strong barriers, but the species still hybridize infrequently in the wild, and around 40% of the genome is influenced by introgression. We tested the hypothesis that genetic barriers between the species are maintained by inversions or other mechanisms that reduce between-species recombination rate. We constructed fine-scale recombination maps for Panamanian populations of both species and their hybrids to directly measure recombination rate within and between species, and generated long sequence reads to detect inversions. We find no evidence for a systematic reduction in recombination rates in F1 hybrids, and also no evidence for inversions longer than 50 kb that might be involved in generating or maintaining species barriers. This suggests that mechanisms leading to global or local reduction in recombination do not play a significant role in the maintenance of species barriers between *H. melpomene* and *H. cydno*.

KEY WORDS: Chromosomal evolution, evolutionary genomics, insects, speciation.

Impact Summary

It is now possible to study the process of species formation by sequencing the genomes of multiple closely related species. *Heliconius melpomene* and *H. cydno* are two butterfly species that have diverged over the past two million years. These species have different color patterns, mate preferences, and host plants, traits that involve variants of multiple genes spread across the genome. However, the species still hybridize infrequently in the wild and exchange large parts of their genomes. Typically, when genomes are exchanged, chromosomes are recombined and gene combinations are broken up, preventing species from forming. Theory predicts that gene variants that define species might be linked together because of structural differences in their genomes, such as chromosome inversions, that will not be broken up when the species hybridize. We sequenced large crosses of butterflies to show that there are almost certainly no megabase-long chromosome regions that are not broken up during hybridization, and while we find evidence for some small chromosome inversions (on the order of tens of kilobases in size), it is unlikely that these are necessary to keep gene combinations together. This suggests that hybridization is rare enough and mate preference is strong enough that inversions are not necessary to maintain the species barrier.

Introduction

It is now widely appreciated that the evolution and maintenance of new species is constrained by genetic as well as ecological and geographical factors (Seehausen et al. 2014). A classic problem for speciation is that if combinations of divergently selected alleles arise in populations that remain in contact, recombination is expected to break down the associations between alleles and prevent speciation from proceeding (Felsenstein 1981). A large body of work has invoked genetic mechanisms that couple species-specific alleles and so reduce the homogenizing effects of gene flow (Smadja and Butlin 2011; Nachman and Payseur 2012), including assortative mating, one-allele mechanisms (Felsenstein 1981), tight physical linkage, pleiotropy, and multiple (or "magic") traits (Servedio et al. 2011). Here, we focus on the role of chromosomal inversions in suppressing recombination of divergently selected alleles in hybrids.

Inversions have frequently been implicated in speciation (White 1978; King 1993; Ayala and Coluzzi 2005; Hoffmann and Rieseberg 2008; Kirkpatrick 2010). Traits associated with reproductive isolation are often linked to inversions (e.g., Noor et al. 2001; Ayala et al. 2013; Fishman et al. 2013) and genetic divergence between species can increase within inverted regions through reduction of gene flow (Navarro and Barton 2003b; Jones et al. 2012; McGaugh and Noor 2012; Lohse et al. 2015). Theory predicts that inversions can spread by reducing recombination between locally adapted alleles (Butlin 2005; Kirkpatrick and Barton 2006; Feder and Nosil 2009; Ortíz-Barrientos et al. 2016), which can either establish or reinforce species barriers by capturing loci for isolating traits such as mating preferences and epistatic incompatibilities (Dagilis and Kirkpatrick 2016) or allow adaptive cassettes to spread between species via hybridization (Kirkpatrick and Barrett 2015). Reduced recombination within and around inversions has been confirmed in several species (Stevison et al. 2011; Farré et al. 2013), although it is unlikely that gene flow is entirely suppressed within inversions, due to double crossovers and gene conversion (Korunes and Noor 2017), factors addressed in some recent models (Guerrero et al. 2012; Feder et al. 2014).

Several authors have predicted that inversions can enable the formation and maintenance of species barriers in sympatry or parapatry by favoring the accumulation of barrier loci in the presence of gene flow (Noor et al. 2001; Rieseberg 2001; Navarro and Barton 2003a; Faria and Navarro 2010), as opposed to older models where inversions have direct effects on hybrid fertility or viability (discussed in Rieseberg 2001). Especially striking is the fact that most sympatric *Drosophila* species pairs differ by one or more inversions, whereas allopatric pairs are virtually all homosequential (Noor et al. 2001). In the particular case of *Drosophila pseudoobscura* and *D. persimilis*, three chromosomes differ by large, fixed inversions and a fourth chromosome has many varied arrangements (Machado et al. 2007; Noor et al. 2007), genome differentiation is greater within and near inversions (Noor et al. 2007; McGaugh and Noor 2012) and sterility factors are associated with inversions in a sympatric species pair, but with collinear regions in

an allopatric pair (Brown et al. 2004). In rodents, sympatric sister species typically have more autosomal karyotypic differences than allopatric sister species (Castiglia 2014). Sympatric sister species of passerine birds are significantly more likely to differ by an inversion than allopatric sister species, with the number of inversion differences best explained by whether the species ranges overlap (Hooper 2016). Although inversions are not the only mechanism by which recombination rate can be modified during speciation, and more recently attention has been drawn to the potential role of genic recombination modifiers (Ortíz-Barrientos et al. 2016), the very strong effect of inversions on recombination rate, and the fact that they are completely linked to the locus at which recombination is reduced, means that they are perhaps the most likely mechanism of recombination rate evolution during speciation.

We set out to test the role of inversions in the maintenance of species barriers in *Heliconius* butterflies. The 46 species of *Heliconius* have been the focus of a wide range of speciation research (Merrill et al. 2011a; Supple et al. 2013; Kozak et al. 2015; Merrill et al. 2015). Chromosomal inversions are known to play an important role in the maintenance of a complex color pattern polymorphism in *Heliconius numata* (Joron et al. 2011). However, no other *Heliconius* inversions have been identified with traditional methods, as *Heliconius* chromosomes typically appear as dots in chromosome squashes, at least in male tissues (Brown et al. 1992).

Here, we systematically searched for inversions between populations of two *Heliconius* species, *H. melpomene rosina* and *H. cydno chioneus*, which are sympatric in the lowland tropical forests of Panama. These species differ by many traits (Jiggins 2008) including color pattern (Naisbit et al. 2003), mate preference (Jiggins et al. 2001; Naisbit et al. 2001; Merrill et al. 2011b), host plant choice (Merrill et al. 2013), pollen load, and microhabitat (Estrada and Jiggins 2002). Hybrid color pattern phenotypes are attacked more frequently than parental forms, indicating disruptive selection against hybrids (Merrill et al. 2012). Assortative mating between the species is strong, and genetic differences in mate preference are linked to different color pattern loci (Merrill et al. 2011b). Matings between *H. cydno* females and *H. melpomene* males produce sterile female offspring, but male offspring are fertile, and female offspring of backcrosses show a range of sterility phenotypes (Naisbit et al. 2002). Hybrids are extremely rare in the wild, but many natural hybrids have been documented in museum collections (Mallet et al. 2007) and examination of present-day genomic sequences indicate that gene flow has been pervasive, affecting around 40% of the genome (Martin et al. 2013; Arias et al. 2014). Modeling suggests that the species diverged around 1.5 million years ago, with hybridization rare or absent for one million years, followed by a period of more abundant gene flow in the last half a million years (Kronforst et al. 2013; Martin et al. 2015b), suggesting that the species originated in parapatry, but have been broadly sympatric and hybridizing during their recent history. Although the species are closely related, they are not sister species; several other species such as *Heliconius timareta* and *Heliconius heurippa* are more closely related to *H. cydno* than *H. melpomene*.

Models predict that inversions, or other modifiers of recombination, can be established during both sympatric speciation and secondary contact (Noor et al. 2001; Rieseberg 2001; Feder and Nosil 2009; Feder et al. 2011; Feder et al. 2014). Furthermore, the genetic basis for species differences between *H. melpomene* and *H. cydno* is well understood and would seem to favor the establishment of inversions. Wing pattern differences are controlled by a few loci of major effect (Naisbit et al. 2003), some of which consist of clusters of linked elements. There is also evidence for linkage between genes controlling wing pattern and those underlying assortative mating (Merrill et al. 2011b). The existing evidence for clusters of linked loci of major effect would therefore seem to favor the evolution of mechanisms to reduce recombination between such loci, and hold species differences in tighter association.

We therefore set out to investigate patterns of recombination and test for the presence of inversions between *H. melpomene rosina* and *H. cydno chioneus*. *H. melpomene melpomene* has a high-quality genome assembly with 99% of the genome placed on chromosomes and 84% ordered and oriented (Heliconius Genome Consortium 2012; Davey et al. 2016). Whole genome resequencing has shown that F_{ST} between *H. melpomene melpomene* and *H. melpomene rosina* is consistently low across the genome, with only a few small, narrow peaks of divergence, but F_{ST} between *H. melpomene rosina* and *H. cydno chioneus* is substantially higher and heterogeneous (Martin et al. 2013), and many gene duplications have been identified between the two species (Pinharanda et al. 2017).

However, *H. melpomene* and *H. cydno* have not yet been examined for evidence of large differences in genome structure such as inversions. To test for this, we constructed fine-scale linkage maps for *H. melpomene*, *H. cydno*, and *H. cydno* x *H. melpomene* hybrids to test for the presence of reduced recombination in hybrids and inverted regions between the species (Fig. 1). Our linkage maps are based on tens of thousands of new single nucleotide polymorphisms (SNPs) discovered and genotyped using RAD Sequencing data from just under 1000 individuals from 24 crosses. We also generated long-read sequencing data and new genome assemblies for both species to test for inversions on smaller scales. This is the first systematic survey of genome structure and recombination at a fine scale in a lepidopteran species, and also one of very few such surveys of both parent species and their hybrids (Ortíz-Barrientos et al. 2016), which we hope will be a valuable test case for the role of inversions in speciation.

Methods

LINKAGE MAPS

Full details of methods for our crosses, library preparations, sequencing, and linkage map construction can be found in Supporting Information. In brief, for the within-species crosses of *H. melpomene rosina* and *H. cydno chioneus*, wild males were mated to virgin stock females and linkage maps were constructed from F1 offspring, whereas for hybrids, *H. cydno* stock females were mated to wild *H. melpomene* males, and F1 males were backcrossed to *H. cydno* stock females, with linkage maps constructed from backcross offspring (Fig. S1). Grandparents, parents, and offspring were RAD sequenced using the PstI restriction enzyme on Illumina HiSeq 2500 and 4000 machines using 100 bp paired end reads, except for one *H. melpomene* and one *H. cydno* trio which were whole genome sequenced with 125 bp paired end Illumina HiSeq 2500 sequencing (previously reported in Malinsky et al. 2016), and 58 hybrid individuals that were sequenced on a HiSeq 2000 using 50 bp single-end sequencing. RAD sequences were demultiplexed with Stacks (Catchen et al. 2013) and Illumina RAD and whole genome reads were aligned to version 2 of the *H. melpomene* genome (Hmel2; Davey et al. 2016) with Stampy (Lunter and Goodson 2011), Picard (http://broadinstitute.github.io/picard/), and GATK (dePristo et al. 2011) and genotype posteriors called with SAMtools mpileup (Li 2011).

Linkage maps were constructed from genotype posteriors using Lep-MAP. Within-species linkage maps for *H. melpomene* and *H. cydno* were built with Lep-MAP2 (Rastas et al. 2016) and some additional bespoke scripts. Due to the more complex cross structure of backcross populations, smaller cross sizes, and lower sequence quality for some crosses, different methods and thresholds were used to construct linkage maps for the *H. cydno* x *H. melpomene* hybrid crosses, now incorporated into Lep-MAP3 (https://sourceforge.net/projects/lep-map3/). Most notably, separate linkage maps were built for each large within-species cross, but only one linkage map was constructed for all hybrids, given the small size of the backcross families. The hybrid linkage map was then divided into four separate maps for each pair of grandparents. Full details can be found in Supporting Information.

In brief, SNPs were filtered to ensure each genotype was supported by multiple reads in the majority of individuals, excluding SNPs with rare alleles and segregation distortion. Missing parental genotypes were called based on related parent and offspring genotypes. Markers were identified by clustering together SNPs with almost identical patterns and filtering candidate markers with low support. Markers were separated into linkage groups, setting parameters empirically to identify 21 linkage groups for the expected 21 *H. melpomene*/*H. cydno* chromosomes, and markers for each

Collinear

Inversion

Reduced hybrid recombination

Misassembly

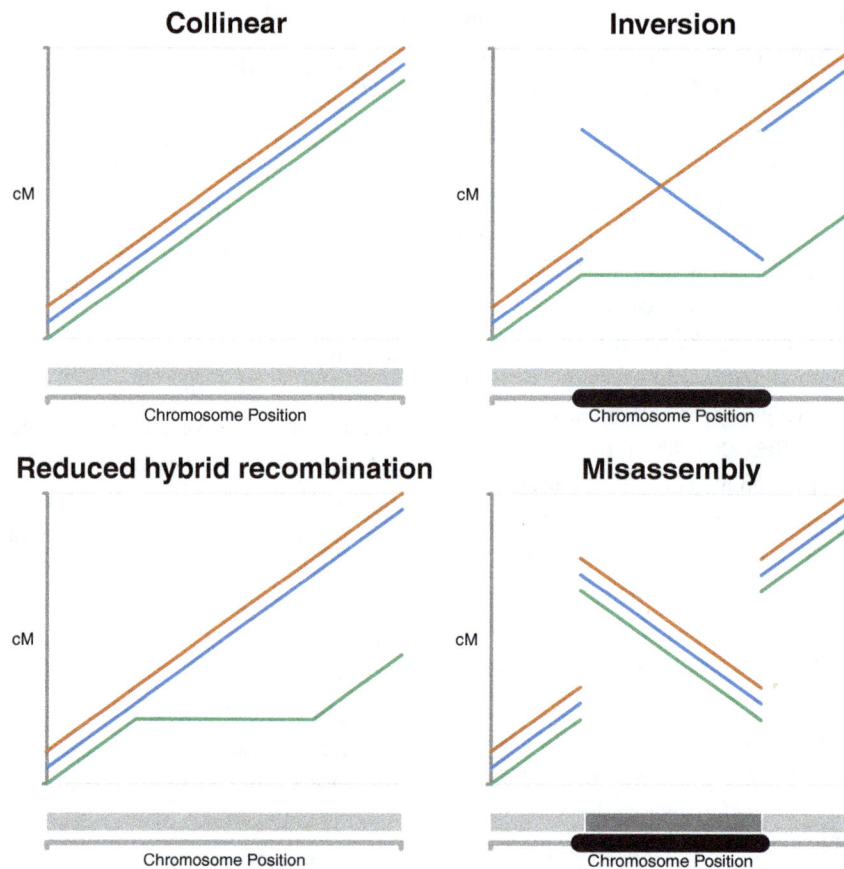

Figure 1. Diagram of expected patterns for collinear, inverted, reduced hybrid recombination, and misassembled genome regions. *Heliconius melpomene*, red; *H. cydno*, blue; *H. cydno* x *H. melpomene* hybrids, green. Gray strip, *H. melpomene* contigs (dark/light gray shows different contigs). Black lozenge, inverted region.

linkage group were ordered. As females do not recombine, maternal markers were easy to identify and unchanging, so we could also make use of thousands of SNPs where both parents were heterozygous, by removing the maternal alleles and so converting the SNPs to paternal-only markers (Jiggins et al. 2005). Initial marker orderings were manually reviewed and edited, and all SNPs were reassigned to the final set of cleaned markers to improve coverage of the genome.

GENOME SCAFFOLDING

Hmel2 scaffolds were manually ordered according to the linkage maps for each of the three *Heliconius melpomene* crosses wherever possible. A small number of misassemblies in Hmel2 were corrected, with scaffolds being split and reoriented where necessary. Not all scaffolds could be ordered based on the linkage maps alone, so Pacific Biosciences reads were also used. PacBio reads were aligned to Hmel2 scaffolds using BWA mem with -x pacbio option (Li 2013). Scaffolds were ordered by manual inspection of spanning reads between scaffolds, identified and summarized by script find_pacbio_scaffold_overlaps.py. Chromosomal positions were assigned by inserting dummy 100 bp gaps between each

pair of remaining scaffolds. Although PacBio sequencing could fill gaps between scaffolds, we chose not to do this for these analyses to avoid disrupting Hmel2 linkage map and annotation feature coordinates.

RECOMBINATION RATE MEASUREMENT AND PERMUTATION TESTING

CentiMorgan values were calculated using the recombination fraction alone, as the maps were sufficiently fine-scale that mapping functions were not necessary (Ziegler and König 2001); see Supporting Information note on crossover detection for further details. Per-cross maps (Fig. S3) and map statistics (Table 1) were calculated for F1 parents within-species and for grandparents for hybrids. Marey maps (Figs. 2 and S3) and total map lengths were calculated using centiMorgan values. Chromosomes were tested for reductions in chromosome-wide recombination rate in the hybrids compared to *H. melpomene* or *H. cydno* using a bootstrapped Kolmogorov–Smirnov test suitable for discrete data with ties such as recombination counts (ks.boot in the R Matching package; Sekhon 2011), using a one-tailed test for reduced rates in hybrids with 10,000 bootstrap samples, declaring significance at a 0.05

Table 1. Cross information for each species. Summary values for each species shown in bold; mean map lengths and sequencing depths shown in italics.

Species	Cross	Offspring	Total map length (cM)	Mean offspring sequencing depth (reads per RAD locus)
Heliconius melpomene	1	111	1048	22
	2	122	1065	23
	3	102	1135	31
	Total/*Mean*	**335**	*1083*	*25*
Heliconius cydno	1	95	1076	15
	2	77	1076	17
	3	125	1070	17
	Total/*Mean*	**297**	*1074*	*16*
H. melpomene × *H. cydno* hybrids	1	170	1090	19
	2	88	1069	30
	3	68	1158	21
	4	5	1040	28
	Total/*Mean*	**331**	*1089*	*25*

false discovery rate with control for multiple testing (42 tests, with two comparisons for each of 21 chromosomes).

Fine-scale recombination rates (Fig. S4) were calculated in windows of 1 Mb with 100 kb steps, counting individual crossovers in each window (see Supporting Information note on crossover detection for further details). One megabase windows were tested for differences in recombination rate by calculating null distributions of rate differences by permutation of species labels across all offspring, testing at a 0.05 false discovery rate over 270,000 permutations, controlling for multiple tests with three comparisons for each of 2549 windows. Ninety-five percent confidence intervals in Figure S4 were calculated by bootstrap, sampling offspring for each species by replacement 10,000 times and calculating centiMorgan values, plotting 2.5 and 97.5% quantiles for each window.

INVERSION DISCOVERY

PBHoney (in PBSuite version 15.8.24 (English et al. 2014)) was used to call candidate inversions between *H. melpomene* and *H. cydno*, using alignments of PacBio data to ordered Hmel2 scaffolds made with BWA mem with -x pacbio option (Li 2013), retaining only primary alignments, and accepting alignments with minimum mapping quality of 30 in Honey.py tails, running separately on each of four samples (*H. cydno* females, *H. cydno* males, *H. melpomene* females, *H. melpomene* males). Break point candidate sets were compiled together into one file and scaffold positions converted to chromosome positions using script compile_tails.py. PBHoney was run with default options, requiring a minimum of three overlapping reads from three unique zero-mode waveguides to call a breakpoint candidate. As the *H. cydno* male sample had low coverage, we also ran PBHoney requiring a min-

imum of two reads from one zero-mode waveguide and included these tentative candidates where they overlapped with candidates from other samples called with the default settings.

PBHoney was tested for false positives by simulating PacBio reads with pbsim 1.0.3 (Ono et al. 2013), generating a sample profile using the *H. melpomene* female sample and simulating 15 "SMRT cells" at 5x coverage each. Simulated data were then aligned with BWA and inversions called with PBHoney as above.

Trio assemblies were aligned to the ordered Hmel2 genome using NUCmer from the MUMmer suite (Kurtz et al. 2004; version 3.23), followed by show-coords with show-Tlcd options, to produce tab-separated output including scaffold lengths, percentage identities, and directions of hits.

Script detect_inversion_gaps.py was used to integrate the PBHoney inversion candidates with the linkage maps, trio alignments, and *H. melpomene* annotation (from Hmel2). As these data are being used to rule out inversions in regions without recombinations, PBHoney inversion candidates were rejected if at least one recombination for the same species as the candidate was contained within the inversion. PBHoney candidates were also rejected if there was a trio scaffold alignment spanning the candidate inversion, with spanning defined as extending more than half the length of the candidate inversion in either direction. Finally, candidates shorter than 1 kb were rejected, as linkage disequilibrium between SNPs separated by 1 kb or less in *H. melpomene* is significantly higher than background levels (Martin et al. 2016) and so inversions below this size are unlikely to be required to maintain linkage. The retained inversion candidates were then combined into groups by overlap.

Each group of overlapping inversion candidates was classified as follows (Fig. 3; Table 4; Figs. S11–S17): *Split reads and*

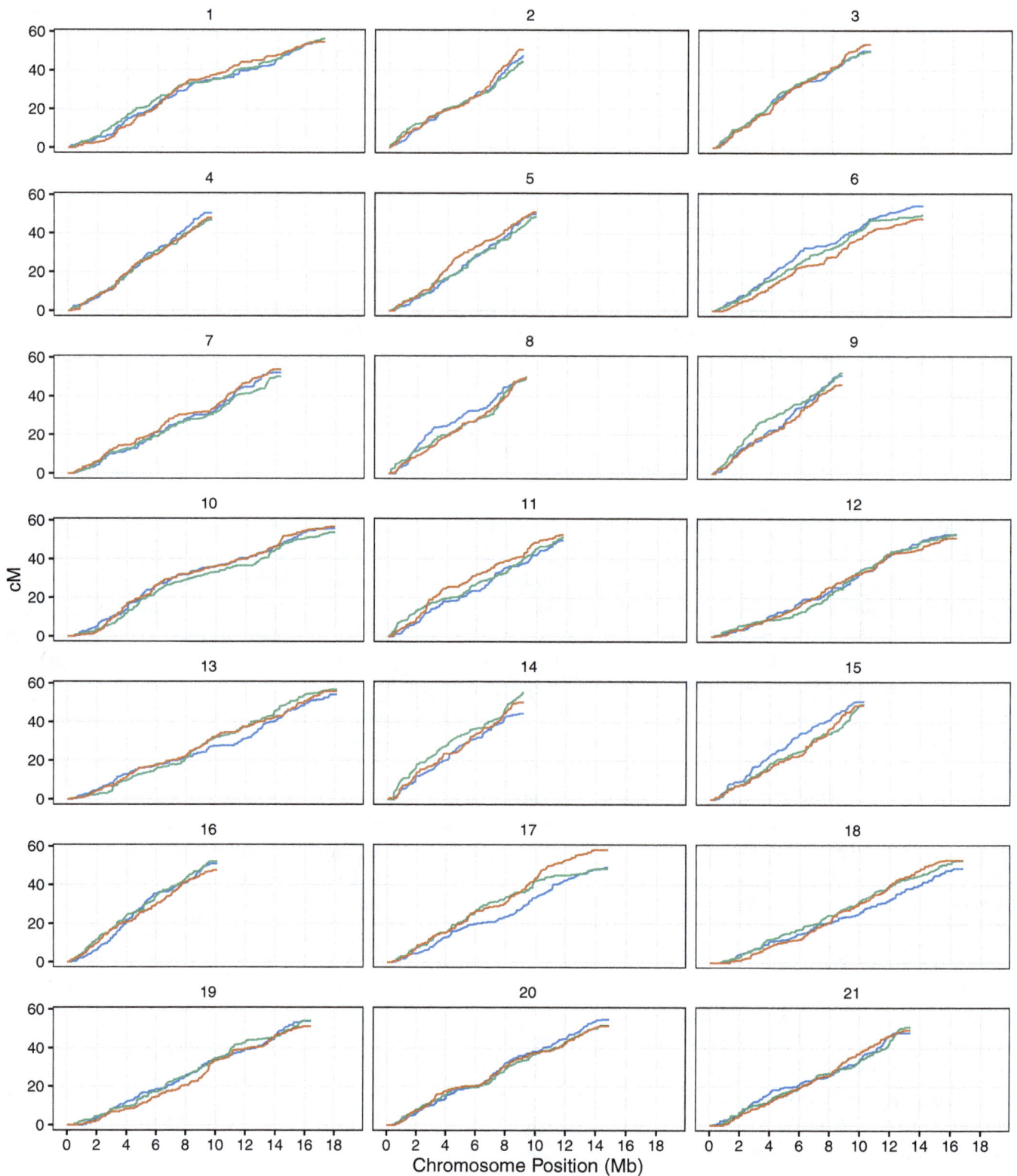

Figure 2. Marey maps of within- and between-species recombination. *Heliconius melpomene*, red; *H. cydno*, blue; *H. cydno* × *H. melpomene* hybrids, green. Chromosomes 1–21 of *H. melpomene* genome assembly version 2 (Hmel2) with improved scaffold ordering shown against cumulative centiMorgan (cM) values.

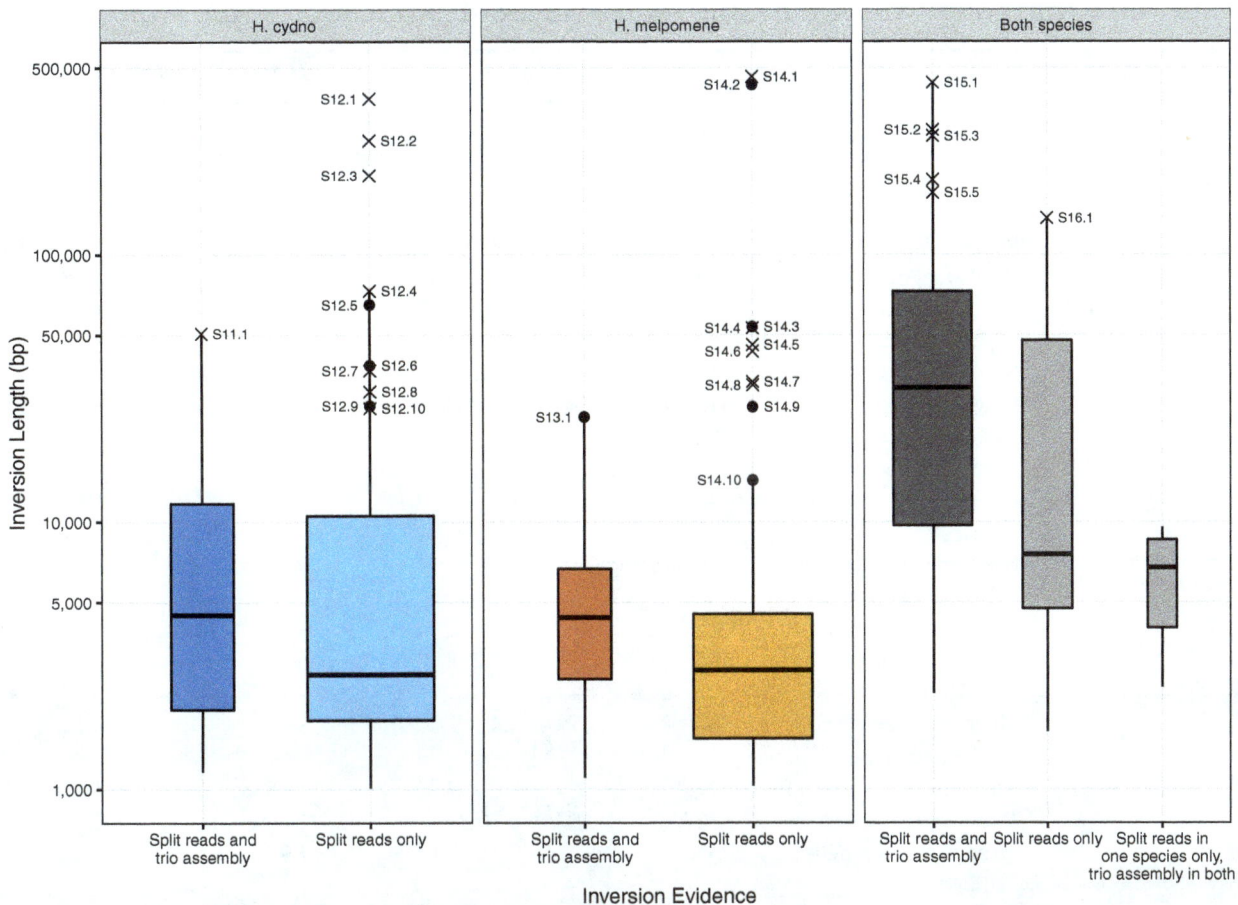

Figure 3. Lengths of candidate inversion groups classified by species and status. See Methods for status definitions. *Heliconius cydno*, blue; *H. melpomene*, red; both species, gray. Dark boxes, evidence from both split reads and trio assemblies; lighter boxes, evidence from either split reads or trio assemblies. Boxes span first and third quartiles; midline shows mean; width represents number of inversions in each category; whiskers extend to the highest value within 1.5 times of the height of the boxes from the edge of the box. Outlier points are shown with crosses if contig gaps fall near inversion breakpoints, circles if not. Labels refer to pages of Figures S11–S17 where full details of each inversion are given.

trio assembly, group has at least one PBHoney inversion candidate and at least one trio scaffold with forward and reverse alignments either side of an inversion breakpoint; *Split reads only*, group has at least one PBHoney inversion candidate in at least one sex, but no matching inverted trio scaffolds; *Split reads in one species*, *trio assembly in both*, group has at least one PBHoney inversion candidate in at least one sex of only one species, but trio assembly has inverted scaffolds in at least one sex in both species. These classifications do not cover whether there are multiple contigs across the candidate inversion (see Table 4; Figs. S11–S17) or whether there are trio scaffolds with alignments that span whole PBHoney inversion candidates or single candidate breakpoints (see Figs. S11–S17).

Martin et al. (2013, 2015a), using all four *H. melpomene rosina*, four *H. cydno chioneus*, four *H. melpomene* French Guiana, and two *H. pardalinus* samples from Martin et al. (2013). Samples were aligned to Hmel2 using bwa mem version 0.7.12 using default parameters and genotypes were called GATK version 3.4 HaplotypeCaller using default parameters except for setting heterozygosity to 0.02. For each candidate inversion, 11 windows equal to the size of the inversion were generated, one for the inversion itself and five either side of the inversion, except where candidates were at the ends of chromosomes. Statistics were calculated for each window with scripts popgenWindows.py and ABBABABAwindows.py in GitHub repository genomics_general (https://github.com/simonhmartin/genomics_general).

POPULATION GENETICS STATISTICS

To look for evidence of variation in gene flow, F_{ST}, d_{XY}, and f_d were calculated within and around candidate inversions following

H. erato ANALYSIS

The *H. erato* version 1 genome assembly was downloaded from LepBase (http://ensembl.lepbase.org/Heliconius_erato_v1)

Table 2. Summary of Pacific Biosciences sequencing and trio assemblies used to identify inversions.

		Heliconius melpomene		*Heliconius cydno*	
		Females	Males	Females	Males
Pacific Biosciences sequences	SMRT cells	15	10	15	10
	Reads	3,138,554	2,079,617	3,022,815	1,218,186
	Reads mapped	3,006,793	1,985,294	2,745,477	1,089,370
	Reads mapped percentage	95.8	95.5	90.8	89.4
	Bases	16,172,976,632	11,157,098,567	15,277,034,979	4,457,967,153
	Bases mapped	15,842,062,694	10,864,609,062	14,285,479,974	4,170,115,456
	Bases mapped percentage	98	97.4	93.5	93.5
	Depth mode for mapped bases	43	27	38	10
	Bases mapped for PBHoney (primary alignments + tails)	10,848,364,007	7,275,050,641	9,173,633,875	2,725,704,499
	Based mapped for PBHoney %	67	65.2	60	61.1
	Mode of base depth for PBHoney bases	37	24	33	8
Trio assemblies	Scaffolds	49,035	46,134	32,548	34,566
	Total length	267.8	276.8	257.9	270.3
	Mean scaffold length (kb)	5.4	6.0	7.9	7.8
	Scaffold N50 (kb)	16.9	20.1	27.0	25.7
	Max scaffold length (kb)	140	165	234	267

and aligned to the ordered Hmel2 scaffolds with LAST version 744 (Kiełbasa et al. 2011). Scaffolds and linkage maps were compared with bespoke scripts Hmel2_Herato_maf.py, compile_Herato_maps.py, and Hmel2_Herato_dotplot.R.

Results

SUMMARY OF SEQUENCED CROSSES, LINKAGE MAPS DATA, IMPROVED ORDERING OF *H. melpomene* ASSEMBLY

We raised crosses within *H. melpomene* (three F_1 crosses, 335 offspring), within *H. cydno* (3 F_1 crosses, 297 offspring) and between *H. cydno* and *H. melpomene* (18 backcrosses of 18 separate F_1 hybrid fathers to 18 separate *H. cydno* females from four pairs of grandparents, 331 offspring; see Table 1 and Fig. S1 for cross designs and Tables S1–S3 for full sample information) and generated PstI RAD sequencing data for a total of 963 offspring as well as whole genome sequencing for parent–offspring trios from *H. melpomene* cross 2 and *H. cydno* cross 1 (Tables S1–S3). Linkage maps were constructed from tens of thousands of SNPs discovered and genotyped in RAD sequencing and whole genome trio

sequencing data (Fig. S2, Tables S4 and S5); separate maps were constructed for each within-species cross, but, due to the varying size and complexity of the hybrid crosses and heterogeneity of hybrid sequencing data, one single linkage map was constructed for all hybrid crosses using more conservative filters, and the single map was divided into separate F1 crosses *post hoc* (Table 1; Fig. S2; see Methods and Supporting Information for full details). We also generated Pacific Biosciences long-read data for pools of male and female larvae from *H. melpomene* cross 2 and *H. cydno* cross 1 (Table 2).

To improve the accuracy of our recombination rate measurements, we first used the new linkage maps and Pacific Biosciences long-read data for *Heliconius melpomene* to improve the scaffolding of version 2 of the *H. melpomene* genome assembly (Hmel2; 795 scaffolds, 275.2 Mb total length, 641 scaffolds placed on chromosomes (274.0 Mb), 2.1 Mb scaffold N50 length; Davey et al. 2016). This resulted in an updated genome assembly with 13 complete chromosomes, the remaining eight chromosomes having one long central scaffold with short unconnected scaffolds at either end (272.6 Mb placed on 21 chromosomes in 38 scaffolds, including 17 minor scaffolds at chromosome ends

Table 3. Physical and genetic map information for each chromosome and species.

Chromosome	Physical length (bp)	Predicted recombination rate (cM/Mb)	Heliconius melpomene		Heliconius cydno		H. cydno x H. melpomene	
			Genetic length (cM)	Rate (cM/Mb)	Genetic length (cM)	Rate (cM/Mb)	Genetic length (cM)	Rate (cM/Mb)
1	17,206,585	5.8	54.6	3.17	54.5	3.17	56.2	3.27
2	9,045,316	11.1	50.7	5.61	47.5	5.25	44.4	4.91
3	10,541,528	9.5	53.7	5.10	50.2	4.76	49.9	4.73
4	9,662,098	10.3	48.1	4.97	50.5	5.23	46.9	4.85
5	9,908,586	10.1	51.0	5.15	50.2	5.06	48.7	4.91
6	14,054,175	7.1	47.8	3.40	54.5	3.88	49.9	3.55
7	14,308,859	7.0	53.7	3.76	52.2	3.65	50.2	3.51
8	9,320,449	10.7	49.3	5.28	49.8	5.35	49.6	5.32
9	8,708,747	11.5	46.3	5.31	50.8	5.84	52.3	6.00
10	17,965,481	5.6	56.7	3.16	55.9	3.11	53.8	3.00
11	11,759,272	8.5	52.5	4.47	49.8	4.24	51.4	4.37
12	16,327,298	6.1	51.0	3.13	52.9	3.24	52.9	3.24
13	18,127,314	5.5	55.8	3.08	54.2	2.99	56.8	3.13
14	9,174,305	10.9	50.2	5.47	44.4	4.84	55.3	6.03
15	10,235,750	9.8	49.0	4.78	50.8	4.97	49.3	4.81
16	10,083,215	9.9	47.5	4.71	50.8	5.04	52.0	5.16
17	14,773,299	6.8	58.2	3.94	49.2	3.33	48.3	3.27
18	16,803,890	6.0	53.1	3.16	48.8	2.91	52.9	3.15
19	16,399,344	6.1	51.0	3.11	53.9	3.29	54.1	3.30
20	14,871,695	6.7	51.3	3.45	54.9	3.69	51.7	3.48
21	13,359,691	7.5	49.6	3.71	48.1	3.60	51.1	3.82
Genome	**272,636,897**	**7.7**	**1081.2**	**3.97**	**1074.1**	**3.94**	**1077.5**	**3.95**
Chromosome		**8.2**	**51.5**	**4.2**	**51.1**	**4.2**	**51.3**	**4.2**

totaling 1.3 Mb; 294 additional scaffolds [2.6 Mb] were not placed on chromosomes and unused in further analyses). This updated reference genome assembly (referred to as ordered Hmel2) was used for all further analyses.

We transferred our existing linkage maps to the new *H. melpomene* chromosomal assembly. Density of SNPs in the final map varies by species and chromosome position (Tables S4 and S5; Figs. S2 and S3; mean paternal SNP density for *H. melpomene*, 6101.1 bp; *H. cydno*, 9043.8 bp; hybrids, 13,642.4 bp). The variation is largely due to variation in sequencing depth and PstI site occurrence, which are both related to GC content (Fig. S3; Benjamini and Speed 2012; see Supporting Information note for full discussion). However, SNP density is not correlated with recombination rate, final map lengths, or crossover resolution, and final map lengths are consistent across all crosses (see below), so we do not believe variation in SNP density has affected our results (see Supporting Information note for further details).

Crossing over has previously been shown to be absent in *Heliconius* females (Turner and Sheppard 1975; Jiggins et al. 2005; Pringle et al. 2007; Davey et al. 2016), and we could find no evidence to the contrary in any of our crosses (Fig. S2), so we

focus on paternal crossovers throughout (see Supporting Information note for a discussion and defense of this point). The paternal linkage maps have a mean genetic length of 51 cM and mean recombination rate of 4.2 cM/Mb per chromosome for both species and hybrids (Table 3). Mean crossovers per offspring across 21 chromosomes were 10.8 in *H. melpomene* (SD 2.4, from 335 offspring) and 10.7 in *H. cydno* (SD 2.2, from 297 offspring). This is consistent with an expectation of one crossover per chromosome per offspring and a 50% chance of inheritance of one of the 2 recombined gametes (from 4 total gametes).

DIFFERENCES IN RECOMBINATION RATE BETWEEN SPECIES AND HYBRIDS

To identify potential genomic regions that may influence the maintenance of the species barrier, we examined our linkage maps for evidence of reduced recombination in the hybrids compared to the within-species crosses at the genome-wide scale, the chromosome scale, and at fine scale (1 Mb windows). Figure 2 shows Marey maps (Chakravarti 1991) for each of the 21 *Heliconius melpomene* chromosomes for *H. melpomene*, *H. cydno*, and *H. cydno* x *H. melpomene* hybrids, with crossovers from all crosses

Table 4. Classification of candidate inversions.

Species	Evidence	Candidate inversions	Breakpoint near contig boundaries (%)	Supporting Information figure
Heliconius cydno	Split reads and trio assembly	13	3 (23%)	S11
	Split reads only	52	17 (33%)	S12
	Total	**65**	**20 (31%)**	
Heliconius melpomene	Split reads and trio assembly	9	4 (44%)	S13
	Split reads only	46	15 (33%)	S14
	Total	**55**	**19 (35%)**	
Both species	Split reads and trio assembly	42	39 (92%)	S15
	Split reads only	17	11 (64%)	S16
	Split reads in one species, trio assembly in both	6	3 (50%)	S17
	Total	**65**	**53 (82%)**	
Grand total		**185**	**92 (50%)**	

per species combined (see Fig. S4 and Table 1 for per-cross Marey maps and map lengths). Mean broad scale recombination rates and total genome-wide map lengths were almost identical across *H. melpomene*, *H. cydno*, and the hybrids (Tables 1 and 3; mean genome-wide recombination rates were all 3.9 cM; mean chromosome-wide recombination rates were all 4.2 cM; total map lengths were *H. melpomene*, 1081 cM; *H. cydno*, 1074.1 cM; hybrids, 1077 cM).

Some differences in chromosome-scale recombination rate between the species maps are visible; for example, on chromosome 17, the *H. melpomene* map is 9.1 cM longer than *H. cydno*; on chromosome 6, *H. cydno* is 6.8 cM longer than *H. melpomene* (Fig. 2; Table 3). However, we are primarily interested in recombination suppression in hybrids, and only chromosome 2 has a significantly reduced chromosome-wide recombination rate in the hybrid crosses, and only when compared to *H. melpomene*, not *H. cydno* (one-tailed Kolmogorov–Smirnov tests; see Methods).

At the fine scale, measuring recombination rate in sliding 1 Mb windows across chromosomes, regions with reduced recombination in the hybrids can be observed (Fig. S5; for example, chromosome 17, 11–13 Mb and chromosome 19, 13.5–14 Mb), but none of these regions are statistically significant (permutation test for fine-scale variation in recombination in 1 Mb sliding windows at a 5% false discovery rate; see Methods).

RECOMBINATION MAPS SHOW NO MAJOR INVERSIONS BETWEEN SPECIES

We also examined our recombination maps for evidence of inversions between species (Fig. 1). There are no regions of any map with a detectable reversed region in *H. cydno* or the hybrids with respect to *H. melpomene* (Fig. 2). This is true for the species

maps and for all individual cross maps (Fig. S4). This indicates there are no large fixed inversions between *H. melpomene* and *H. cydno*.

Known or predicted chromosome inversions involved in the maintenance of species barriers are typically megabases long, and models indicate that inversions may have to be very large to become fixed in a population (Feder et al. 2014). Our maps are sufficiently fine scale to rule out the presence of inversions on the megabase scale (*H. melpomene* mean gap between markers, 115 kb, median 87 kb, maximum 1.38 Mb; *H. cydno* mean 135 kb, median 101 kb, maximum 1.14 Mb; see Figs. S6 and S7, and Supporting Information notes on our ability to detect and resolve crossovers). Simulation of random inversions indicates that our existing maps give us power to detect ~98% of 500 kb inversions, ~90% of 250 kb inversions and ~75% of 100 kb inversions (Fig. S8). These sizes are smaller than most inversions known to be associated with adaptive traits or species barriers, which are typically on the megabase scale; however, they are on the order of the sizes of the known inversions involved in within-species polymorphism in *H. numata* (see Introduction). The recombination maps alone do not rule out the presence of an inversion in any remaining gap between markers within *H. melpomene* or *H. cydno*.

DETECTION OF SMALL INVERSIONS WITH LONG SEQUENCE READS AND TRIO ASSEMBLIES

To test for the presence of smaller fixed inversions between *H. melpomene* and *H. cydno* that were undetectable using our recombination maps, we generated Pacific Biosciences long-read sequence data for pools of male and female larvae from one each of the *H. melpomene* and *H. cydno* crosses used to generate recombination maps (Table 2; Figs. S9 and S10). We called candidate inversions from the long-read data using PBHoney to identify reads with clipped alignments, realign the clipped read ends, and detect such split reads with inverted alignments.

We also generated Illumina short-read assemblies of the maternal and paternal genomes of one offspring from the same crosses used to generate the linkage maps and PBHoney candidates. These assemblies were constructed using a trio assembly method that separates maternal and paternal reads from one offspring and constructs haplotypic assemblies of each parental genome, providing longer and more accurate contigs compared to standard Illumina assemblies of heterogenous genomes such as those of *Heliconius* species (Malinsky et al. 2016; Table 2). We aligned these trio assemblies to the *H. melpomene* genome and assessed whether the resulting alignments supported or conflicted with candidate split read inversions.

In total, 1494 raw PBHoney split read candidates were identified across the four samples (two sexes for each of two species; Tables 2 and S6). As we consider our linkage maps to be reliable, and we are concerned with regions of the genome where our linkage maps do not contain recombinations, we rejected 438 split read candidates (30%) that spanned recombinations in the linkage maps (Table S6), of which 294 (20%) were longer than 1 Mb, with 36 (2.5%) longer than 10 Mb. The remaining candidates were all in regions that may contain crossovers but where crossover location could not be resolved, or in regions where multiple SNPs showed that there were no crossovers and so recombination could not be used to detect inversions (see Supporting Information note for discussion).

A further 344 candidates (23%) were removed because the candidate was spanned by a trio scaffold from the same species by 50% of the inversion length on either side (Table S6). These rejected candidates are likely to be mostly false positives; when we simulated PacBio reads directly from the ordered Hmel2 reference genome, PBHoney called 49 "false-positive" inversions. Alternatively, they may be generated by polymorphic inversions that are not present in the two parental haplotypes in the trio assemblies, but are present in at least one of the other two parental haplotypes and so detectable in the PacBio data, but as we expect only fixed inversions to contribute to species barriers, we have not considered these candidates any further.

A further 199 of the 1494 candidates (13%) were removed because they were shorter than 1 kb (Table S6) on the grounds that there is already above-background linkage disequilibrium between SNPs separated by 1 kb or less in *H. melpomene* (Martin et al. 2016). The remaining 463 split read candidate inversions from the four samples were merged into 185 candidate groups based on their overlaps. We expect fixed inversions to be present in both sexes for each species, but the four samples were sequenced with variable coverage, with particularly low coverage for the

H. cydno males (Table 2). Given this, 173 additional candidates with less robust support that overlapped with the 185 merged groups were included in the dataset (Table S6). Each of the merged groups was then classified based on their presence in either or both species and their support by split read and trio assembly evidence (Table 4; Figs. 3 and S11–S17; Table S7; see Methods for full criteria).

Despite the high rate of likely false positives, PBHoney does appear to detect some genuine inverted sequences relative to the reference genome. Where candidate inversions are identified in the same location from both *H. melpomene* and *H. cydno* sequence data, it is likely that these candidates are accurately reflecting a misassembly in the reference genome. This is especially the case where the inversion breakpoints fall at contig boundaries in the assembly, as local misassembly can prevent neighboring contigs from being assembled. There were 59 candidate groups where PBHoney found overlapping inversions in both *H. cydno* and *H. melpomene*, 50 (85%) of which span multiple contigs, with most inversion breakpoints falling at or near to the end of a contig (Table 4; Figs. 3 and S15–S17). This indicates either that some whole contigs are inverted, or that the ends of contigs have inverted regions that need to be reassembled (which perhaps explains the failure to fill the contig gaps during assembly). In contrast, candidate inversions specific to one or other species were less likely to span multiple contigs (20 of 65 *H. cydno* candidates (31%), and 19 of 56 *H. melpomene* candidates (35%); Figs. S11–S14). We suggest that while some of these species-specific inversions could be explained by misassemblies and incomplete PacBio coverage across both species, many of them could be genuine inversions.

CANDIDATE INVERSIONS ASSESSED USING TRIO ASSEMBLIES AND POPULATION GENETICS

As the false positive rate for PBHoney is high, we made further use of the trio assemblies to find support for the remaining PBHoney candidate inversion groups (Tables 4 and S7, Figs. S11–S17). Thirteen *H. cydno* and nine *H. melpomene* groups were further supported by trio scaffolds aligning with inverted hits within inversion breakpoints (Fig. 3, "Split reads and trio assembly"; Figs. S11 and S13). Of these, eight *H. cydno* and three *H. melpomene* candidates did not have inversion breakpoints near contig boundaries, suggesting that they are less likely to be due to genome misassemblies. If these inversions are species-specific, as indicated by the PBHoney output, we expect support for the reference genome order in the species that does not possess the inversion candidate. Indeed, six of these *H. cydno* and all three *H. melpomene* candidates have trio scaffolds of the other species spanning the whole inversion or one of the breakpoints, supporting the inversion as being species-specific (Figs. S11.2, S11.4, S11.6, S11.8, S11.10, S11.11; Fig S13.6, S13.8, S13.9). Hence, we have likely detected a small number of species-specific inversions. However, the longest of these candidates is Figure S11.2 at 20,247 bp, far shorter than any known inversion relevant for speciation and shorter than is expected to become fixed in simulations (Feder et al. 2014). Furthermore, this is only slightly larger than the distance at which linkage disequilibrium in *H. melpomene* reaches background levels (~10 kb; Martin et al. 2016), such that any effect of reduced recombination would be slight in population genetic terms. We conclude that there are a small number of likely species-specific inversions, but that these are too small to play a role in speciation via reduced recombination. Notably, none of these candidate inversions were located near loci known or suspected to determine species differences in wing pattern or any other trait with known locations (Nadeau et al. 2014, Davey et al. 2016; we have transferred the *H. melpomene* loci to positions in the ordered Hmel2 genome in Table S8 and also included a table of all candidate inversion positions for comparison in Table S7; the BD region has been narrowed based on the results of Wallbank et al. 2016).

We also calculated F_{ST}, d_{XY}, and f_d (Cruickshank and Hahn 2014; Martin et al. 2015a) across inverted regions (Figs. S11–S17) to look for evidence of variation in gene flow at the inversion relative to surrounding regions. An inversion acting as a species barrier typically produces a signal of elevated F_{ST} and reduced admixture (here estimated using f_d; Aulard et al. 2002; Deng et al. 2008; Huynh et al. 2011; Nachman and Payseur 2012; Fontaine et al. 2015; Love et al. 2016), and an inversion enabling the spread of an adaptive cassette between species (Kirkpatrick and Barrett 2015) might produce a signal of elevated f_d. However, we see very little evidence for deviations in these statistics within the handful of candidate inversions compared to the surrounding regions, with only one *H. cydno* inversion (Fig. S11.4, 11,719 bp long) showing a noticeable localized increase in F_{ST} and small increase in d_{XY}. This region contains no annotated features, although of course this does not rule out some functional importance of this region.

Some candidates with only split read evidence, many in only one sex, are hundreds of kilobases long (outliers labeled in Fig. 3, particularly those marked with circles, where breakpoints are not near contig boundaries), which, if real, may be relevant to speciation. However, given the large number of false positives produced by PBHoney, the lack of supporting evidence from trio assemblies, and the lack of clear, localized deviations in F_{ST}, d_{XY}, and f_d signals at these candidates, it is unlikely these candidates, even if they are real, are substantial species barriers.

THE *H. melpomene* AND *H. erato* GENOMES ARE MOSTLY COLLINEAR, BUT DO CONTAIN INVERTED REGIONS

We used the recently completed *H. erato* genome assembly (Van Belleghem et al. 2017) to investigate the incidence of inversions between more divergent genomes in the *Heliconius* genus.

Heliconius melpomene and *H. erato* diverged 10 million years ago (Kozak et al. 2015; Fig. S18), considerably more than the ~1.5 million years between *H. melpomene* and *H. cydno*. Despite the substantial divergence time, the chromosomes of the two species are collinear throughout at the large scale, with a few exceptions. There are many regions of the *H. erato* genome assembly that are inverted relative to the ordered Hmel2 assembly, but they fall within regions where the *H. erato* or *H. melpomene* linkage maps were not informative and so may be due to genome misassemblies. For example, *H. erato* scaffolds Herato0201, Herato0202, and Herato0203 on chromosome 2, and the first 300 kb of *H. melpomene* chromosome 3, may be misoriented rather than genuinely inverted.

However, three large inverted and/or translocated regions are well supported by linkage map markers in both species, and so are likely to be genuine inversions (Fig. S18; chromosome 2, *H. erato* 7–10 Mb, *H. melpomene* 4–7 Mb; chromosome 6, *H. erato* 16–18 Mb, *H. melpomene* 12–13 Mb; chromosome 20, *H. erato* 13–15 Mb, *H. melpomene* 11–12 Mb). The chromosome 2 rearrangement is particularly striking, spanning four *H. erato* scaffolds (Herato0211, Herato0212, Herato0213, and Herato0214) and multiple linkage map markers in both species. On current scaffold ordering, this rearrangement appears to be an inversion followed by a translocation (for scaffold Herato0214), but it is likely to be a single inversion; as scaffolds Herato0212, Herato0213, and Herato0214 are all found at the same marker on the linkage map, it may be that these scaffolds need to be reoriented and reordered, inserting scaffold Herato0212 at the start of the inversion in Herato0211 and inverting Herato0214. Nevertheless, this large region deserves further attention, especially as some pairs of *H. erato* subspecies appear to have elevated F_{st} in the center of chromosome 2 (see Fig. 2 of Van Belleghem et al. 2017). It is unclear whether this inversion is polymorphic only in *H. erato* or whether it is present in other *Heliconius* species.

Discussion

We have systematically tested the hypothesis that inversions causing reduced recombination rates in hybrids might maintain species barriers with gene flow (Ortíz-Barrientos et al. 2016). High-density linkage maps and high-coverage long-read sequence data give us considerable power to both measure recombination rate and detect structural rearrangements. We find evidence for some small inversions, but not for inversion differences between *H. melpomene* and *H. cydno* at a scale that is likely to influence the speciation process.

Our data have some limitations that might have prevented us from identifying genuine inversions between *H. melpomene* and *H. cydno*. First, we have only sequenced crosses from three or four pairs of parents per species, and so may have missed polymorphic inversions absent from our sampling of wild individuals. However, any inversion important for speciation is expected to be fixed between the species, so it should have been detected even in small samples. Second, our ability to detect inversions and differences in recombination is limited by the size of our crosses (roughly 300 individuals for each species and for the hybrids), and the maps contain regions of the genome up to a maximum of 1.3 Mb without crossovers that might conceivably harbor inversions (see Results, recombination maps show no major inversions between species); further crosses could improve resolution in these areas. Third, we have used RAD sequencing data to measure recombinations, which is limited to ∼10 kb resolution (using the PstI restriction enzyme); some of the smaller candidate inversions could be confirmed by developing further markers within them at narrower resolution, but this would not change our conclusion that inversions are unlikely to be involved in speciation between *H. melpomene* and *H. cydno*.

One important aspect of our experimental design is that we have measured recombination in hybrids as well as investigating gene order in the parental species. This gives power to detect reduced hybrid recombination rate more generally as well as specifically the presence of inversions. We have found no evidence for significantly reduced recombination in hybrids, at the broad (chromosome) scale or megabase scale, suggesting that genic modifiers of recombination are unlikely to have widespread effects in these species. Larger crosses would give greater resolution to this test, and might detect smaller regions of reduced recombination. Nevertheless, we can decisively rule out the presence of any multimegabase rearrangements among these samples.

We complemented the linkage maps with PacBio sequencing and trio assemblies to detect candidate inverted regions at a smaller scale. This approach also has challenges and generated a high rate of false positives. One potential source of difficulties is reliance on alignment to the *H. melpomene* reference genome assembly. The existing assembly has 25% transposable element content (Lavoie et al. 2013) and is likely missing around 6% of true genome sequence, mostly due to collapsed repeats (Davey et al. 2016). Inversion breakpoints are typically repeat-rich, which increases the likelihood that reads or scaffolds will not align correctly, and that the breakpoint regions could be misassembled or absent in the reference genome and in the trio assemblies. This problem may be worse for *H. cydno*, where more divergent sequence may align incorrectly or not align at all, and unique *H. cydno* sequence will not be present in the reference (an additional ∼5% of *H. cydno* sequence did not map to the *H. melpomene* genome compared to *H. melpomene* samples; Table 2). These issues may explain the high observed rate of false positives in our data.

Nonetheless, the detection of likely genome misassemblies indicates that our methods do indeed have the power to detect real rearrangements. These are supported by multiple lines of evidence in both species and fall near contig boundaries. These misassemblies could be due to whole inverted contigs, or to misassembled inverted regions at the ends of contigs, which may be preventing the contigs being joined by spanning reads. Misassemblies demonstrate that our methods are capable of detecting large rearrangements in the sampled reads relative to the genome assembly.

In contrast, our candidate species-specific inversions are typically smaller than the misassemblies, and are mostly not supported by multiple lines of evidence. Indeed, we can find no compelling fixed candidate inversions supported by both the split read and trio assembly datasets that also show evidence of an increase in F_{ST} or d_{XY}, except for the 11.7 kb inversion shown in Figure S11.4, which is probably too small to substantially increase linkage across this locus beyond that expected by normal decay of linkage disequilibrium (Martin et al. 2016). It is possible that some of the candidates with less robust evidence are genuine, given the limitations described above, but on the existing evidence we cannot identify any inversions that are likely to be involved in maintaining species barriers between *H. melpomene* and *H. cydno*.

Although existing models identify situations where chromosome inversions can spread to fixation between two species and maintain a species barrier, they do not show that inversions always spread during speciation with gene flow. For example, in the model of Feder et al. (2014), inversions only fix when the strength of selection on the loci captured by the inversion is considerably lower than migration between the species. Similarly, Dagilis and Kirkpatrick (2016), modeling the spread of inversions that capture a mate preference locus and one or more epistatic hybrid viability genes, show that inversions are unlikely to spread where pre- and post-zygotic reproductive isolation is already strong. In a recent review, Ortíz-Barrientos et al. (2016) also highlight that during reinforcement, assortative mating and recombination modifiers such as inversions are antagonistic; if strong assortative mating arises first, there is only weak selection for reduced recombination.

We considered *H. melpomene* and *H. cydno* to be good candidates for the spread of inversions because there are linked loci causing reproductive isolation, because hybridization has been ongoing for much of their history, because an inversion is known to maintain color pattern polymorphism in *H. numata* (Joron et al. 2011), and because they are a parallel case to that of *D. pseudoobscura* and *D. persimilis*, where inversions do appear to maintain the species barrier (Noor et al. 2007). Comparisons between sympatric and allopatric populations of the two *Heliconius* species have shown that almost a third of the genome is admixed in sympatry and that hybridization has been ongoing for a long time (Martin et al. 2013), perhaps at a low rate.

However, strong selection on species differences and assortative mating are not conducive to the spread of inversions. Aposematic warning patterns are strongly selected (Mallet and Barton 1989) with F_1 hybrids twice as likely to be attacked as parental phenotypes (Merrill et al. 2012), and prezygotic isolation in the form of mate preference is almost complete (Jiggins et al. 2001). Therefore, inversions may not be necessary for divergent loci to accumulate between the species. Thus, in this case, the evolution of strong assortative mating may have been favored by reinforcement selection and close physical linkage between preference and wing-patterning loci (Merrill et al. 2011b), and it is likely that the species barrier between *H. melpomene* and *H. cydno* has persisted with gene flow, but without the suppression of recombination by chromosome inversions.

An alternative and complementary explanation is that the rate of production of inversions may simply be low in *Heliconius*. This is suggested by the low background rate of fixation of inversions in *Heliconius* genomes. We have shown that, not only is there little evidence for substantial, fixed inversions between *H. melpomene* and *H. cydno*, but also that *H. melpomene* and *H. erato*, which last shared a common ancestor over 10 million years ago, have largely collinear genomes, and it is also known that there is substantial chromosomal synteny across the Nymphalids (Ahola et al. 2014). The association of multiple inversions with the wing pattern polymorphism in *H. numata* is all the more remarkable given the low background rate of inversions in these butterflies. This contrasts with, for example, the many fixed or polymorphic inversions in the genomes of *Drosophila* (Krimbas and Powell 1992), *Anopheles* (Ayala et al. 2014), and primates (Samonte and Eichler 2002), and especially with the solid case for the influence of inversions on speciation between sympatric populations of *D. pseudoobscura* and *D. persimilis* (Noor et al. 2001; Machado et al. 2007; Noor et al. 2007), where major fixed inversions occur on most chromosomes. Although *H. melpomene* and *H. cydno* have similarly divergent genomes overall compared to the *Drosophila* species pair, we do not find evidence for a similar role for inversions in maintaining the species barrier. Although inversions are clearly involved in speciation in many taxa studied to date, they appear to be absent in *H. melpomene* and *H. cydno* and in the flycatcher species pair *Ficedula albicollis* and *F. hypoleuca* (Ellegren et al. 2012), so the possibility of speciation without inversions should be kept in mind. We conclude that species barriers can persist during speciation with gene flow without substantial suppression of between-species recombination.

AUTHOR CONTRIBUTIONS

JWD and CJ conceived and designed the project. JWD carried out *Heliconius melpomene* crosses, analyzed the data, and wrote the manuscript. SLB extracted DNA samples and prepared RAD libraries. PMR and JWD wrote linkage mapping software and built linkage maps. AP carried out

H. cydno crosses and extracted DNA samples. SHM prepared population samples and wrote population genetics software. RMM carried out hybrid crosses and prepared RAD libraries. JWD, RMM, SHM, RD, and CDJ designed experiments and sequencing strategies. WOM managed insectaries in Panama and contributed to crossing designs. All authors read and commented on the manuscript.

ACKNOWLEDGMENTS

We are grateful to A. Tapia, L. Evans, M.-C. Melo, J. Scott, B. v. Schooten, A. Gonzalez-Karlsson, M. Abanto, T. Thurman, and all staff at the Smithsonian Tropical Research Institute for support and assistance with insectary work. The RAD sequencing protocol was developed by P. Etter at the University of Oregon and further enhanced by P. Fuentes-Utrilla and C. Eland at Edinburgh Genomics. PacBio sequencing was carried out by K. Oliver at the Sanger Institute, supported by European Research Council grant number 339873 and Wellcome Trust grant number 098051. We thank BGI and Edinburgh Genomics for RAD sequencing, M. Malinsky for whole genome trio sequencing, and J. Barna for computing support. Analyses were performed using the Darwin Supercomputer of the University of Cambridge High Performance Computing Service (http://www.hpc.cam.ac.uk/), provided by Dell, Inc. and funded using Strategic Research Infrastructure Funding from the Higher Education Funding Council for England, and funding from the Science and Technology Facilities Council.

This work was primarily funded by European Research Council Advanced Grant SpeciationGenetics 339873 to CDJ. JWD was funded by a Herchel Smith Postdoctoral Research Fellowship and a Smithsonian Tropical Research Institute Fellowship. SHM was funded by a research fellowship from St. John's College, Cambridge. RMM was funded by a Biotechnology and Biological Sciences Research Council doctoral training grant and a Junior Research Fellowship from King's College, Cambridge. The authors have no conflicts of interest to declare.

LITERATURE CITED

Ahola, V., R. Lehtonen, P. Somervuo, L. Salmela, P. Koskinen, P. Rastas, et al. 2014. The Glanville fritillary genome retains an ancient karyotype and reveals selective chromosomal fusions in Lepidoptera. Nat. Commun. 5:4737. Available at https://doi.org/10.1038/ncomms5737.

Arias, C. F., C. Salazar, C. Rosales, M. R. Kronforst, M. Linares, E. Bermingham, and W. O. McMillan. 2014. Phylogeography of *Heliconius cydno* and its closest relatives: disentangling their origin and diversification. Mol. Ecol. 23:4137–4152.

Aulard, S., J. R. David, and F. Lemeunier. 2002. Chromosomal inversion polymorphism in Afrotropical populations of *Drosophila melanogaster*. Genet. Res. Camb. 79:49–63.

Ayala, D., R. F. Guerrero, and M. Kirkpatrick. 2013. Reproductive isolation and local adaptation quantified for a chromosome inversion in a malaria mosquito. Evolution 67:946–958.

Ayala, D., A. Ullastres, and J. González. 2014. Adaptation through chromosomal inversions in *Anopheles*. Front. Genet. 5:129.

Ayala, F. J., and M. Coluzzi. 2005. Chromosome speciation: humans, *Drosophila*, and mosquitoes. Proc. Natl. Acad. Sci. USA. 102:6535–6542.

Benjamini, Y., and T. P. Speed. 2012. Summarizing and correcting the GC content bias in high-throughput sequencing. Nucleic Acids Res. 40:e72.

Brown, K. M., L. M. Burk, L. M. Henagan, and M. A. F. Noor. 2004. A test of the chromosomal rearrangement model of speciation in *Drosophila pseudoobscura*. Evolution 58:1856–1860.

Brown, K. S., T. C. Emmel, P. J. Eliazar, and E. Suomalainen. 1992. Evolutionary patterns in chromosome numbers in neotropical Lepidoptera. Hereditas 117:109–125.

Butlin, R. K. 2005. Recombination and speciation. Mol. Ecol. 14:2621–2635.

Castiglia, R. 2014. Sympatric sister species in rodents are more chromosomally differentiated than allopatric ones: implications for the role of chromosomal rearrangements in speciation. Mammal Rev. 44:1–4.

Catchen, J., P. A. Hohenlohe, S. Bassham, A. Amores, and W. A. Cresko. 2013. Stacks: an analysis tool set for population genomics. Mol. Ecol. 22:3124–3140.

Chakravarti, A. 1991. A graphical representation of genetic and physical maps: the Marey map. Genomics 11:219–222.

Cruickshank, T. E., and M. W. Hahn. 2014. Reanalysis suggests that genomic islands of speciation are due to reduced diversity, not reduced gene flow. Mol. Ecol. 23:3133–3157.

Dagilis, A. J., and M. Kirkpatrick. 2016. Prezygotic isolation, mating preferences, and the evolution of chromosomal inversions. Evolution 70:1465–1472.

Davey, J. W., M. Chouteau, S. L. Barker, L. Maroja, S. W. Baxter, F. Simpson, et al. 2016. Major improvements to the *Heliconius melpomene* genome assembly used to confirm 10 chromosome fusion events in 6 million years of butterfly evolution. G3 6:695–708.

Deng, L., Y. Zhang, J. Kang, T. Liu, H. Zhao, Y. Gao, et al. 2008. An unusual haplotype structure on human chromosome 8p23 derived from the inversion polymorphism. Hum. Mutat. 29:1209–1216.

DePristo, M. A., E. Banks, R. Poplin, K. V. Garimella, J. R. Maguire, C. Hartl, et al. 2011. A framework for variation discovery and genotyping using next-generation DNA sequencing data. Nat Genet 43:491–498.

Ellegren, H., L. Smeds, R. Burri, P. I. Olason, N. Backström, T. Kawakami, et al. 2012. The genomic landscape of species divergence in Ficedula flycatchers. Nature 491:756–760.

English, A. C., W. J. Salerno, and J. G. Reid. 2014. PBHoney: identifying genomic variants via long-read discordance and interrupted mapping. BMC Bioinform. 15:180.

Estrada, C., and C. D. Jiggins. 2002. Patterns of pollen feeding and habitat preference among *Heliconius* species. Ecol. Entomol. 27:448–456.

Faria, R., and A. Navarro. 2010. Chromosomal speciation revisited: rearranging theory with pieces of evidence. Trends Ecol. Evol. 25:660–669.

Farré, M.,Micheletti, D., &A. Ruiz-Herrera. 2013. Recombination rates and genomic shuffling in human and chimpanzee–a new twist in the chromosomal speciation theory. Evol. 30:853–864.

Feder, J. L., and P. Nosil. 2009. Chromosomal inversions and species differences: when are genes affecting adaptive divergence and reproductive isolation expected to reside within inversions? Evolution 63:3061–3075.

Feder, J. L., R. Gejji, T. H. Q. Powell, and P. Nosil. 2011. Adaptive chromosomal divergence driven by mixed geographic mode of evolution. Evolution 65:2157–2170.

Feder, J. L., P. Nosil, and S. M. Flaxman. 2014. Assessing when chromosomal rearrangements affect the dynamics of speciation: implications from computer simulations. Front Genet. 5:295.

Felsenstein, J. 1981. Skepticism towards Santa Rosalia, or why are there so few kinds of animals? Evolution 35:124–138.

Fishman, L., A. Stathos, P. M. Beardsley, C. F. Williams, and J. P. Hill. 2013. Chromosomal rearrangements and the genetics of reproductive barriers in *Mimulus* (monkey flowers). Evolution 67:2547–2560.

Fontaine, M. C., J. B. Pease, A. Steele, R. M. Waterhouse, D. E. Neafsey, I. V. Sharakhov, et al. 2015. Extensive introgression in a malaria vector species complex revealed by phylogenomics. Science 347:1258524-1–1258524-6.

Guerrero, R. F., F. Rousset, and M. Kirkpatrick. 2012. Coalescent patterns for chromosomal inversions in divergent populations. Phil. Trans. R. Soc. B 367:430–438.

Heliconius Genome Consortium. 2012. Butterfly genome reveals promiscuous exchange of mimicry adaptations among species. Nature 487:94–98.

Hoffmann, A. A., and L. H. Rieseberg. 2008. Revisiting the impact of inversions in evolution: from population genetic markers to drivers of adaptive shifts and speciation? Annu. Rev. Ecol. Syst. 39:21–42.

Hooper, D. M. 2016. Range overlap drives chromosome inversion fixation in passerine birds. bioRxiv. https://doi.org/10.1101/053371.

Huynh, L. Y., D. L. Maney, and J. W. Thomas. 2011. Chromosome-wide linkage disequilibrium caused by an inversion polymorphism in the white-throated sparrow (*Zonotrichia albicollis*). Heredity 106:537–546.

Jiggins, C. D. 2008. Ecological speciation in mimetic butterflies. BioScience 58:541–548.

Jiggins, C. D., R. E. Naisbit, R. L. Coe, and J. L. B. Mallet. 2001. Reproductive isolation caused by colour pattern mimicry. Nature 411:302–305.

Jiggins, C. D., J. Mavarez, M. Beltrán, W. O. McMillan, J. S. Johnston, and E. Bermingham. 2005. A genetic linkage map of the mimetic butterfly *Heliconius melpomene*. Genetics 171:557–570.

Jones, F. C., M. G. Grabherr, Y. F. Chan, P. Russell, E. Mauceli, J. Johnson, et al. 2012. The genomic basis of adaptive evolution in threespine sticklebacks. Nature 484:55–61.

Joron, M., L. Frezal, R. T. Jones, N. L. Chamberlain, S. F. Lee, C. R. Haag, et al. 2011. Chromosomal rearrangements maintain a polymorphic supergene controlling butterfly mimicry. Nature 477:203–206.

Kiełbasa, S. M., R. Wan, K. Sato, P. Horton, and M. C. Frith. 2011. Adaptive seeds tame genomic sequence comparison. Genome Res 21:487–493.

King, M. 1993. Species evolution: the role of chromosome change. Cambridge Univ. Press, Cambridge, U.K.

Kirkpatrick, M. 2010. How and why chromosome inversions evolve. PLOS Biol 8:e1000501.

Kirkpatrick, M., and B. Barrett. 2015. Chromosome inversions, adaptive cassettes and the evolution of species' ranges. Mol. Ecol. 24:2046–2055.

Kirkpatrick, M., and N. Barton. 2006. Chromosome inversions, local adaptation and speciation. Genetics 173:419–434.

Korunes, K. L., and M. A. F. Noor. 2017. Gene Conversion and Linkage: effects on genome evolution and speciation. Mol. Ecol. 26:351–364.

Kozak, K. M., N. Wahlberg, A. F. E. Neild, K. K. Dasmahapatra, J. L. B. Mallet, and C. D. Jiggins. 2015. Multilocus species trees show the recent adaptive radiation of the mimetic *Heliconius* butterflies. Syst. Biol. 64:505–524.

Krimbas, C. B. and J. R. Powell. 1992. Drosophila inversion polymorphism. CRC Press, Boca Raton, FL.

Kronforst, M. R., M. E. B. Hansen, N. G. Crawford, J. R. Gallant, W. Zhang, R. J. Kulathinal, et al. 2013. Hybridization reveals the evolving genomic architecture of speciation. Cell Rep. 5:666–677.

Kurtz, S., A. Phillippy, A. L. Delcher, M. Smoot, M. Shumway, C. Antonescu, and S. L. Salzberg. 2004. Versatile and open software for comparing large genomes. Genome Biol. 5:R12.

Lavoie, C. A., R. N. Platt, P. A. Novick, B. A. Counterman, and D. A. Ray. 2013. Transposable element evolution in *Heliconius* suggests genome diversity within Lepidoptera. Mobile DNA 4:21.

Li, H. 2011. A statistical framework for SNP calling, mutation discovery, association mapping and population genetical parameter estimation from sequencing data. Bioinformatics 27:2987–2993.

———. 2013. Aligning sequence reads, clone sequences and assembly contigs with BWA-MEM. arXiv 1303.3997v2

Lohse, K., M. Clarke, M. G. Ritchie, and W. J. Etges. 2015. Genome-wide tests for introgression between cactophilic *Drosophila* implicate a role of inversions during speciation. Evolution 69:1178–1190.

Love, R. R., A. M. Steele, M. B. Coulibaly, S. F. Traorè, S. J. Emrich, M. C. Fontaine, and N. J. Besansky. 2016. Chromosomal inversions and ecotypic differentiation in *Anopheles gambiae*: the perspective from whole-genome sequencing. Mol. Ecol. 25:5889–5906.

Lunter, G., and M. Goodson. 2011. Stampy: a statistical algorithm for sensitive and fast mapping of Illumina sequence reads. Genome Res. 21: 936–939.

Machado, C. A., T. S. Haselkorn, and M. A. F. Noor. 2007. Evaluation of the genomic extent of effects of fixed inversion differences on intraspecific variation and interspecific gene flow in *Drosophila pseudoobscura* and *D. persimilis*. Genetics 175:1289–1306.

Malinsky, M., J. T. Simpson, and R. Durbin. 2016. trio-sga: facilitating de novo assembly of highly heterozygous genomes with parent-child trios. bioRxiv. https://doi.org/10.1101/051516

Mallet, J. L. B., and N. H. Barton. 1989. Strong natural selection in a warning-color hybrid zone. Evolution 43:421–431.

Mallet, J. L. B., M. Beltrán, W. Neukirchen, and M. Linares. 2007. Natural hybridization in heliconiine butterflies: the species boundary as a continuum. BMC Evol. Biol. 7:28.

Martin, S. H., K. K. Dasmahapatra, N. J. Nadeau, C. Salazar, J. R. Walters, F. Simpson, et al. 2013. Genome-wide evidence for speciation with gene flow in *Heliconius* butterflies. Genome Res. 23:1817–1828.

Martin, S. H., J. W. Davey, and C. D. Jiggins. 2015a. Evaluating the use of ABBA-BABA statistics to locate introgressed loci. Mol. Biol. Ecol. 32:244–257.

Martin, S. H., A. Eriksson, K. M. Kozak, and A. Manica. 2015b. Speciation in Heliconius butterflies: minimal contact followed by millions of generations of hybridisation. bioRxiv. https://doi.org/10.1101/015800

Martin, S. H., M. Möst, W. J. Palmer, C. Salazar, W. O. McMillan, F. M. Jiggins, and C. D. Jiggins. 2016. Natural Selection and Genetic Diversity in the Butterfly *Heliconius melpomene*. Genetics 203:525–541.

McGaugh, S. E., and M. A. F. Noor. 2012. Genomic impacts of chromosomal inversions in parapatric *Drosophila* species. Phil. Trans. R. Soc. B 367:422–429.

Merrill, R. M., Z. Gompert, L. M. Dembeck, M. R. Kronforst, W. O. McMillan, and C. D. Jiggins. 2011a. Mate preference across the speciation continuum in a clade of mimetic butterflies. Evolution 65:1489–1500.

Merrill, R. M., B. Van Schooten, J. A. Scott, and C. D. Jiggins. 2011b. Pervasive genetic associations between traits causing reproductive isolation in *Heliconius* butterflies. Proc. R. Soc. B 278:511–518.

Merrill, R. M., R. W. R. Wallbank, V. Bull, P. C. A. Salazar, J. L. B. Mallet, M. Stevens, and C. D. Jiggins. 2012. Disruptive ecological selection on a mating cue. Proc. R. Soc. B. 279:4907–4913.

Merrill, R. M., R. E. Naisbit, J. L. B. Mallet, and C. D. Jiggins. 2013. Ecological and genetic factors influencing the transition between host-use strategies in sympatric *Heliconius* butterflies. J. Evol. Biol. 26:1959–1967.

Merrill, R. M., K. K. Dasmahapatra, J. W. Davey, D. D. Dell'Aglio, J. J. Hanly, B. Huber, et al. 2015. The diversification of *Heliconius* butterflies: what have we learned in 150 years? J. Evol. Biol. 28:1417–1438.

Nachman, M. W., and B. A. Payseur. 2012. Recombination rate variation and speciation: theoretical predictions and empirical results from rabbits and mice. Phil. Trans. R. Soc. B 367:409–421.

Nadeau, N. J., M. Ruiz, P. Salazar, B. Counterman, J. A. Medina, H. Ortiz-Zuazaga, et al. 2014. Population genomics of parallel hybrid zones in the

mimetic butterflies, *H. melpomene* and *H. erato*. Genome Res. 24:1316–1333.

Naisbit, R. E., C. D. Jiggins, and J. L. Mallet. 2001. Disruptive sexual selection against hybrids contributes to speciation between *Heliconius cydno* and *Heliconius melpomene*. Proc. R. Soc. B 268:1849–1854.

Naisbit, R. E., C. D. Jiggins, M. Linares, C. Salazar, and J. L. B. Mallet. 2002. Hybrid sterility, Haldane's rule and speciation in *Heliconius cydno* and *H. melpomene*. Genetics 161:1517–1526.

Naisbit, R. E., C. D. Jiggins, and J. L. B. Mallet. 2003. Mimicry: developmental genes that contribute to speciation. Evol. Dev. 5:269–280.

Navarro, A., and N. H. Barton. 2003a. Accumulating postzygotic isolation genes in parapatry: a new twist on chromosomal speciation. Evolution 57:447–459.

———. 2003b. Chromosomal speciation and molecular divergence—accelerated evolution in rearranged chromosomes. Science 300:321–324.

Noor, M. A. F., K. L. Grams, L. A. Bertucci, and J. Reiland. 2001. Chromosomal inversions and the reproductive isolation of species. Proc. Natl. Acad. Sci. USA 98:12084–12088.

Noor, M. A. F., D. A. Garfield, S. W. Schaeffer, and C. A. Machado. 2007. Divergence between the *Drosophila pseudoobscura* and *D. persimilis* genome sequences in relation to chromosomal inversions. Genetics 177:1417–1428.

Ono, Y., K. Asai, and M. Hamada. 2013. PBSIM: PacBio reads simulator—toward accurate genome assembly. Bioinformatics 29:119–121.

Ortíz-Barrientos, D., J. Engelstädter, and L. H. Rieseberg. 2016. Recombination rate evolution and the origin of species. Trends Ecol. Evol. 31:226–236.

Pinharanda, A., S. H. Martin, S. L. Barker, J. W. Davey, and C. D. Jiggins. 2017. The comparative landscape of duplications in *Heliconius melpomene* and *Heliconius cydno*. Heredity 118:78–87.

Pringle, E. G., S. W. Baxter, C. L. Webster, A. Papanicolaou, S. F. Lee, and C. D. Jiggins. 2007. Synteny and chromosome evolution in the Lepidoptera: evidence from mapping in *Heliconius melpomene*. Genetics 177:417–426.

Rastas, P., F. C. F. Calboli, B. Guo, T. Shikano, and J. Merilä. 2016. Construction of ultradense linkage maps with Lep-MAP2: stickleback F2 recombinant crosses as an example. Genome Biol. Evol. 8:78–93.

Rieseberg, L. H. 2001. Chromosomal rearrangements and speciation. Trends Ecol. Evol. 16:351–358.

Samonte, R. V., and E. E. Eichler. 2002. Segmental duplications and the evolution of the primate genome. Nat. Rev. Genet. 3:65–72.

Seehausen, O., R. K. Butlin, I. Keller, C. E. Wagner, J. W. Boughman, P. A. Hohenlohe, et al. 2014. Genomics and the origin of species. Nat. Rev. Genet. 15:176–192.

Sekhon, J. S. 2011. Multivariate and propensity score matching software with automated balance optimization: the matching package for R. J. Stat. Softw. 42:1–52.

Servedio, M. R., G. S. Van Doorn, M. Kopp, A. M. Frame, and P. Nosil. 2011. Magic traits in speciation: "magic" but not rare? Trends Ecol. Evol. 26:389–397.

Smadja, C. M., and R. K. Butlin. 2011. A framework for comparing processes of speciation in the presence of gene flow. Mol. Ecol. 20:5123–5140.

Stevison, L. S., K. B. Hoehn, and M. A. F. Noor. 2011. Effects of inversions on within- and between-species recombination and divergence. Genome Biol. Evol. 3:830–841.

Supple, M., R. Papa, B. Counterman, and W. O. McMillan. 2013. The genomics of an adaptive radiation: insights across the *Heliconius* speciation continuum. Adv. Exp. Med. Biol. 781:249–271.

Turner, J. R. G., and P. M. Sheppard. 1975. Absence of crossing-over in female butterflies (*Heliconius*). Heredity 34:265–269.

Van Belleghem, S. M., P. Rastas, A. Papanicolaou, S. H. Martin, C. F. Arias, M. A. Supple, et al. 2017. Complex modular architecture around a simple toolkit of wing pattern genes. Nat. Ecol. Evol. 1.

Wallbank, R. W. R., S. W. Baxter, C. Pardo-Diaz, J. J. Hanly, S. H. Martin, J. L. B. Mallet, et al. 2016. Evolutionary novelty in a butterfly wing pattern through enhancer shuffling. PLOS Biol. 14:e1002353.

White, M. J. D. 1978. Modes of speciation. Freeman & Co., San Francisco, CA.

Ziegler, A., and I. R. König. 2001. Genetic distance and mapping functions. eLS. https://doi.org/10.1038/npg.els.0005399

5

Experimental evolution reveals that sperm competition intensity selects for longer, more costly sperm

Joanne L. Godwin,[1] Ramakrishnan Vasudeva,[1] Łukasz Michalczyk,[2] Oliver Y. Martin,[3] Alyson J. Lumley,[1] Tracey Chapman,[1] and Matthew J. G. Gage[1,4]

[1]School of Biological Sciences, University of East Anglia, Norwich Research Park, Norwich NR4 7TJ, United Kingdom

[2]Institute of Zoology, Jagiellonian University, Kraków, Poland

[3]ETH Zürich, Institute of Integrative Biology, Zürich, Switzerland

[4]E-mail: m.gage@uea.ac.uk

It is the differences between sperm and eggs that fundamentally underpin the differences between the sexes within reproduction. For males, it is theorized that widespread sperm competition leads to selection for investment in sperm numbers, achieved by minimizing sperm size within limited resources for spermatogenesis in the testis. Here, we empirically examine how sperm competition shapes sperm size, after more than 77 generations of experimental selection of replicate lines under either high or low sperm competition intensities in the promiscuous flour beetle *Tribolium castaneum*. After this experimental evolution, populations had diverged significantly in their sperm competitiveness, with sperm in ejaculates from males evolving under high sperm competition intensities gaining 20% greater paternity than sperm in ejaculates from males that had evolved under low sperm competition intensity. Males did not change their relative investment into sperm production following this experimental evolution, showing no difference in testis sizes between high and low intensity regimes. However, the more competitive males from high sperm competition intensity regimes had evolved significantly longer sperm and, across six independently selected lines, there was a significant association between the degree of divergence in sperm length and average sperm competitiveness. To determine whether such sperm elongation is costly, we used dietary restriction experiments, and revealed that protein-restricted males produced significantly shorter sperm. Our findings therefore demonstrate that sperm competition intensity can exert positive directional selection on sperm size, despite this being a costly reproductive trait.

KEY WORDS: Anisogamy, directional selection, sexual selection, stabilising selection, Tribolium.

Impact Summary

When sexual reproduction evolved, it is theorized that a phenomenon known as "disruptive selection" drove the evolution of gametes into two different types (= anisogamy), making them equally adapted in their own ways for reproduction. Eggs were primarily shaped by the requirement to resource the embryo, and therefore needed to be large (making them few in number). Sperm was primarily shaped by the requirement to locate and fertilize eggs and, when there was competition for fertilizations, needed to be plentiful (making them small in size). For sperm, therefore, widespread numerical competition for fertilizations is thought to have made them numerous and tiny. We now know, however, that sperm are extremely diverse in size and shape, so their evolution may be more complicated than a simple drive to maximize numbers. To better understand these important cells, we used laboratory evolution with an insect model to measure how competition shapes sperm size. We maintained lines of beetles for over five years under identical conditions except, at every adult generation, we created mating regimes presenting very high or very low intensities of sperm competition. After ~80 generations, we found that sperm competitiveness had diverged significantly: sperm of males from the high competition regime achieved 20% higher fertilization success in competition than sperm of males evolving under the low competition regime. Importantly, we also found that these more competitive males had evolved significantly longer sperm, indicating that sperm competition can select for qualitative aspects of sperm form and function, and that competition is not just a numbers game. To assess whether sperm elongation places demands on males, we used dietary restriction experiments and found obvious costs to producing longer sperm because protein-restricted males developed shorter sperm. Our results demonstrate that competition in the struggle to reproduce can increase sperm size, despite this carrying costs, and therefore that the selective forces controlling the evolution of many, tiny sperm are more complex than originally assumed.

Introduction

Our understanding of the evolution of sperm form and function has its roots in anisogamy theory, where numerical competition for fertilizations was proposed to shape the fundamental phenotype of a male gamete that was produced in vast numbers, achieved via minimizing sperm size and increasing testicular investment (Parker et al. 1972, 1997; Parker 1982; Lessells et al. 2009; Parker and Pizzari 2010). It was this logic that answered the question "why are there so many tiny sperm?" (Parker et al. 1972; Parker 1982; Pizzari and Parker 2009). There is good evidence that numerical superiority of sperm (the "raffle principle") is indeed shaped by sperm competition (e.g., Wedell et al. 2002; Parker and Pizzari 2010; Kelly and Jennions 2011), and we also recognize that varying selection from sperm competition is a near-universal force among sexual reproducers (Taylor et al. 2014). However, we also recognize that sperm cells are not always minimally sized for the production of maximal numbers. In fact, spermatozoa are the most morphologically diverse eukaryotic cell types known (Pitnick et al. 2009a); even sperm size alone varies more than 8000-fold, from the diminutive gametes of the male braconid parasitoid wasp *Cotesia congregata* (7 μm; Uzbekov et al. 2017) to the giant sperm of *Drosophila bifurca* (58,290 μm; Pitnick et al. 1995). Most of this profound variation remains unexplained, but the diversity in form and function suggests that the selective forces shaping sperm form and function are more complex than a basic drive to win fertilization competitions by maximizing sperm number and minimizing sperm size.

There is increasing attention to the possibility that much of the huge variation in spermatozoa could be the result of postcopulatory sexual selection, when the forces of competition and choice act on gamete biodiversity (Snook 2005; Rowe et al. 2015;

Lüpold et al. 2016). We now know that fertilization outcome can depend on variation in form, function, and identity of competing sperm (Pizzari and Parker 2009), and that females have evolved mechanisms at the cryptic level of the gamete to control fertilization and influence paternity (Pitnick et al. 2009b). The evolution of spermatozoa will therefore be subject to focused selection from a complexity of interacting forces arising both from cryptic female choice, and competition between rival males and their sperm. Previously, results from cross-species studies were mainly used to infer how postcopulatory sexual selection had shaped sperm cell phenotypes, with evidence that increasing sperm competition intensity can have positive, negative, or neutral effects on sperm size variation (reviewed in Snook 2005). Findings have therefore been mixed, with one possible reason being due to differences between species in the additional impact of female-controlled influences on sperm competition, cryptic choice, and fertilization. By analyzing both sperm number and sperm size simultaneously, for example, in the context of varying selection from female size and tract dilution, Lüpold and Fitzpatrick (2015) showed that there is positive selection from sperm competition intensity on both sperm number and sperm size across mammals. However, responses to selection were greater for sperm number, and greater still in those larger species where sperm dilution in the competitive arena of the female tract was more likely to exist (Lüpold et al. 2016). Comparative studies can also be sensitive to assumptions about phylogenetic relatedness. Previous analyses of the relationship between sperm competition intensity and sperm size across mammals have found both positive and neutral relationships, which could result from the use of different phylogenies (Gomendio and Roldan 1991; Gage and Freckleton 2003; Tourmente et al. 2011), and analyses across passerine birds have revealed significant relationships between sperm competition and sperm dimensions, but these differed depending in Family, being positive in the Fringillidae, and negative in the Sylviidae (Immler and Birkhead 2007).

Here, we use long-term experimental evolution within a single species to measure how sperm length evolves across independently replicated lineages. The flour beetle, *Tribolium castaneum*, is promiscuous and females store sperm (Fedina and Lewis 2008; Michalczyk et al. 2010; Michalczyk et al. 2011; Lumley et al. 2015). Sperm form and function are therefore expected to be key targets of postmating sexual selection. Manipulation of adult sex ratios was used to create Male-biased (90 males to 10 females) intense sperm competition populations, contrasting with Female-biased (10 males to 90 females) relaxed sperm competition populations. After 77–83 generations of this experimental evolution, we measured how ejaculate competitiveness, overall male investment to spermatogenesis, and sperm size had evolved. We hypothesized that, if sperm competitiveness responds to divergent intensities of sperm competition, the "raffle principle" within anisogamy theory (Parker et al. 1972; Pizzari and Parker 2009) should drive sperm size to either decrease, or remain at some biological minimum within investment to spermatogenesis in the testis, maximizing numerical superiority of sperm within competitions for fertilizations. On the other hand, if competition selects for qualitative adaptations in sperm for winning fertilizations (Lüpold et al. 2016), elevated postcopulatory sexual selection could increase sperm size. Additionally, to test the hypothesis that any evolution of sperm length is constrained by the existence of costs, we also employed dietary restriction experiments that limited the amount of resource available for spermatogenesis, and then assessed the relative impact upon sperm elongation. Therefore, as well as measuring how experimental variation in postcopulatory sexual selection shapes sperm competitiveness and sperm length evolution, this study also empirically examines the costs of sperm elongation for males.

Methods

EXPERIMENTAL EVOLUTION UNDER DIVERGENT SPERM COMPETITION INTENSITIES

Was conducted with beetles of the widely used Georgia 1 (GA1) "wild-type" strain, originating from the Beeman Lab (United States Department of Agriculture). Populations were maintained at standard conditions of 30°C and 60% humidity, within ad libitum medium consisting of 90% organic white flour, 10% brewer's yeast, and a thin layer of oats to aid traction. Divergent operational sex ratios during the adult life stage were used to apply a Male-biased (90 males to 10 females), intense sperm competition regime, contrasted against a Female-biased (10 males to 90 females), relaxed sperm competition regime. Male and female mating potential and promiscuity assays demonstrate that our male- versus Female-biased regimes create extreme divergence in sperm competition intensities; full details are in Michalczyk et al. (2011) and Lumley et al. (2015). The regimes were structured so that theoretical effective population size (Wright 1931) was equalized to minimize the opportunity for differences in genetic drift and/or inbreeding to influence either the male- or Female-biased regimes. By structuring our adult regimes as either 90:10 or 10:90, we generated adult N_e throughout of 36; post hoc genetic testing using a suite of microsatellites confirmed that we had retained equal levels of heterozygosity between the regimes (Lumley et al. 2015). Three independent lines within each regime were maintained. For each independent line, at every generation, pupae were sexed to ensure virginity. Adults were then placed in fresh medium for seven days for mating, sperm competition, and oviposition, after which they were removed and eggs/larvae left to develop in standardized conditions until pupae were ready for beginning the next generation.

SPERM COMPETITIVENESS

Figure 1 of ejaculates from individual males evolved under contrasting sperm competition intensity regimes was assessed following 77 generations of experimental evolution. Competition took place between an experimental male and a marker mutant "Reindeer" (Rd) male, allowing paternity to be assigned (Lewis et al. 2005; Tregenza et al. 2009; Fig. 1A). The Rd mutation is dominant and homozygous within the Rd population, therefore all offspring sired by a Rd male have distinctive swollen antennae, while offspring of the sex ratio treatment regime males have wild-type (WT), filiform antennae. All adults were virgin and 10–12 days posteclosion when used, and experimentally evolved males were isolated at the pupal stage to equalize any developmental effects. Females were paired with an Rd male for 24 h in a 1 cm diameter 7 mL vial containing 2 g fresh medium, allowing frequent mate encounter and ample opportunities for mating and full sperm storage. In *T. castaneum*, even a single mating is sufficient for females to fertilize 700 eggs across four months of oviposition (Bloch Qazi et al. 1996), and both males and females mate frequently (Fedina and Lewis 2008). We can therefore assume that virgin Rd males paired for 24 h with single virgin females in these conditions will fill the limited spermathecal and bursal storage sites with Rd sperm (Bloch Qazi et al. 1996), so that the subsequent competitor males must win fertilizations within females that have been fully inseminated. After 24 h, Rd males were removed and females were presented with an unmated experimental focal male from either the male- or Female-biased sperm competition intensity regime, pairing Male-biased males with Male-biased females, and vice versa. Pairs were left in an empty 1 cm diameter 7 mL vial to mate for 1 h, after which males were removed. After the second mating, females were transferred to Petri dishes of fresh medium, and left to oviposit across two 10-day blocks before being removed. Thus, the outcomes of sperm competitions lasting 20 days were recorded, which typically accounts for ~50% of the total offspring production by females from such a mating period before they run out of functional sperm (M. J. G. Gage, unpubl. data). Eggs/larvae were left to develop in individual Petri dishes, with the number of each phenotype (Rd and wild type) counted as adults. Females from the respective experimental evolution regime were used in sperm competition experiments to maintain any coevolved male–female effects within regimes, and to avoid the possibility for differential female effects when paired with males of different experimental evolution populations. To balance within-line versus between-line coevolutionary effects, males were competed with females from their own line and with females from the other two independent replicate lines within the regime, applying a balanced design to equalize within- versus between-line effects (Fig. S1). There was no evidence of any within- versus between-line influences on differential fertilization success, under either

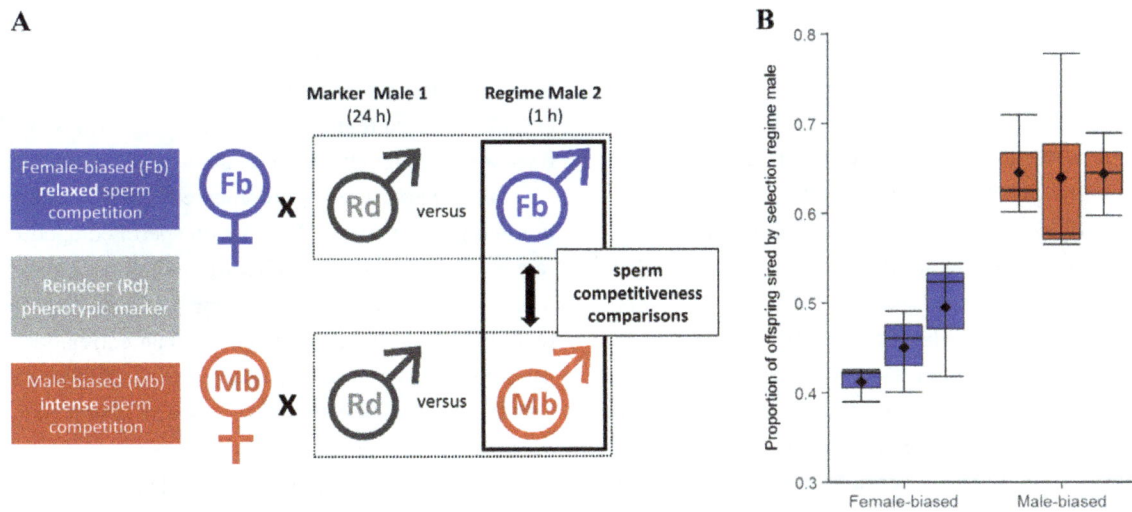

Figure 1. Experimental design and outcome of a sperm competition assay comparing ability of sperm from relaxed Female-biased (Fb), and intense Male-biased (Mb), sperm competition regime males to compete for fertilizations against sperm from marker Reindeer (Rd) males. (A) Experimental design of sperm competition assay. (B) Proportion of offspring sired by males from contrasting Female-biased and Male-biased sperm competition regimes, following sperm competition through 20 days of oviposition. Male-biased males sired a significantly greater proportion of offspring (negative binomial GLMM: $\chi^2_{(1)}$ = 5.58, P = 0.02). Data grouped by independent line (n = 13–17 replicates × 3 crosses per line; average number of offspring scored across n = 282 sperm competitions = 158 (± 3 SEM), range 24–295).

intense or relaxed evolutionary histories of sperm competition (Fig. S2).

SPERM LENGTH FOLLOWING EXPERIMENTAL EVOLUTION

Was measured in the same replicate lines and regimes after 83 generations of experimental evolution, using similar 10-day to 12-day posteclosion unmated males, sexed and isolated at the pupal stage. Sperm were recovered using microdissection from females, soon after mating and spermatophore deposition. Pairs of beetles (n = 9–10 males × 3 independent populations = 29–30 total males per sex ratio regime treatment) were placed in 1 × 1 cm² mating arenas at standard conditions for 30 min to allow spermatophore transfer. Females were then removed and decapitated, and the reproductive tract isolated by extruding the ovipositor and gently pulling to detach it from the abdomen. This was placed in a drop of buffer (0.9% NaCl) and the spermatophore isolated. Finally, the spermatophore was transferred to a 10 µL drop of fresh buffer on a clean microscope slide, teased open, and the slide flooded with further buffer to disperse the sperm. Slides were left to dry, then dipped in distilled water to remove buffer salt residue, and redried. Cleared slides were viewed with phase contrast, at 60× magnification and images were captured. Total sperm length was measured by creating a segmented line that traced the entire length of the cell using the "ImageJ" image analysis package (Schneider et al. 2012; inset Fig. 2A). Thirty sperm per male were measured (n = 30 sperm × 9–10 males × 3 lines = 870–900 total sperm in either treatment). Sperm were measured by two

investigators. To assess repeatability, 20 sperm were measured by both investigators, and the intraclass correlation coefficient (ICC) calculated (Lessells and Boag 1987; Nakagawa and Schielzeth 2010). Repeatability was found to be very high, ICC = 0.99 (± 0.005), with close correlations between pairs of measurements (r = 0.99), and a Bland-Altman plot revealed consistent agreement between the investigators across the parameter range (Bland and Altman 1986; Bartlett and Frost 2008).

TESTIS SIZE VARIATION FOLLOWING EXPERIMENTAL EVOLUTION

Was measured in the same replicate lines after 100/107 generations. Ten-day to 12-day posteclosion unmated males, sexed and isolated at the pupal stage, were frozen and elytra length (n = 10 males × 3 independent populations = 30 males per sperm competition regime) and testes volume (n = 15 males × 3 independent populations = 45 males per sperm competition regime) were measured. Testes were dissected out and images of the follicles, which make up the testes in *T. castaneum*, were captured. "ImageJ" (Schneider et al. 2012) was then used to measure the circumference of the follicle, and volume calculated by assuming a spherical shape. Where possible, all 12 follicles per male were measured, however, where this was not possible due to damage to fragile follicles during dissection, total testes volume was calculated as mean measured follicle size multiplied by 12. To assess the accuracy of this estimate, uniformity of follicle size was investigated by calculating the ICC of two randomly selected follicles per male. Uniformity was found to be very high (ICC =

0.99). Hence, our measure gave an accurate estimate of total testis volume. Mean number of measured follicles was 8.82 ($\pm$ 0.22) per male.

DIETARY RESTRICTION EFFECTS ON BODY SIZE, TESTES VOLUME, AND SPERM LENGTHS

Were measured in *T. castaneum* males taken from a standard stock population of Krakow super strain (KSS) created by Ł. Michalczyk in 2008. Eggs were collected and randomly assigned to either control medium (90% organic white flour and 10% brewer's yeast) or a protein-restricted medium in which brewer's yeast was not added (0% yeast). Ten-day to 12-day posteclosion unmated males, sexed and isolated at the pupal stage, were frozen and elytra length and testes volume ($n = 32$ males per nutritional treatment), and sperm length ($n = 15$ sperm x 32 males per nutritional treatment) were measured as previously described.

STATISTICAL ANALYSES

Were conducted in "R" (R Core Team 2015), with "plyr" (Wickham 2015), "pastecs" (Grosjean et al. 2014), "car" (Fox et al. 2015), and "stats" (R Core Team 2015) packages used for data exploration, descriptive statistics, and testing assumptions. Figures were created using "ggplot2" (Wickham and Chang 2015). For all boxplots, a horizontal line indicates the median, boxes indicate the interquartile range (IQR), whiskers indicate points within 1.5 IQR, and any data not included in the box and whiskers are shown as outliers (small filled points). An additional point (filled diamond) was added to display the mean. Linear and generalized linear mixed effects models (LMM and GLMM, respectively), with appropriate error distributions, were used to include random effects to account for nesting in the data. Mixed models were fitted by maximum-likelihood and likelihood ratio tests and AIC values were used to compare models with and without the factor of interest (Crawley 2013). All models were implemented in "lme4" (Bates et al. 2015) unless otherwise stated.

Sperm competitiveness, measured as the proportion of offspring sired by males from experimentally evolved regimes, was compared by constructing a GLMM, with a negative binomial error structure to account for overdispersion in the data, using the "glmmADMB" (Fournier et al. 2012) and "R2admb" (Bolker et al. 2015). The response variable was entered as a paired variable containing the number of offspring sired by each male to retain information on sample size within the model (Crawley 2013). A maximal model was fitted with "sperm competition intensity" (Female- or Male-biased) and "cross" (within or between regime replicate) entered as fixed effects. Cross was then dropped from the fixed effects as it did not significantly improve the explanatory power of the model. Female line (A, B, C), nested within male line (A, B, C), were entered as random effects to check for any

within- versus between-line compatibility, and account for nesting in the data. To check for differential offspring mortality effects, total offspring production was also compared between female- and Male-biased regimes by constructing a LMM with the same fixed and random effects structure.

Total sperm length was compared between experimental evolution regimes using an LMM with sperm competition intensity regime (female- or Male-biased) entered as a fixed effect and replicate male (a–j), nested within replicate line (A, B, C) as random effects. In addition, a Spearman correlation, carried out using the "Hmisc" package (Harrell et al. 2016), was used to assess the association between mean sperm length and mean sperm competitiveness across treatments ($n = 6$).

Sperm length variance was calculated as a standardized coefficient of variation (CV) both within males (CV_{wm}) and between males (CV_{bm}) for each independent treatment replicate. The within male CV for a population is a mean of 10 individual male CVs. The between-male CV for a population was calculated using the mean sperm length of each male. A linear mixed effects model was fitted to compare within male CV ($n = 9$–10×3 lines = 29–30 total CV_{wm} per treatment) between regimes. A maximal model was fitted with experimental evolution regimes (female- or Male-biased) entered as the fixed effect, and line (A, B, C) entered as a random effect. An unpaired Wilcoxon rank-sum test was used to compare between male CVs ($n = 1 \times 3$ populations per treatment).

Testes size and elytra length were compared between sexual selection regimes using LMMs with sperm competition intensity regime (female- or Male-biased) entered as a fixed effect and replicate line (A, B, C) entered as random effects.

Male morphometric and reproductive traits were compared between control and protein-restricted dietary treatments using *t* tests (absolute testes volume and within male sperm length variance [CV]) or the nonparametric unpaired Wilcoxon rank-sum test (elytra length). In addition, relative testes volume was compared between dietary treatments using analysis of covariance (ANCOVA) with elytra length incorporated as a covariate to account for allometry in the growth of body parts. Finally, a linear mixed model was used to compare sperm length, with diet as a fixed effect, and male (1 to 32) as a random effect to account for nesting of the data.

Results

SPERM COMPETITIVENESS

After 77 generations of experimental evolution under either intense or relaxed competition for fertilizations, sperm competitiveness had significantly diverged between regimes. Sperm from Male-biased males won significantly greater numbers of fertilizations across 20 days of competition and oviposition than did

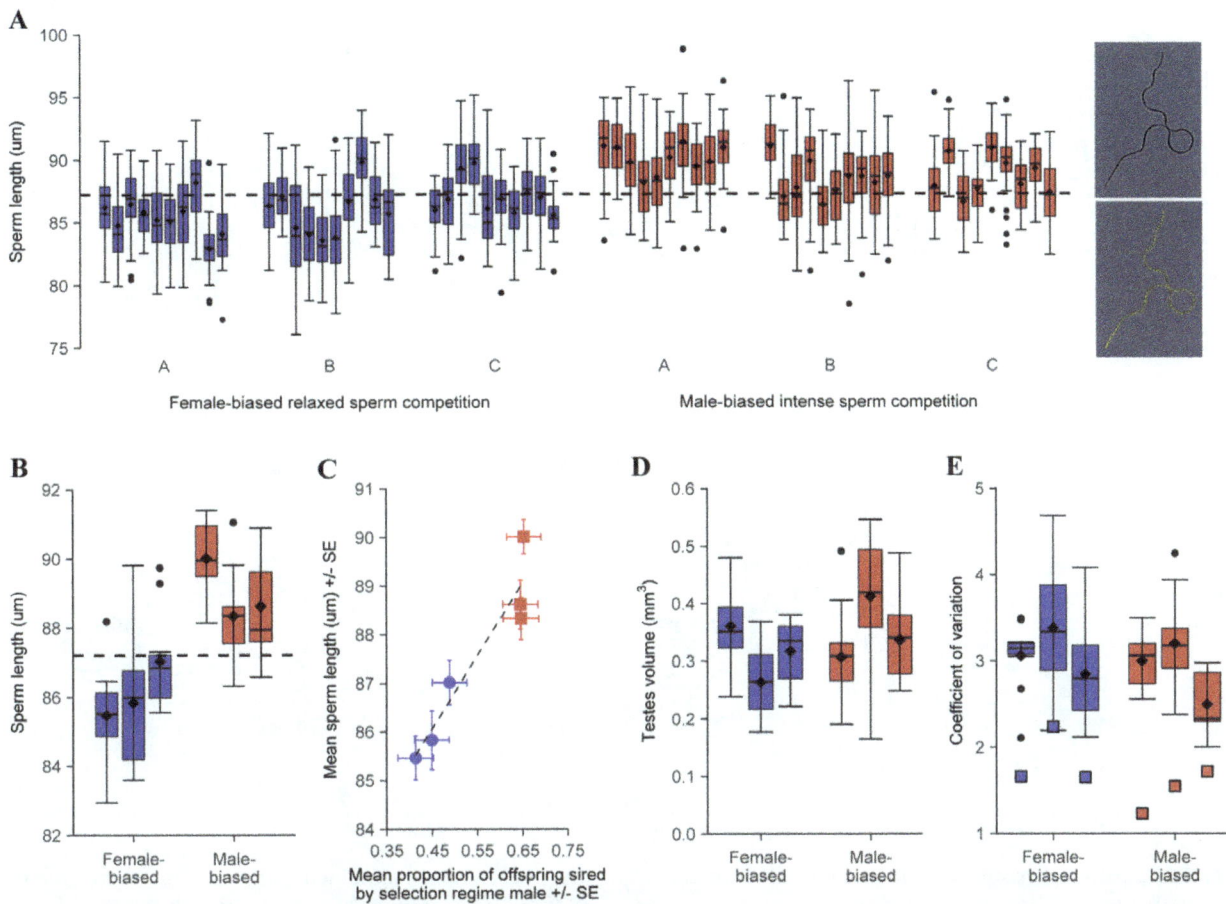

Figure 2. Sperm length and variance of males from relaxed Female-biased (blue) versus intense Male-biased (red) sperm competition histories. (A) Sperm length per male ($n = 30$ sperm x 9–10 males x 3 lines = 870–900 total sperm in either treatment), and sperm micrograph and measuring technique using "ImageJ" image analysis package (see Methods). The difference between regimes was highly significant (LMM $\chi^2_{(1)} = 6.69$, $P < 0.01$). Dashed line shows mean sperm length of ancestral GA1 males (86.8 μm) to compare increases and decreases within either selection regime. (B) Sperm length grouped by independent line and dashed line for ancestral mean. (C) Correlation between sperm length and sperm competitive ability across Female-biased (blue circles), and Male-biased (red squares) independent lines ($r = 0.94$). (D) Testes volume ($n = 15$ males per population) also did not differ between regimes (LMM: $\chi^2_{(1)} = 1.13$, $P = 0.29$). (E) Within- and between-male coefficients of variation (CVs) in sperm length. Neither within-male CVs (boxplots; $n = 30$ sperm x 9–10 males per population), nor between-male CV (filled square markers) calculated using mean sperm length of each male ($n = 9$–10 males per population) differed significantly between regimes (within; LMM $\chi^2_{(1)} = 0.79$, $P = 0.37$, between; $W = 7$, $P = 0.40$).

sperm from Female-biased males (negative binomial GLMM: $\chi^2_{(1)} = 5.58$, $P = 0.02$; Fig. 1B). Scoring an average of 158 ($\pm$ 3 SEM) offspring per competition, and with regime males mated second, Male-biased, intense sperm competition regime males won 20% more fertilizations (65% $\pm$ 4 SEM, $n = 143$ competitions) than males from the Female-biased, relaxed sperm competition regime (45% $\pm$ 4 SEM, $n = 139$ sperm competitions). Removing those competitions where either of the males gained zero or 100% paternity ($n = 12$ of 143 Male-biased competitions, and 13 of 139 Female-biased competitions) to control for the possibility that failed matings explained the paternity biases, did not change the results, with paternity share still showing a 17% difference between regimes (negative binomial GLMM: $\chi^2_{(1)} = 4.72$, $P < 0.05$). Likewise, there was no indication that differen-

tial offspring mortality explained the male- versus Female-biased paternity differences, because there was no significant difference between the regime crosses in the numbers of adult offspring that were produced (LMM: $\chi^2_{(1)} = 2.95$, $P = 0.09$). Moreover, the direction of any difference was conservative to the difference in paternity, with Male-biased regime sperm competitions (where more offspring were sired by the experimental evolution males) yielding slightly fewer total offspring ($n = 151 \pm 8$ SEM) than the Female-biased trials (166 ± 6 SEM).

SPERM LENGTH

In addition to superior competitive ability, sperm produced by males from the Male-biased, intense competition regime were significantly longer than those of males derived from the

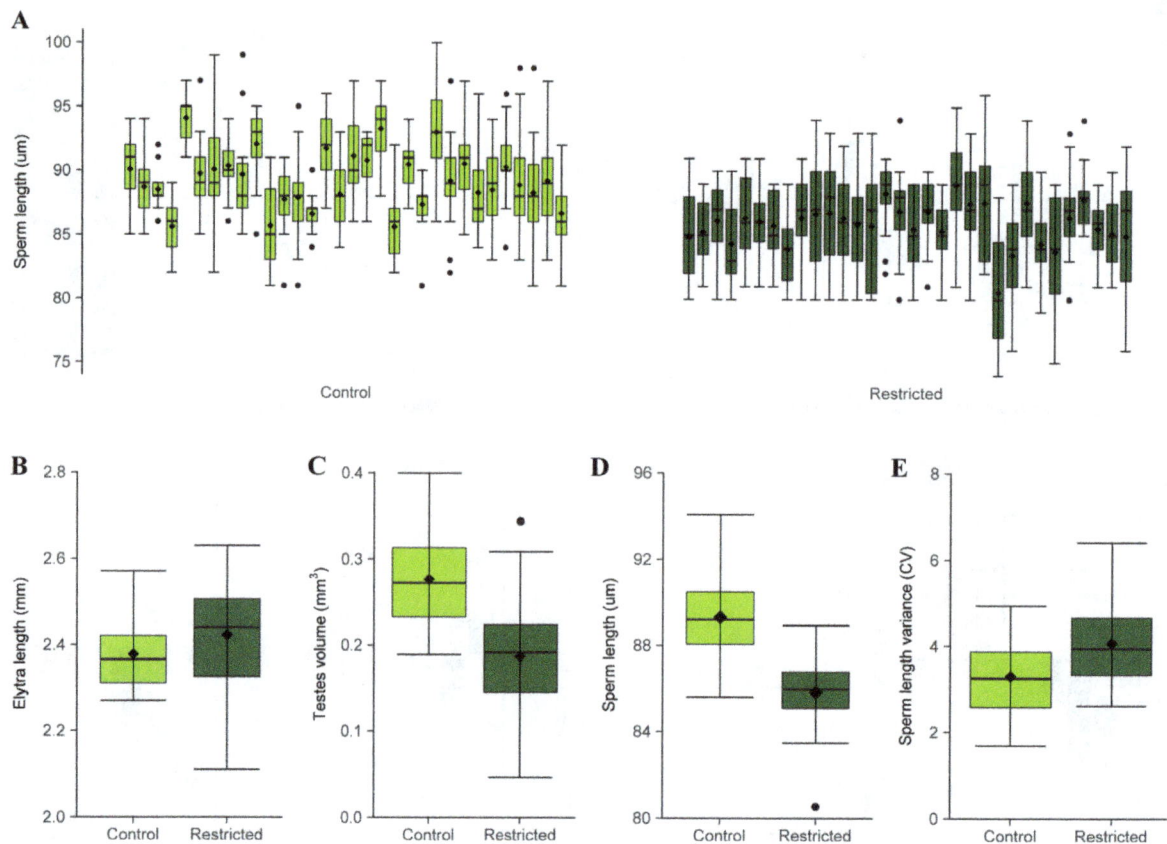

Figure 3. Comparison of morphometric and reproductive traits in males reared and maintained under control (light green) versus protein-restricted (dark green) diets ($n = 32$ males per dietary treatment). (A) Sperm length per male ($n = 15$ sperm per male). (B) Elytra length did not differ between diets (Wilcoxon rank sum test: $W = 383$, $P = 0.08$). (C) Testes volume was significantly greater in control males (absolute testes volume; $t_{62} = 5.74$, $P < 0.01$, relative testes volume; ANCOVA $F_{1,61} = 33.58$, $P < 0.01$). (D) Mean sperm length ($n = 15$ sperm $\times$ 32 males $= 480$ total sperm in either treatment) was significantly longer in control males (LMM: $\chi^2_{(1)} = 204.09$, $P < 0.01$). (E) Within-male coefficient of variation (CV) in sperm length was significantly higher in protein-restricted males ($t_{62} = -3.40$, $P < 0.01$).

Female-biased, relaxed competition regime (LMM: $\chi^2_{(1)} = 6.69$, $P < 0.01$; Fig. 2A and B). Mean ($\pm$ SEM) sperm length of Male-biased males was 89.0 μm ($\pm$ 0.57) compared to 86.1 μm ($\pm$ 0.40) in Female-biased males. Comparisons of the experimentally evolved sperm length distributions against their ancestral stock population revealed that average sperm length had both increased and decreased in the high and low sperm competition regimes, respectively (see Fig. 2A and B). In addition, there was a significant positive association across the six independent selection lines between average sperm length and mean competitive ability (Spearman correlation: $r = 0.94$, $n = 6$, $P < 0.01$; Fig. 2C).

SPERM LENGTH VARIANCE

In addition to evidence for directional selection, we explored whether sperm competition intensity had also exerted stabilizing selection so that sperm length had evolved to a narrower optimum by comparing CVs between regimes (Lifjeld et al. 2010). Despite evidence for significant divergence in sperm competitiveness and

sperm length, we found that neither within-male CVs (LMM: $\chi^2_{(1)} = 0.79$, $P = 0.37$), nor between-male CVs (Wilcoxon rank sum test: $W = 7$, $P = 0.4$) differed between sperm competition selection regimes (Fig. 2E).

TESTES SIZE AND BODY SIZE

Relative investment into spermatogenesis was similar for males from both male- and Female-biased regimes, with no differences in testes volume (LMM: $\chi^2_{(1)} = 1.13$, $P = 0.29$; Fig. 2D) or elytra length (LMM: $\chi^2_{(1)} = 0.40$, $P = 0.53$) between males from contrasting regimes.

DIETARY RESTRICTION AND SPERM LENGTH

Males reared and maintained under protein-restricted conditions without supplementary yeast showed reduced investment in spermatogenesis, with significantly smaller absolute ($t_{62} = 5.74$, $P < 0.01$) and relative (ANCOVA: $F_{1,61} = 33.58$, $P < 0.01$) testis sizes than controls (Fig. 3C). Importantly, these males suffering dietary constraints to spermatogenic investment produced sperm cells that were significantly shorter than those from males reared on a standard 10% yeast diet (LMM: $\chi^2_{(1)} = 204.09$, $P < 0.01$; Fig. 3A and D). Sperm size variance was also greater within males suffering dietary restriction ($t_{62} = -3.40$, $P < 0.01$), supporting the idea that environmental stress reduced male ability to produce more uniform, as well as more elongate, sperm cell phenotypes (Fig. 3E).

Discussion

After more than six years of experimental evolution under controlled but widely differing intensities of sexual selection and sperm competition (Lumley et al. 2015), we found replicated evidence for significant divergence in both sperm competitiveness and sperm size. In competition with sperm from standardized marker males, ejaculates from intense Male-biased selection histories won an average of 65% of the fertilizations, whereas sperm from relaxed Female-biased selection histories only achieved 45% fertilization success (Fig. 1B). In parallel with the divergence in sperm competitiveness, we also found significant differences in sperm length under strong versus weak selection from sperm competition, and both increases and decreases relative to the ancestral population average (Fig. 2A and B). Contrary to basic expectations from the "raffle principle" (Parker et al. 1972; Parker 1982; Pizzari and Parker 2009), therefore, sperm had therefore become significantly larger in males exposed to selection from high levels of sperm competition (Fig. 2A–C). By contrast with sperm length, we found no divergence in testis size (or body size) following histories of selection under intense versus relaxed sperm competition (Fig. 2D), indicating that males under both regimes made similar overall levels of investment to spermatogenesis, possibly as a result of equal and parallel selection from sperm competition and mating frequency, respectively (e.g., Reuter et al. 2008; Crudgington et al. 2009).

Given similar investment into spermatogenesis, the divergence in sperm competitiveness and sperm length between our selection regimes indicates that more intense sperm competition can select for qualitative improvements within individual sperm cell phenotypes, and not necessarily a basic drive to increase sperm numbers. Longer sperm may achieve greater mobility or velocity (Lüpold et al. 2009; Fitzpatrick et al. 2010), providing a competitive advantage if races for fertilization are important (Gage et al. 2004). However, despite the seemingly intuitive relationship between flagellum length and speed, evidence linking the two is ambiguous (Simpson et al. 2014; reviewed in Snook 2005), and may depend upon physical complexities that affect hydrodynamics of very small flagellated cells (Humphries et al. 2008). Notably, in experiments with externally fertilizing myobatrachid frogs, it is slower swimming (potentially longer lived) sperm that win more competitive fertilizations (Dziminski et al. 2009) and,

across myobatrachid species and Chinese anurans, there are significant positive relationships between level of sperm competition and sperm length (Byrne et al. 2003; Zeng et al. 2014). Increasing flagellum length will theoretically deliver additional mechanical thrust (Katz and Drobnis 1990), but that may be translated into different swimming patterns. For example, thrust and torque, not speed, may be advantageous in species where females store sperm from multiple males at high densities in narrow tract tubules, with selection on sperm to resist displacement and secure optimal storage sites for fertilization (Immler et al. 2011; Lüpold et al. 2012). Both bursal and spermathecal storage sites are invariably densely packed with sperm following mating opportunities in *T. castaneum*, with the position in the bursa being important for proximate fertilization success and storage in the spermathecal tubules possibly playing roles over longer oviposition periods (Droge-Young et al. 2016). Whatever the specific mechanistic advantage that longer sperm have for fertilization success, our findings provide clear evidence that sperm competition can directionally select for increased investment in sperm size, revealing the importance of qualitative aspects of sperm cell phenotypes that are relevant for models explaining the evolution and maintenance of anisogamy (Parker 1982; Snook 2005; Lüpold et al. 2016).

Very few previous studies have examined the response of sperm length to experimental evolution of mating pattern and sperm competition. Studies using *Drosophila, Callosobruchus*, and house mice (38–81, 90, and eight generations, respectively) found no significant responses by sperm length in regimes experiencing enforced monogamy, compared with regimes allowing polyandry (Pitnick et al. 2001; Gay et al. 2009; Firman and Simmons 2010). Also, in a well-replicated study employing experimental evolution and controlling for effective population sizes in *Drosophila pseudoobscura*, the lengths and component dimensions of both short and long sperm morphs showed no change after more than 40 generations of selection under elevated promiscuity versus monogamy (Crudgington et al. 2009). This lack of response could not be explained by low genetic variability in either sperm or female tract length, both of which show moderate potential to evolve under direct selection (Snook et al. 2010; Moore et al. 2013). However, in *Caenorhabditis* nematodes, which produce amoeboid sperm, LaMunyon and Ward (2002) compared selfing hermaphrodites (where sperm competition is absent) versus sexually crossing lines (where males compete) and found that sperm evolved to be bigger in the context of sperm competition across 60 generations. Finally, in a study across 20 generations, monogamy did not change sperm lengths of *Macrostomum* flatworms, but length of the sperm bristles did elongate in polygamous lines (Janicke et al. 2016).

Building on cross-species analyses (e.g., Lüpold et al. 2016), evidence for larger sperm advantages in sperm competition have been previously confirmed where natural sperm size variation exists between individual males within a species (reviewed in Snook 2005). In *Caenorhabditis* nematodes and *Rhizoglyphus* bulb mites, which produce amoeboid sperm, males with larger sperm won more fertilizations within sperm competitions (Radwan 1996; LaMunyon and Ward 1998). However, studies using natural variation in flagellated sperm length in other insect, fish, and mammal species found no longer sperm advantages (Gage and Morrow 2003; Simmons et al. 2003; Gage et al. 2004), although in birds, male zebra finches producing longer sperm win more fertilizations under sperm competition (Bennison et al. 2015). *Drosophila* experiments have provided important insights, with experimental evidence for both qualitative and quantitative advantages for sperm length and number in sperm competition in *D. melanogaster* (Miller and Pitnick 2002; Patterini et al. 2006). Importantly, the qualitative advantages achieved by longer sperm in *Drosophila* competitions are known to interact with the structure of the female reproductive tract, such that long-sperm advantages only become evident when dimensions of the female tract are also enlarged (Miller and Pitnick 2002).

The *Drosophila* studies revealing that sperm qualitative advantages prevail in the context of variation by the female reproductive tract, demonstrate the existence of "cryptic female choice," where particular sperm quality phenotypes must co-adapt in males under selection from competition for fertilizations in a particular female-controlled environment (Miller and Pitnick 2002; Snook 2005). We measured the relative competitiveness of our selection regime males against standard male competitors (carrying the Reindeer marker) for females of the same regime (Fig. 1). Having run sperm competition trials both within and between either selection regime's three lines (Fig. S1), we found no evidence of line coevolution (Fig. S2). However, our 20% differences in overall sperm competitiveness between male- and Female-biased males may also have been influenced by female postcopulatory processes (Miller and Pitnick 2002; Snook 2005). If longer sperm are costly to maintain, then cryptic female choice could logically drive sperm elongation through sexual selection on male condition via a process where sperm phenotypes act as gametic equivalents of the peacocks' tail. This possibility has been recently confirmed through comparative analyses revealing that sperm size can exaggerate via Fisherian runaway sexual selection, mediated by cryptic female choice (Lüpold et al. 2016). If sperm cells act as postcopulatory signals of male quality under such sexual selection, then the trait must be honest and costly to develop and maintain (Lüpold et al. 2016). Our experiment comparing testis and sperm length development within males reared on protein-rich versus protein-poor diets demonstrates that sperm elongation is indeed costly in *T. castaneum* and is dependent on male condition (Fig. 3), allowing sperm size to represent a reliable signal for postcopulatory sexual selection (Lüpold et al. 2016). Previous work in *T. castaneum* has shown that diet restriction also con-

strains male fertility and sperm competitiveness (Sbilordo et al. 2011).

In addition to demonstrating that sperm competition can exert directional selection on sperm length, our experiments allow us to test directly the hypothesis that sperm competition generates stabilizing selection on sperm size. If sperm competition levels are high, and there is an optimally competitive sperm length phenotype, then selection could stabilize morphological variation more tightly around the optimum (Parker and Begon 1993; Lifjeld et al. 2010). Such stabilization could act at the population level, driving down between-male variation in sperm size to the population optimum (Morrow and Gage 2001), or within individual males, driving up quality control within spermatogenesis and reducing production errors to maximize the number of optimally competitive sperm produced (Lifjeld et al. 2010). The converse situation may also apply, where very relaxed postcopulatory sexual selection could allow more variant sperm morphology, perhaps exemplified by species showing exceptionally degenerative sperm morphology for their taxa, and also associated with very low levels of sperm competition (van der Horst and Maree 2014; Stewart et al. 2016). A number of comparative studies using passerine birds (Immler et al. 2008; Kleven et al. 2008; Lifjeld et al. 2010; Laskemoen et al. 2013), murine rodents (Varea- Sánchez et al. 2014), and social insects (Fitzpatrick and Baer 2011) have found evidence that decreased sperm length variance, both between and within males, exists where sperm competition levels are higher. Building on these findings, Lifjeld et al. (2010) have proposed that sperm length variance itself could represent an objective index of the intensity of sperm competition sustained by a species. However, we found no evidence, following controlled experimental evolution on a single ancestral population across 83 generations of exposure to divergent levels of sperm competition, that increased sperm competition stabilizes and reduces sperm length variation (Fig. 2E). Our findings therefore do not support the universal use of sperm length variance as an indicator of the level of sperm competition in a population (Lifjeld et al. 2010). However, the cross-species evidence for sperm competition and stabilizing selection on sperm comes from wild systems (Immler et al. 2008; Kleven et al. 2008; Lifjeld et al. 2010; Varea- Sánchez et al. 2014), where environmental variation and a greater range of stresses exist. Under natural conditions, developmental stability within spermatogenesis may be more difficult to achieve, making the existence of sperm size variance in the absence of selection from sperm competition the default condition. Our results showing increased sperm length variance in males exposed to dietary stress, compared with low variance in optimal and benign culture conditions support this idea (Fig. 3E), as does previous work in this system showing increased sperm length variance under genetic stress from inbreeding (Michalczyk et al. 2011).

In conclusion, we applied experimental evolution through variation in adult mating pattern to successfully and significantly diverge sperm competitiveness of independent replicate lines. Although testis and body size measures showed that selection did not change male relative investment into spermatogenesis through experimental evolution, we found that exposure to more intense sperm competition regimes caused males to evolve longer sperm, and there are obvious costs to such sperm elongation under diet restriction. Our findings therefore demonstrate positive directional selection from sperm competition on costly sperm size, revealing that postcopulatory sexual selection can generate qualitative selection on sperm form and function.

AUTHOR CONTRIBUTIONS
Experimental evolution lines were initiated in 2005 by ŁM, OYM, and MJGG and maintained since by AJL and JJG MJGG, JLG, and ŁM conceived, designed, and analyzed the study, with input from all authors. RV conducted the diet experiments. MJGG and JLG wrote the paper, with contributions from all authors.

ACKNOWLEDGMENTS
We thank the Natural Environment Research Council (projects NE/K013041/1 and NE/J024244/1), University of East Anglia, and Leverhulme Trust for funding this work.

LITERATURE CITED
Bartlett, J. W., and C. Frost. 2008. Reliability, repeatability and reproducibility: analysis of measurement errors in continuous variables. Ultrasound Obst. Gyn. 31:466–475.

Bates, D., M. Maechler, B. Bolker, S. Walker, R. H. B. Christensen, H. Singmann, B. Dai, et al. 2015. lme4: linear mixed-effects models using "Eigen" and S4. Available at https://cran.r-project.org/web/packages/lme4

Bennison, C., N. Hemmings, J. Slate, and T. R. Birkhead. 2015. Long sperm fertilize more eggs in a bird. Proc. R. Soc. B Biol. 282: p.20141897.

Bland, J., and D. Altman. 1986. Statistical methods for assessing agreement between two methods of clinical measurement. Lancet 1:307–310.

Bloch Qazi, M. C., J. T. Herbeck, and S. M. Lewis. 1996. Mechanisms of sperm transfer and storage in the red flour beetle (Coleoptera: Tenebrionidae). Ann. Ent. Soc. Am. 89:892–897.

Bolker, B., H. Skaug, and J. Laake. 2015. R2admb: "ADMB" to R interface functions. Available at https://cran.r-project.org/web/packages/R2admb

Byrne, P. G., L. W. Simmons, and J. D. Roberts. 2003. Sperm competition and the evolution of gamete morphology in frogs. Proc. R. Soc. B Biol. 270:2079–2086.

Crawley, M. J. 2013. The R book. 2nd ed. John Wiley & Sons, Chichester, U.K.

Crudgington, H. S., S. Fellows, N. S. Badcock, and R. R. Snook. 2009. Experimental manipulation of sexual selection promotes greater male mating capacity but does not alter sperm investment. Evolution 63:926–938.

Droge-Young, E. M., J. M. Belote, G. S. Perez, and S. Pitnick. 2016. Resolving mechanisms of short-term competitive fertilization success in the red flour beetle. J. Insect Physiol. 93–94:1–10.

Dziminski, M. A., J. D. Roberts, M. Beveridge, and L. W. Simmons. 2009. Sperm competitiveness in frogs: slow and steady wins the race. Proc. R. Soc. B Biol. 276:3955–3961.

Fedina, T. Y., and S. M. Lewis. 2008. An integrative view of sexual selection in Tribolium flour beetles. Biol. Rev. 83:151–171.

Firman, R. C., and L. W. Simmons. 2010. Experimental evolution of sperm quality via postcopulatory sexual selection in house mice. Evolution 64:1245–1256.

Fitzpatrick, J. L., and B. Baer. 2011. Polyandry reduces sperm length variation in social insects. Evolution 65:3006–3012.

Fitzpatrick, J. L., F. Garcia-Gonzalez, and J. P. Evans. 2010. Linking sperm length and velocity: the importance of intra-male variation. Biol. Lett. 6:797–799.

Fournier, D. A., H. J. Skaug, J. Ancheta, J. Ianelli, A. Magnusson, M. N. Maunder et al. 2012. AD Model Builder: using automatic differentiation for statistical inference of highly parameterized complex nonlinear models. Optim. Method. Softw. 27:233–249.

Fox, J., S. Weisberg, D. Adler, D. Bates, G. Baud-Bovy, S. Ellison, et al. 2015. Car: companion to applied regression. Available at https://cran.r-project.org/web/packages/car

Gage, M. J. G., and R. P. Freckleton. 2003. Relative testis size and sperm morphometry across mammals: no evidence for an association between sperm competition and sperm length. Proc. R. Soc. Lond. B Biol. 270:625–632.

Gage, M. J. G., and E. H. Morrow. 2003. Experimental evidence for the evolution of numerous, tiny sperm via sperm competition. Curr. Biol. 13:754–757.

Gage, M. J. G., C. P. Macfarlane, S. Yeates, R. G. Ward, J. B. Searle, and G. A. Parker. 2004. Spermatozoal traits and sperm competition in Atlantic salmon: relative sperm velocity is the primary determinant of fertilization success. Curr. Biol. 14:44–47.

Gay, L., D. J. Hosken, R. Vasudev, T. Tregenza, and P. E. Eady. 2009. Sperm competition and maternal effects differentially influence testis and sperm size in Callosobruchus maculatus. J. Evol. Biol. 22:1143–1150.

Gomendio, M., and E. R. S. Roldan. 1991. Sperm competition influences sperm size in mammals. Proc. R. Soc. Lond. B Biol. 243:181–185.

Grosjean, P., F. Ibanez, and M. Etienne. 2014. pastecs: package for analysis of space-time ecological series. Available at https://cran.r-project.org/web/packages/pastecs

Harrell, F. E., et al. 2016. Hmisc: Harrell miscellaneous. Available at https://cran.r-project.org/package=Hmisc

van der Horst, G., and L. Maree 2014. Sperm form and function in the absence of sperm competition. Mol. Reprod. Dev. 81, 204–216.

Humphries, S., J. P. Evans, and L. W. Simmons. 2008. Sperm competition: linking form to function. BMC Evol. Biol. 8:1–11.

Immler, S., and T. R. Birkhead. 2007. Sperm competition and sperm midpiece size: no consistent pattern in passerine birds. Proc. R. Soc. Lond. B Biol. 274:561–568.

Immler, S., S. Calhim, and T. R. Birkhead. 2008. Increased post-copulatory sexual selection reduces the intra-male variation in sperm design. Evolution 62:1538–1543.

Immler, S., S. Pitnick, G. A. Parker, K. L. Durrant, S. Lüpold, S. Calhim, et al. 2011. Resolving variation in the reproductive trade-off between sperm size and number. Proc. Natl. Aacd. Sci. 108:5325–5330.

Janicke, T., P. Sandner, S. A. Ramm, D. B. Vizoso, and L. Schärer. 2016. Experimentally evolved and phenotypically plastic responses to enforced monogamy in a hermaphroditic flatworm. J. Evol. Biol. 29:1713–1727.

Katz, D. F., and E. Z. Drobnis. 1990. Analysis and interpretation of the forces generated by spermatozoa. In B. D. Bavister, J. Cummins, and E. R. S. Roldan Fertilization in mammals. Serono Symposia, Norwell, Massachusetts, USA. 125–137.

Kelly, C. D., and M. D. Jennions. 2011. Sexual selection and sperm quantity: meta-analyses of strategic ejaculation. Biol. Rev. 86:863–884.

Kleven, O., T. Laskemoen, F. Fossoy, R. J. Robertson, and J. T. Lifjeld. 2008. Intraspecific variation in sperm length is negatively related to sperm competition in passerine birds. Evolution 62:494–499.

LaMunyon, C. W., and S. Ward. 1998. Larger sperm outcompete smaller sperm in the nematode *Caenorhabditis elegans*. Proc. R. Soc. B Biol. 265:1997–2002.

———. 2002. Evolution of larger sperm in response to experimentally increased sperm competition in *Caenorhabditis elegans*. Proc. R. Soc. B Biol. 269, 1125–1128.

Laskemoen, T., T. Albrecht, A. Bonisoli-Alquati, J. Cepak, F. de Lope, I. G. Hermosell, et al. 2013. Variation in sperm morphometry and sperm competition among barn swallow (*Hirundo rustica*) populations. Behav. Ecol. Sociobiol. 67:301–309.

Lessells, C., and P. Boag. 1987. Unrepeatable repeatabilities—a common mistake. Auk 104:116–121.

Lessells, C. M., R. R. Snook, D. J. Hosken, T. R. Birkhead, D. J. Hosken, and S. Pitnick. 2009. The evolutionary origin and maintenance of sperm: selection for a small, motile gamete mating type. Pp. 43–67 in T. R. Birkhead, D. J. Hosken, and S. S. Pitnick, eds. Sperm biology: an evolutionary perspective. Academic Press, London.

Lewis, S. M., A. Kobel, T. Fedina, and R. W. Beeman. 2005. Sperm stratification and paternity success in red flour beetles. Physiol. Entomol. 30:303–307.

Lifjeld, J. T., T. Laskemoen, O. Kleven, T. Albrecht, and R. J. Robertson. 2010. Sperm length variation as a predictor of extrapair paternity in passerine birds. PLoS One. 5:e13456.

Lumley, A. J., Ł. Michalczyk, J. J. N. Kitson, L. G. Spurgin, C. A. Morrison, J. L. Godwin, et al. 2015. Sexual selection protects against extinction. Nature 522:470–473.

Lüpold, S., and J. L. Fitzpatrick. 2015. Sperm number trumps sperm size in mammalian ejaculate evolution. Proc. R. Soc. B Biol. 282: p.20152122

Lüpold, S., S. Calhim, S. Immler, and T. R. Birkhead. 2009. Sperm morphology and sperm velocity in passerine birds. Proc. R. Soc. B Biol. 276:1175–1181.

Lüpold, S., M. K. Manier, K. S. Berben, K. J. Smith, B. D. Daley, S. H. Buckley et al. 2012. How multivariate ejaculate traits determine competitive fertilization success in *Drosophila melanogaster*. Curr. Biol. 22:1667–1672.

Lüpold, S., M. K. Manier, N. Puniamoorthy, C. Schoff, W. T. Starmer, S. H. B. Luepold et al. 2016. How sexual selection can drive the evolution of costly sperm ornamentation. Nature 533:535–538.

Michalczyk, Ł., O. Y. Martin, A. L. Millard, B. C. Emerson, and M. J. G. Gage. 2010. Inbreeding depresses sperm competitiveness, but not fertilization or mating success in male *Tribolium castaneum*. Proc. R. Soc. B Biol. 277:3483–3491.

Michalczyk, Ł., A. L. Millard, O. Y. Martin, A. J. Lumley, B. C. Emerson, and M. J. G. Gage. 2011. Experimental evolution exposes female and male responses to sexual selection and conflict in *Tribolium castaneum*. Evolution 65:713–724.

Miller, G. T., and S. Pitnick. 2002. Sperm-female coevolution in *Drosophila*. Science 298, 1230–1233.

Moore, A. J., L. D. Bacigalupe, and R. R. Snook. 2013. Integrated and independent evolution of heteromorphic sperm types. Proc. R. Soc. B Biol. 280(1769). p.20131647.

Morrow, E. H., and M. J. Gage. 2001. Consistent significant variation between individual males in spermatozoal morphometry. J. Zool. 254:147–153.

Nakagawa, S., and H. Schielzeth. 2010. Repeatability for Gaussian and non-Gaussian data: a practical guide for biologists. Biol. Rev. 85:935–956.

Parker, G. A. 1982. Why are there so many tiny sperm? Sperm competition and the maintenance of two sexes. J. Theor. Biol. 96:281–294.

Parker, G. A., and M. E. Begon. 1993. Sperm competition games: sperm size and number under gametic control. Proc. R. Soc. Lond. B Biol. 253:255–262.

Parker, G. A., and T. Pizzari. 2010. Sperm competition and ejaculate economics. Biol. Rev. 85:897–934.

Parker, G. A., R. R. Baker, and V. G. F. Smith. 1972. The origin and evolution of gamete dimorphism and the male-female phenomenon. J. Theor. Biol. 36:529–553.

Parker, G. A., M. A. Ball, P. Stockley, and M. J. G. Gage. 1997. Sperm competition games: a prospective analysis of risk assessment. Proc. R. Soc. Lond. B Biol. 264:1793–1802.

Pattarini, J. M., W. T. Starmer, A. Bjork, and S. Pitnick. 2006. Mechanisms underlying the sperm quality advantage in *Drosophila melanogaster*. Evolution 60:2064–2080.

Pitnick, S., G. S. Spicer, and T. A. Markow. 1995. How long is a giant sperm? Nature 375:109–109.

Pitnick, S., G. T. Miller, J. Reagan, and B. Holland. 2001. Males' evolutionary responses to experimental removal of sexual selection. Proc. R. Soc. B Biol. 268:1071–1080.

Pitnick, S., D. J. Hosken, and T. R. Birkhead. 2009a. Sperm morphological diversity. Pp. 69–149 *in* T. R. Birkhead, D. J. Hosken, and S. S. Pitnick, eds. Sperm biology: an evolutionary perspective. Academic Press, London.

Pitnick, S., M. F. Wolfner, and S. S. Suarez. 2009b. Ejaculate-female and sperm-female interactions. Pp. 247–304 *in* T. R. Birkhead, D. J. Hosken, and S. S. Pitnick, eds. Sperm biology: an evolutionary perspective. Academic Press, London.

Pizzari, T., and G. A. Parker. 2009. Sperm competition and sperm phenotype. Pp. 207–245 *in* T. R., Birkhead, D. J., Hosken, and S. S., Pitnick, eds. Sperm biology: an evolutionary perspective. Academic Press, London.

R Core Team. 2015. R: a language and environment for statistical computing. R Foundation for Statistical Computing, Vienna.

Radwan, J. 1996. Intraspecific variation in sperm competition success in the bulb mite: a role for sperm size. Proc. R. Soc. B Biol. 263:855–859.

Reuter, M., J. R. Linklater, L. Lehmann, K. Fowler, T. Chapman, and G. D. D. Hurst. 2008. Adaptation to experimental alterations of the operational sex ratio in populations of *Drosophila melanogaster*. Evolution 62:401–412.

Rowe, M., T. Albrecht, A. Johnsen, E. R. A. Cramer, T. Laskemoen, J. T. Weir, et al. 2015. Postcopulatory sexual selection is associated with accelerated evolution of sperm morphology. Evolution 69:1044–1052.

Schneider, C. A., W. S. Rasband, and K. W. Eliceiri. 2012. NIH image to ImageJ: 25 years of image analysis. Nat. Methods 9:671–675.

Sbilordo, S. H., V. M. Grazer, M. Demont, and O. Y. Martin. 2011. Impacts of starvation on male reproductive success in *Tribolium castaneum*. Evol. Ecol. Res. 13:347–359

Simmons, L. W., J. Wernham, F. García-González, and D. Kamien. 2003. Variation in paternity in the field cricket *Teleogryllus oceanicus*: no detectable influence of sperm numbers or sperm length. Behav. Ecol. 14:539–545.

Simpson, J. L., S. Humphries, J. P. Evans, L. W. Simmons, and J. L. Fitzpatrick. 2014. Relationships between sperm length and speed differ among three internally and three externally fertilizing species. Evolution 68:92–104.

Snook, R. R. 2005. Sperm in competition: not playing by the numbers. Trends Ecol. Evol. 20:46–53.

Snook, R. R., L. D. Bacigalupe, and A. J. Moore. 2010. The quantitative genetics and coevolution of male and female reproductive traits. Evolution 64:1926–1934.

Stewart, K. A., R. Wang, and R. Montgomerie. 2016. Extensive variation in sperm morphology in a frog with no sperm competition. BMC Evol. Biol. 16:29.

Taylor, M. L., T. A. R. Price, and N. Wedell. 2014. Polyandry in nature: a global analysis. Trends Ecol. Evol. 29:376–383.

Tourmente, M., M. Gomendio, and E. R. Roldan. 2011. Sperm competition and the evolution of sperm design in mammals. BMC Evol. Biol. 11:12.

Tregenza, T., F. Attia, and S. S. Bushaiba. 2009. Repeatability and heritability of sperm competition outcomes in males and females of *Tribolium castaneum*. Behav. Ecol. Sociobiol. 63:817–823.

Uzbekov, R., J. Burlaud-Gaillard, A. S. Garanina, and C. Bressac. 2017. The length of a short sperm: elongation and shortening during spermiogenesis in *Cotesia congregata* (Hymenoptera, Braconidae). Arthropod Struct. Dev. 46:265–273.

Varea-Sánchez, M., L. Gómez Montoto, M. Tourmente, and E. R. S. Roldan. 2014. Postcopulatory sexual selection results in spermatozoa with more uniform head and flagellum sizes in rodents. PLoS One 9 (9):e108148.

Wedell, N., M. J. Gage, and G. A. Parker. 2002. Sperm competition, male prudence and sperm-limited females. Trends Ecol. Evol. 17:313–320.

Wickham, H. 2015. plyr: tools for splitting, applying and combining data. Available at https://cran.r-project.org/web/packages/plyr

Wickham, H., and W. Chang. 2015. ggplot2: an implementation of the grammar of graphics. Available at https://cran.r-project.org/web/packages/ggplot2

Wright, S. 1931. Evolution in Mendelian populations. Genetics 16:97–159.

Zeng, Y., S.-L. Lou, W.-B. Liao, and R. Jehle. 2014. Evolution of sperm morphology in anurans: insights into the roles of mating system and spawning location. BMC Evol. Biol. 14:104.

6

Interpreting differentiation landscapes in the light of long-term linked selection

Reto Burri[1,2]

[1] Department of Population Ecology, Friedrich Schiller University Jena, Dornburger Strasse 159, D-07743 Jena, Germany

[2] E-mail: burri@wildlight.ch

Identifying genomic regions underlying adaptation in extant lineages is key to understanding the trajectories along which biodiversity evolves. However, this task is complicated by evolutionary processes that obscure and mimic footprints of positive selection. Particularly, the long-term effects of linked selection remain underappreciated and difficult to account for. Based on patterns emerging from recent research on the evolution of differentiation across the speciation continuum, I illustrate how long-term linked selection affects the distribution of differentiation along genomes. I then argue that a comparative population genomics framework that exploits emergent features of long-term linked selection can help overcome shortcomings of traditional genome scans for adaptive evolution, but needs to account for the temporal dynamics of differentiation landscapes.

KEY WORDS: Adaptation, comparative population genomics, genome scans, speciation.

Impact Summary

The quest for genomic regions involved in adaptation or speciation is still most frequently based on the screening of genomes for regions exhibiting accentuated differentiation. This framework assumes accentuated differentiation to predominantly evolve through positive selection of beneficial variants or selection against gene flow. However, together with longstanding theory, recent research challenges this assumption, suggesting that accentuated differentiation may evolve through processes unrelated to adaptation or speciation of extant lineages. Here, I outline the common patterns emerging from recent research to highlight yet underappreciated caveats and opportunities for the inference of genomic regions underlying adaptation and speciation. In particular, I argue that the repeated evolution of highly correlated differentiation landscapes in independent lineages is a predicted consequence of long-term selection at linked sites, which is likely to include important contributions of purifying selection. Meanwhile, accentuated differentiation at early stages of differentiation may more likely reflect the action of positive selection. I then put forward that the correlation among independent differentiation landscapes can be exploited in a comparative population genomics framework to empirically formulate dynamic baselines across the genome against which to discriminate candidate regions involved in adaptation or speciation in extant lineages. I then formulate a verbal model for the temporal dynamics of the correlation among differentiation landscapes of independent lineages. Finally, I argue that by integrating these temporal dynamics, the comparative population genetic framework provides a powerful tool to take into account the long-term effects of linked selection in the quest for genomic regions underlying adaptation or speciation.

Introduction

Since the dawn of high-throughput sequencing, the inference of genomic regions of accentuated differentiation (Box 1)—variously referred to as differentiation islands, divergence islands, or speciations islands (Box 2)—has become a central focus of research on local adaptation and speciation. The past years have seen an unprecedented quest for such regions (Haasl & Payseur 2016) that assumed accentuated differentiation to evolve trough processes related to adaptation or speciation, in particular positive selection of beneficial variants (Maynard Smith & Haigh 1974; Kaplan et al. 1989) or selection against gene flow (Turner et al. 2005) in extant populations or species (in the following referred to as "extant lineages", see Glossary). However, recent research highlights that accentuated differentiation may evolve through processes other than positive selection (e. g., Bank et al. 2014; Cruickshank & Hahn 2014; Haasl & Payseur 2016).

Particularly, several aspects of linked selection may remain underappreciated. First, awareness that purifying selection at linked sites (background selection, BGS) (Charlesworth et al. 1993; Charlesworth 2013) makes important contributions to linked selection and can mimic the footprints of positive selection (Stephan 2010) is still limited. Furthermore, part of the effects of linked selection observed today may have accumulated over extended periods of time and therefore be related to adaptation in ancestral rather than extant lineages (McVicker et al. 2009; Munch et al. 2016; Phung et al. 2016). Approaches that assist disentangling the effects of alternative forms of selection and the timescales at which they acted are called for.

Here, I first showcase the impact of linked selection on the long-term evolution of genetic diversity and hence differentiation (Charlesworth 1998). I then discuss how these effects lead to the evolution of temporally dynamic correlations of differentiation landscapes, and how this process may be influenced by demography and the evolution of genome features. I close by outlining how emergent features of long-term linked selection can be exploited to empirically take into account the long-term effects of linked selection in a comparative population genomics framework that takes into account the temporal dynamics of differentiation landscapes.

Effects of Linked Selection in Heterogeneous Recombination Landscapes

The association of physically linked genetic variants within chromosomes, and recombination—the force that can break it up—have a profound effect on the distribution of genetic diversity along genomes (Cutter and Payseur 2013). Genetic diversity becomes not only reduced at sites targeted by positive selection but also at surrounding sites. During selective sweeps, genetic variants linked to beneficial mutations hitchhike along, while others are lost (Maynard Smith and Haigh 1974). Likewise, selection against gene flow limits the levels of genetic diversity in genomic regions involved in reproductive isolation compared to introgressed proportions of the genome (Turner et al. 2005). Impor-

tantly, in a similar way a reduction of genetic diversity at linked sites is caused by selection against deleterious variants, that is BGS, although at lower rates (Charlesworth et al. 1993; Zeng and Charlesworth 2011). The extent to which these processes affect particular genomic regions, besides selection strength, are determined by local recombination rates (Maynard Smith and Haigh 1974; Kaplan et al. 1989). Specifically, in genomic regions with infrequent recombination linked selection affects physically larger chromosome stretches, and it does so at a higher rate in regions with a high density of functional sites (hereafter "functional density") that constitute targets for selection (Payseur and Nachman 2002). Therefore, variation in recombination rate and in functional densities along the genome are bound to result in a heterogeneous genomic landscape of diversity, with regions of low recombination exhibiting troughs in diversity (and therefore reduced local effective population size, N_e, as diversity $\pi = 4N_e\mu$). Because differentiation is intrinsically affected by diversity (Charlesworth 1998), genomic regions of low recombination inevitably evolve toward accentuated differentiation under any of these forms of selection.

Evolution of Correlated Differentiation Landscapes

Recent research contributes to the emergence of a striking picture of how differentiation landscapes evolved among closely related species. *Helianthus* sunflowers, *Heliconius* butterflies, *Ficedula* flycatchers, crows, greenish warblers (*Phylloscopus trochiloides*), and stonechats (Fig. 1) exhibit highly similar differentiation landscapes among closely related species (Kronforst et al. 2013; Martin et al. 2013; Renaut et al. 2014; Burri et al. 2015; Irwin et al. 2016; Vijay et al. 2016; Van Doren et al. 2017).

This pattern of repeated evolution is a direct prediction of the long-term action of linked selection in heterogeneous recombination landscapes. Over timescales across which the genome features constraining genetic diversity are stable, diversity landscapes are expected to correlate even among independent lineages for two reasons. First, diversity levels along the genome scale with levels of ancestral diversity (Fig. 2A and B). This cumulative impact of linked selection is reflected in a correlation of extant diversity (π) with ancestral diversity (reflected in sequence divergence, d_{XY}, among closely related species, Box 1) (Nachman and Payseur 2012; Burri et al. 2015). The reduction of ancestral diversity through long-term linked selection may be strong enough to even impact sequence divergence between distantly related species, such as human and mouse, and bird species as divergent as 50 my (Phung et al. 2016; Dutoit et al. 2017; Van Doren et al. 2017; Vijay et al. 2017). Second, linked selection within extant lineages maintains the relative levels of diversity among genomic regions. This is reflected most directly in reduced levels of private polymorphisms in low-recombination regions, as observed in

Figure 1. Systems with correlated differentiation landscapes. From top left to bottom right: *Helianthus* sunflowers, collared flycatcher (*Ficedula albicollis*), pied flycatcher (*Ficedula hypoleuca*), *Heliconius melpomene rosina*, hooded crow (*Corvus cornix*), carrion crow (*Corvus corone*), Siberian stonechat (*Saxicola maurus*), *Heliconius cydno galanthus* (picture credits: sunflowers, Takeshi Kawakami; *Heliconius* butterflies, Richard Merrill; birds, Reto Burri)

flycatchers (Burri et al. 2015). Importantly, despite extant diversity in part being explained by ancestral diversity, correlated *differentiation* landscapes can evolve independently; differentiation evolves only through the differential sorting of genetic diversity in independent extant lineages. In conclusion, the evolution of correlated differentiation landscapes is an effect of long-term linked selection exposing conserved genomic features, namely conserved landscapes of recombination and functional densities.

Given that BGS explains a high proportion of the variation in diversity along genomes (Lohmueller et al. 2011; Comeron 2014; Corbett-Detig et al. 2015; Phung et al. 2016), in finite populations that are mutation limited, BGS may arguably play a major role in the evolution of these patterns. Beneficial mutations are rare (Eyre-Walker and Keightley 2007), and the frequency at which positive selection hits the same genomic regions repeatedly is expected to be limited. In contrast, deleterious variants represent the majority of non-neutral mutations. With a steady influx of deleterious mutations, the cumulative time span during which BGS acts is expected to be much longer than that of episodic positive selection, and mutations representing targets for BGS are abundant and widespread across the genome. Consequently, while being pervasively affected by BGS, not all functional regions are necessarily hit by events of positive selection. Therefore, *in extremis*, genomic landscapes correlated among species may evolve even under the action of BGS alone.

Even though beneficial mutations are rare, at vast timescales there is an increasing chance for several of them to occur within

the same genomic region. In such cases, also positive selection may affect the same genomic regions repeatedly. However, it may often have occurred at timescales exceeding lineages' divergence times, and therefore involve events of positive selection other than ones associated with adaptive evolution in extant lineages.

It appears that on the long run low-recombination regions are predestined to evolve toward accentuated differentiation due to the cumulative effects of pervasive BGS and, presumably to a lesser extent, recurring sweeps (McVicker et al. 2009; Munch et al. 2016). As a consequence of this process, conservation of recombination rates and functional densities among lineages will lead to the evolution of correlated diversity and differentiation landscapes even among distantly related lineages.

Unpredictable Emergence of Differentiation Islands at Early Stages of Differentiation

The existence of genomic regions predestined to evolve accentuated differentiation prompts the question whether the same patterns are already present at early stages of differentiation. Indeed, sharing of highly differentiated regions at early stages of differentiation has been reported from diverse organisms (Jones et al. 2012; Renaut et al. 2013; Soria-Carrasco et al. 2014; Fraser et al. 2015). Moreover, in the few studies that investigated multiple timescales, regions shared among older comparisons are on average also more differentiated earlier on. In flycatchers and crows (Burri et al. 2015; Vijay et al. 2016), regions exhibiting

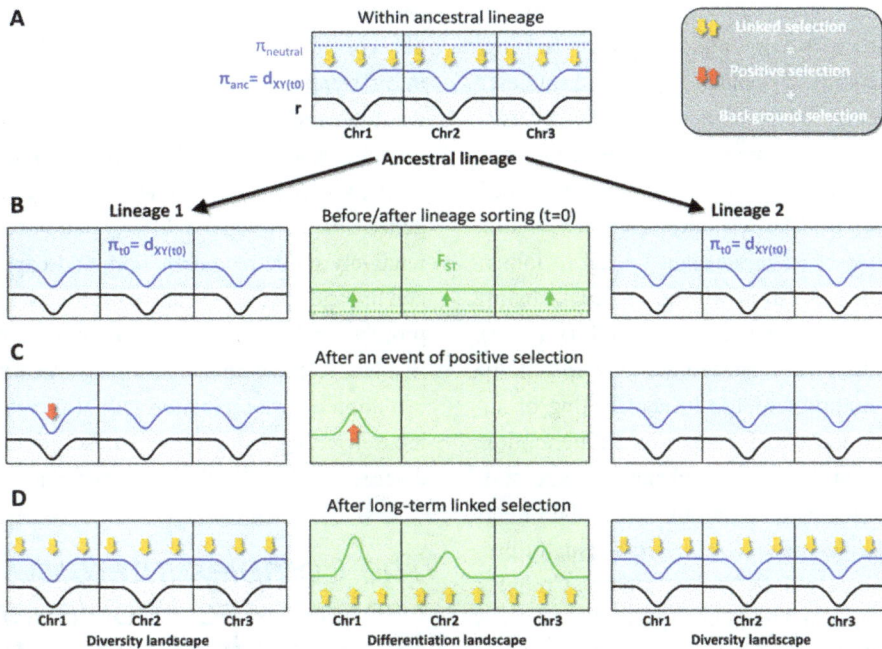

Figure 2. Genetic diversity (π, blue shaded boxes) and differentiation (F_{ST}, green shaded boxes) from early to late stages of differentiation and the role of linked selection in their evolution. Three example chromosomes are depicted. The grey shaded parts of the boxes represent the underlying recombination landscapes. (A) Genetic diversity in a hypothetical population with a heterogeneous recombination landscape before (broken line) and after (solid line) the action of any form of selection. (B) Genetic diversity and differentiation after a population split. Differentiation is shown before (broken line) and after lineage sorting (solid line) with lineage sorting depicted by green arrows. Before lineage sorting, genetic diversity in the descendent populations (π_{t0}) equals ancestral diversity ($d_{XY(t0)} = 4N_{e(ANC)}\mu + 2\mu T = 4N_{e(ANC)}\mu$). After some amount of lineage sorting, differentiation starts building up genome wide. At this timescale, background selection had limited power to reduce diversity and enhance differentiation in low-recombination regions, and no event of positive selection may have occurred. (C) After the occurrence of an event of positive selection in one population, differentiation is enhanced in one chromosome. At this timescale, background selection still had limited power to reduce diversity and enhance differentiation in low-recombination regions. (D) As a long-term effect of linked selection (including both positive selection and BGS) differentiation starts building up more strongly in all low-recombination regions than in the remainder of the genome.

accentuated differentiation among populations do so also between species. However, vice versa, only a subset of regions exhibiting accentuated differentiation between species do so among populations. Similar patterns are found in *Heliconius* butterflies (Kronforst et al. 2013; Martin et al. 2013), and in *Helianthus* sunflowers (Renaut et al. 2014). These observations suggest that the effects of linked selection expose low-recombination regions less homogeneously at early stages of differentiation than in the long run.

Current evidence suggests that at early stages of differentiation BGS has limited power to explain accentuated differentiation. The effects of BGS may be too subtle to drive accentuated differentiation at short timescales. Otherwise, differentiation would be predicted to evolve in identical genomic regions in related lineages (Munch et al. 2016). However, this expectation is not met by observations from early differentiation stages in flycatchers and crows: genomic regions highly differentiated among populations within one flycatcher species or across one crow hybrid zone usually exhibit average differentiation in other

species/hybrid zones (Burri et al. 2015; Vijay et al. 2016). The emergence of differentiation islands (Box 2) at these rather recent timescales, therefore, appears to be unpredictable. In particular the observation that the locations of such regions differ between closely related species in flycatchers and across crow hybrid zones suggest that, apart from the effects of drift connected with lineage splits, the evolution of accentuated differentiation at early stages of differentiation may involve positive selection rather than BGS.

A Temporal Perspective: Dynamic Correlations among Genomic Landscapes Along the Differentiation Continuum

The outlined observations suggest temporal dynamics of differentiation landscapes during which (i) differentiation increasingly reflects underlying recombination rates and functional densities, and correlations among independent differentiation landscapes increase with advancing differentiation, and (ii) the interpretation of accentuated differentiation may depend on the differentiation stage in focus. These temporal dynamics have important bearings on the design and interpretation of population genomic studies.

In populations evolving under neutrality, genetic diversity (π) is unaffected by recombination rate variation (Fig. 2A). However, selection reduces genetic diversity, with strongest effects on low-recombination regions. The resulting heterogeneous diversity landscape is passed on to descendant lineages. Shortly after the lineages split, differentiation will be minimal, mainly driven by the sampling effect connected with the split (Fig. 2B). In a first stage, differentiation will then start building up genome wide, predominantly through differential lineage sorting of ancestral polymorphisms. At these early stages of differentiation, BGS has made limited contributions to enhance lineage sorting in low-recombination regions. Therefore, unless the effects of genetic drift connected with lineage splits are unusually pronounced in particular genome regions (see section The Complex Interplay of Genomic Features and Population-Level Processes in the Evolution of Differentiation Landscapes), the evolution of accentuated differentiation may require positive selection (Fig. 2C). As a function of the waiting times for positive selection to occur, the emergence of regions of strongly accentuated differentiation may be unpredictable, and the correlation of differentiation landscapes among independent comparisons remain limited (Fig. 3A).

As differentiation advances, the effects of linked selection will start accumulating and differentiation landscapes will increasingly reflect the underlying recombination landscape, such as illustrated, for example, in stickleback fish (Roesti et al. 2013). Compared to earlier stages of differentiation, these long-term effects of linked selection will contribute to a fundamentally different picture characterized by consistently accentuated differentiation in low-recombination regions (Fig. 2D). Across timescales for which the recombination landscape is stable—this can be across numerous speciation events, and tens of millions of years in birds (Singhal et al. 2015)—this process will result in strong correlations among differentiation landscapes from independent comparisons (Fig. 3A). In contrast to shorter timescales, the cumulative effects of BGS had more time to play out, and accentuated differentiation may evolve even in the absence of positive selection.

Altogether, this leads to temporally dynamic correlations among differentiation landscapes from independent comparisons (Fig. 3A). In contrast, at population genetic timescales (Fig. 3 left) no such temporal dynamics is expected for the genomic landscapes genetic of diversity (π) and sequence divergence (d_{XY}), because these are passed down over lineage splits (Fig. 3B).

However, in the long run, the correlations among genomic landscapes are eventually bound to decay along two axes: (i) a population genetic timescale, that is, the stage of differentiation *within* each pairwise comparison (Fig. 3 left), and (ii) a phylo-

genetic timescale, that is, the phylogenetic divergence *between* pairwise comparisons (Fig. 3 right). Within the first, the variance in differentiation (F_{ST}, Box 1) is reduced as it reaches unity toward the end stages of differentiation. Stochastic variation may therefore lead to decreasing correlations at this stage (Fig. 3 right). Perhaps more importantly, at the phylogenetic timescale, as divergence time between the differentiation landscape increases, even relatively stable recombination landscapes are bound to diverge, and lineage-specific events of positive selection accumulate. Both processes will contribute to erode the correlation of differentiation landscapes with increasing divergence (Fig. 3). The divergent evolution of recombination will also contribute to erode the correlation among independent landscapes of diversity and sequence divergence (Fig. 3B) over phylogenetic timescales.

The Complex Interplay of Genomic Features and Population-Level Processes in the Evolution of Differentiation Landscapes

The outlined temporal dynamics may vary strongly and depend on the complex interplay of various genomic features and population-level processes.

Recombination rate evolution differs substantially among species (Smukowski and Noor 2011), and may have a strong impact on the evolution of genomic landscapes. First, among taxa in which the amplitude of recombination rate variation is limited, random variation in genetic diversity and differentiation may dominate and differentiation landscapes may correlate poorly (Table 1). Second, the conservation of recombination landscapes, determines the timescale across which the same (low-recombination) regions are exposed to strong effects of linked selection. In taxa with long-term conserved recombination landscapes, such as birds (Singhal et al. 2015), diversity reductions through linked selection in low-recombination regions will be particularly strong, and diversity and differentiation landscapes will be correlated over longer timescales than in taxa with more dynamically evolving recombination landscapes, such as most mammals (e.g., Oliver et al. 2009; Baudat et al. 2010) (Table 1).

How other genomic features affect the evolution of correlated differentiation landscapes may depend on the extent and sign of their correlation with recombination rate and the turnover of their distribution along the genome (Table 1). Mutation rate can vary substantially (e.g., Hodgkinson and Eyre-Walker 2011) and appears to be elevated in regions of high recombination (e.g., Cutter and Payseur 2013; Arbeithuber et al. 2015; Terekhanova et al. 2017). Although such a positive correlation of mutation with recombination might reinforce the effects of linked selection, there is little evidence for such a concerted effect on genetic diversity (Francioli et al. 2015; Ellegren and Galtier 2016). Meanwhile,

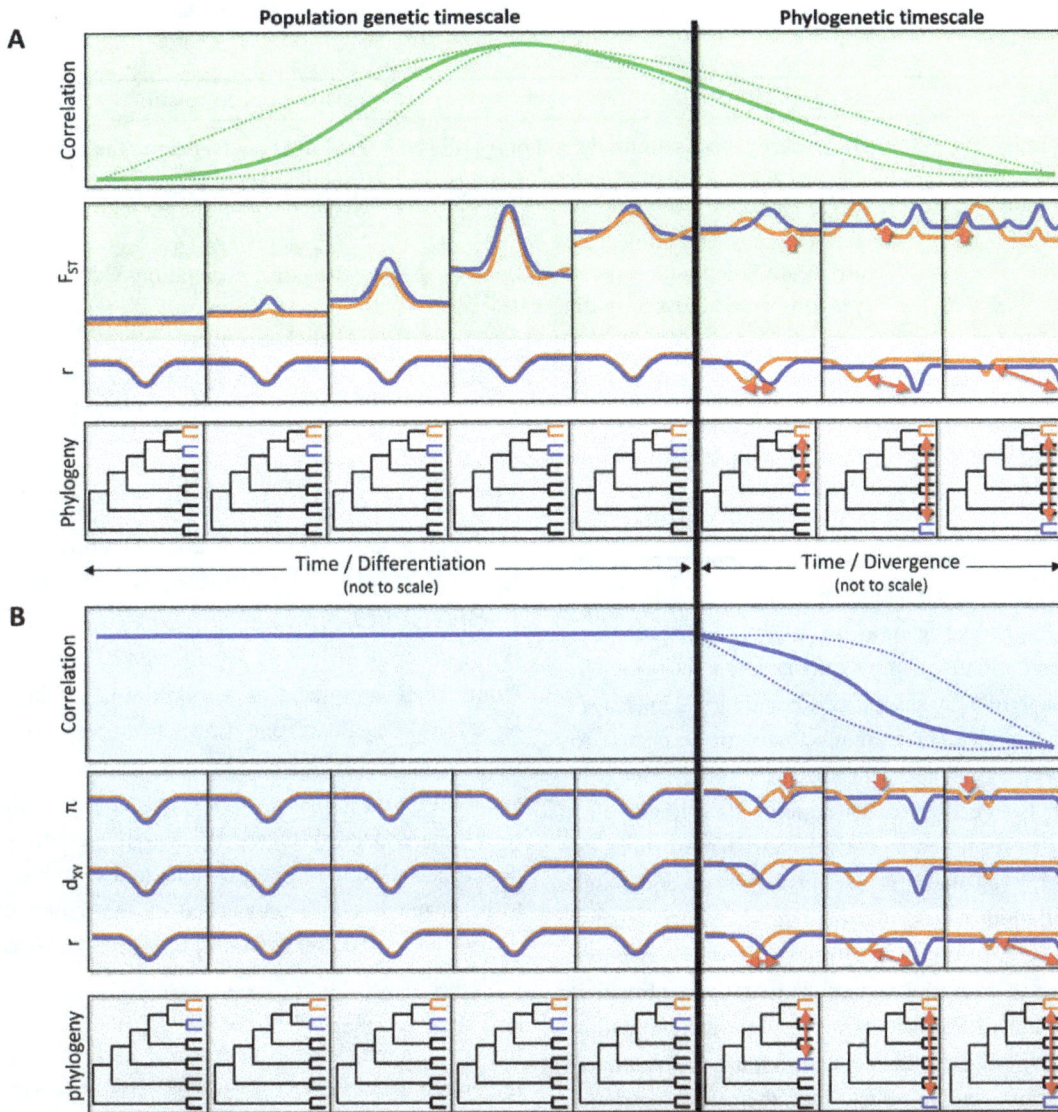

Figure 3. Sketch of temporal dynamics of the correlation among independent differentiation, diversity, and divergence landscapes with increasing differentiation and divergence. **(A)** Temporal dynamics of differentiation landscapes and their correlation. Top panel: Temporal dynamics of the correlation of differentiation landscapes. The form of this correlation is not clear, as indicated by the alternative trajectories (broken lines). Middle panel: Evolution of the underlying differentiation and recombination landscapes of the two comparisons (depicted in blue and orange; blue Comparison 1; orange, Comparison 2). Bottom panel: Phylogenetic topologies indicating the divergence between the two pairwise comparisons. **(B)** Temporal dynamics of the genomic landscapes of diversity (π) and sequence divergence (d_{XY}). The temporal dynamics of the correlations in π and d_{XY} among lineages/comparisons are expected to be similar, for which reason only one temporal dynamics is shown (upper panel). π and d_{XY} are not affected the same way as F_{ST} by lineage sorting after a lineage split and therefore are expected to be correlated throughout the population genetic timescale. Note that selection in extant lineages does not affect d_{XY}. In both panels the temporal dynamics at the population genetic (left) and phylogenetic timescales (right) are shown. Axes are not to scale. Fat red arrows, lineage-specific sweeps; slim red arrows, divergent evolution of the recombination landscapes, and divergence between comparisons.

functional densities are positively correlated with recombination rate to varying degrees (e.g., Flowers et al. 2012; Nam and Ellegren 2012; Kawakami et al. 2014). In taxa in which this antagonistic correlation is strong, it may dampen the effects of linked selection (Cutter and Payseur 2013). Moreover, a quick turnover of hot spots and cold spots of mutation and functional densities would distribute the respective effects more homogeneously along the genome in the long run. The turnover of functional densities may be moderate; even though new genes emerge at high rates, the majority is short lived (Schlötterer 2015). Meanwhile mutation rates

Table 1. Conditions under which genomic features and population-level processes may favor or disfavor the evolution of correlated differentiation landscapes.

	Favors correlation	Disfavors correlation
Recombination rates	Highly heterogeneous along the genome Conserved across long timescales	Homogeneous along the genome Quick turnover
Functional densities	Negative correlation with recombination rates Conserved across long timescales	Positive correlation with recombination rate Quick turnover
Mutation rates	Positive correlation with recombination rate *and* conserved across long timescales	Negative correlation with recombination rate Quick turnover
N_e	High N_e	Low N_e
Demography	Similar demography among lineages	Lineage-specific demographic events

appear to evolve at a faster pace than recombination (Terekhanova et al. 2017), and may therefore have more ephemeral effects on the heterogeneity of genomic landscapes than long-term linked selection in conserved recombination landscapes.

In addition to genome features, population-level processes and parameters, in particular demography and genome-wide N_e, may play a crucial role in shaping differentiation landscapes (Table 1). Even though demographic effects are expected to reduce genetic diversity genome-wide, they may have particularly strong effects on low-recombination regions. First, the higher linkage among sites results in an elevated variance of diversity and differentiation even under neutral evolution. Even though at early differentiation stages, this may result in accentuated differentiation to be mistaken as a footprint of positive selection, by itself it does not increase differentiation in low-recombination regions systematically. Second, however, low-recombination regions affected by long-term linked selection are characterized by reduced N_e. Demographic effects may therefore contribute to systematic increases in differentiation in these regions, and reduce the correlation between differentiation landscapes of taxa with unequal demographic histories.

Finally, correlated genomic landscapes may only evolve in taxa with sufficiently high N_e, in which selection is not overwhelmed by genetic drift (Ohta 2002) (Table 1). In line with this, linked selection appears to reduce genetic diversity below neutral expectations across a wider range of recombination rates in taxa with high N_e (Corbett-Detig et al. 2015). In taxa with very low N_e, effects of linked selection may be negligible altogether, and correlations between genomic landscapes be absent, or even inversed (Van Doren et al. 2017; Burri 2017b)

In conclusion, the alignment of multiple genomic features in space (along the genome) and over time in combination with population-level processes are expected to determine the conditions under which correlated differentiation landscapes may evolve and the timescales over which correlations persist (see section Outstanding Questions).

Accounting for the Long-Term Effects of Linked Selection: A Comparative Population Genomics Framework

While genomic regions of accentuated differentiation have commonly been interpreted as a footprint of ecological adaptation or selection against gene flow, awareness is increasing that such an approach is naïve toward the vast recombination rate variation found in many species, the long-term effects of linked selection associated therewith, and the contribution of BGS therein. Distributions against which to discriminate outliers need to be adapted, and, as suggested by the temporal dynamics of differentiation landscapes outlined above, timescales may need to be accounted for.

At early stages of differentiation, genome-wide differentiation may offer a baseline against which to discriminate candidate regions (but see below). However, later on, owing to the long-term effects of linked selection, this baseline is shifted upwards in low-recombination regions (Fig. 2D). With recombination rate data and functional annotations at hand, it is possible to devise comprehensive models that estimate parameters of BGS along the genome to adapt baselines accordingly and enable the discrimination of positive selection against BGS (McVicker et al. 2009; Elyashiv et al. 2016; Huber et al. 2016). In the absence of such data, alternative approaches are required.

In line with previous suggestions, I advocate that a comparative genomic framework may significantly assist the inference of candidate regions involved in adaptation and/or speciation (Berner and Salzburger 2015). First, this approach provides information on whether constraints imposed by long-term linked selection apply to the taxa of interest. Second, the correlation of genomic landscapes among independent comparisons at advanced stages of differentiation can be exploited to expose the (relative) baseline levels of diversity and differentiation expected to evolve under long-term linked selection.

In the simplest form of this approach, observations across multiple comparisons serve as *qualitative* "evolutionary controls": Differentiation patterns common to all (focal and control) comparisons constitute a baseline hypothesis that assumes the evolution of common patterns through long-term linked selection. Only accentuated differentiation specific to single lineages may be taken to suggest positive selection in extant lineages (in the focal lineage or in the ancestor of the remaining lineages if they share a common ancestor to the exclusion of the focal lineage). Ideally, this criterion is applied even at early stages of differentiation to safeguard against false positives in regions of exceptionally low recombination and high functional density that may evolve accentuated differentiation most rapidly.

This qualitative criterion rigidly treats genomic regions with accentuated differentiation in all comparisons as unrelated to positive selection. However, in genomic regions strongly affected by long-term linked selection the effects of lineage-specific positive selection accumulate on top of already accentuated differentiation. The qualitative criterion may therefore be overly conservative. Instead, one might rather want to determine the varying amplitude to which baseline levels of differentiation need to be shifted along the genome. At differentiation stages at which differentiation landscapes are strongly correlated, the differentiation patterns common to independent comparisons are expected to reflect the long-term effects of linked selection (Munch et al. 2016), and may therefore be used as *quantitative* evolutionary controls. A dynamic baseline across the genome taking the effects of long-term linked selection into account can therefore be directly formulated based on the empirically observed common pattern. Several flavors of this approach can be envisaged (Box 3) and have helped to focus research on refined sets of candidate regions (e.g., Roesti et al. 2012; Roesti et al. 2013; Vijay et al. 2016).

Temporal Dynamics of Differentiation Landscapes in the Comparative Population Genomics Framework

Although both qualitative and quantitative evolutionary controls have found application (e.g., Roesti et al. 2012; Roesti et al. 2013; Renaut et al. 2014; Feulner et al. 2015; Vijay et al. 2016), the outlined temporal dynamics open an issue that has found less attention: the choice of appropriate evolutionary controls. I argue that these need to be chosen carefully with respect to both the stage of differentiation and the phylogenetic divergence.

Mismatched evolutionary controls may result in high false discovery rates. For comparisons at early differentiation stages (Fig. 2B–Dand Fig. 3 left) controls at an advanced stage (Fig. 2D and Fig. 3 right) risk yielding high rates of false negatives; although in some genomic regions accentuated differentiation may be bound to evolve in the long run, at early differentiation stages it may require positive selection. Conversely, for comparisons situated at advanced differentiation stages, comparisons at earlier stages, at which the effects of BGS may not yet be discernible, are prone to yield false positives. Similar problems may be induced by exceedingly divergent evolutionary controls (Fig. 3 right), which are prone to bias inference toward false positives. Due to an increased risk of divergent recombination landscapes, such divergent controls may not adequately reflect the long-term effects of linked selection relevant for the comparison. Consequently, evolutionary controls may yield most conclusive results when chosen at differentiation stages similar to that of comparisons, and in close phylogenetic proximity to the latter.

Nevertheless, finding adequate evolutionary controls may be complex. In many instances differentiation and divergence parameters may not be known when setting out for population genomics experiments. Ideally, experimental designs therefore may incorporate a range of evolutionary controls strategically placed along the differentiation continuum.

Conclusions

Long-term linked selection in heterogeneous recombination landscapes leads to the evolution of highly heterogeneous differentiation landscapes. This process results in temporally dynamic correlations among independent differentiation landscapes of related lineages, within which the interpretation of accentuated differentiation in terms of positive selection in extant lineages is complicated to different extents depending on the timescale studied. The comparative population genetic approach provides a powerful tool to qualitatively or quantitatively adapt baselines along the genome to take into account the long-term effects of linked selection. However, evolutionary controls have to be matched to the stage of differentiation and chosen at appropriate phylogenetic divergence. Empirical research and simulations are now called for to investigate the temporal dynamics of correlations among independent differentiation landscapes (see section Outstanding Questions), and to study the interplay of different forms of selection, recombination rate variation, and demography in their evolution.

Box 1. Measures of differentiation and divergence.

The interchangeable use of differentiation (allele frequency divergence) and divergence (sequence divergence) has led to their distinction as "relative divergence" and "absolute divergence," respectively. This distinction is important, because they describe different properties of genetic variation and are affected differently by evolutionary processes (e.g., Noor and Bennett 2009; Cruickshank and Hahn 2014). Here, for reason of simplicity I use the former, original definition.

F_{ST}

A measure of *differentiation* deeply rooted in population genetics theory (Whitlock 2011). It measures the relative contribution of genetic diversity between lineages to the total observed genetic diversity and is therefore dependent on genetic diversity within lineages (Charlesworth 1998). A particularly simple definition was given by Hudson et al. (1992):

$$F_{ST} = \frac{\pi_{total} - \pi_{within}}{\pi_{total}} = 1 - \frac{\pi_{within}}{\pi_{total}}$$

Where π_{total} and π_{within} are total nucleotide diversity and nucleotide diversity observed within lineages, respectively. In practice, F_{ST} is most often estimated using the unbiased estimator of Weir and Cockerham (1984). See Bhatia et al. (2013) for recommendations regarding F_{ST} estimation. It is worth noting that the π/d_{XY} ratio that has recently found application to interpret differentiation patterns (Irwin et al. 2016; Van Doren et al. 2017) is directly related to F_{ST}, as π_{total} is strongly related to d_{XY}.

d_{XY} (Nei & Li 1979)

A measure of *divergence*. It provides the average number of differences per site observed between two random haplotypes drawn from two different lineages and (in the absence of selection) and is defined as: $d_{XY} = 4N_{e(ANC)}\mu + 2\mu T$, where $N_{e(ANC)}$ represents ancestral N_e, μ the mutation rate (and thus $4N_{e(ANC)}\mu$ genetic diversity in the ancestral lineage), and T divergence time since lineage split. In practice, d_{XY} is estimated as:

$$\frac{1}{n} \sum_{i=1}^{m} p_{i1}q_{i2} + p_{i2}q_{i1}$$

where p_{i1} and q_{i1} are the frequencies of alternative alleles at locus i in lineage 1; p_{i2} and q_{i2} the frequencies of alternative alleles at the same locus in lineage 2; m the number of variable sites; and n the sum of variable and invariable sites (importantly, following the same filtering criteria) within the interval for which d_{XY} is estimated. Contrary to F_{ST}, d_{XY} is not sensitive to selection in extant lineages, but sensitive to selection in the ancestor (Nachman and Payseur 2012). At early stages of differentiation ($T \sim 0$), d_{XY} reflects ancestral diversity, $4N_{e(ANC)}\mu$. As the contribution of $4N_{e(ANC)}\mu$ relative to $2\mu T$ diminishes, d_{XY} converges toward d (see below).

Relative node depth, RND (Feder et al. 2005)

RND is a measure of *divergence* based on d_{XY} that aims at correcting d_{XY} for mutation rate variation along the genome. To this end, it divides d_{XY} of the focal comparison by the average d_{XY} observed between the focal species and an outgroup. However, by doing so also variation in ancestral population size along the genome is corrected for. This can be problematic in species in which the recombination landscape is conserved between outgroup and focal taxa, for instance if RND is used instead of d_{XY} in F_{ST}-d_{XY} contrasts.

d_A (Nei & Li 1979)

A measure of *differentiation*. This measure has been devised to capture pairwise differences that arose since the lineage split and is estimated/defined as:

$$d_A = d_{XY} - \frac{\pi_1 - \pi_2}{2} \sim 2\mu T$$

where π_1 and π_2 are genetic diversities observed in extant lineages 1 and 2, respectively. It has often been mistaken as a measure of divergence (see Cruickshank and Hahn 2014). Owing to the approximation of $4N_{e(ANC)}\mu$ through average diversity in descendant populations, it is dependent on genetic diversity.

d_f

A measure of *differentiation*. It represents the number of fixed differences per site (first referred to as d_f by Ellegren et al. 2012), and therefore is readily mistaken as a measure of divergence (see Cruickshank and Hahn 2014). The evolution of the density of fixed differences is complex (Hey 1991). At early stages of differentiation, the number of fixed sites depends on the fixation of ancestral variants exclusively. d_f is therefore highly dependent on genetic diversity within lineages. On the long run, as fixed differences between lineages predominantly get to represent mutations which arose since lineage split, d_f converges toward d.

d (Nei 1972)

A measure of *divergence* used in the field of molecular evolution that assumes all variation between lineages to be fixed variants arisen by mutation since lineages split. As such, it is defined as $d = 2\mu T$. The estimation of d is usually based on the pairwise divergence between single sequences from two species, assuming that all variation found between sequences is fixed between species.

Box 2. Types of Genomic Islands

Descriptions of genomic landscapes, in particular genomic regions of accentuated differentiation, include a number of terms that have been used interchangeably. Discussions on the existence of genomic islands (Pennisi 2014) appear of limited use. Rather, their interpretation requires a comprehensive understanding of the processes involved in the evolution of heterogeneous genomic landscapes, which is assisted by clear-cut terminology. The island metaphor has its limits in capturing the processes underlying the evolution of genomic landscapes, and interpretations of the latter may require more explicit terminology clearly distinguishing between pattern, process, and interpretation in terms of adaptation and speciation models.

Differentiation islands

Genomic islands exhibiting accentuated genetic differentiation (e.g., F_{ST}, Box 1). Differentiation is sensitive to any process affecting allele frequencies, and differentiation islands can therefore evolve as a consequence of positive selection, gene flow in surrounding regions, or background selection. This term therefore has the broadest application and should be preferably used in the absence of evidence for a particular process underlying the evolution of genomic islands.

Divergence islands

Genomic islands exhibiting accentuated sequence divergence (d_{XY}, Box 1). Divergence is mostly sensitive to selection in ancestral (but importantly not extant) lineages, and gene flow (for useful illustrations see Nachman and Payseur 2012; Cruickshank and Hahn 2014). Both processes reduce divergence, and divergence islands are therefore usually related to gene flow in the surrounding genomic regions.

Speciation islands

Defined as genomic regions that "remain differentiated despite considerable gene flow" (Turner et al. 2005). Therefore, contrary to the previous terms, relating not only to pattern, but to the supposed role in speciation. According to Nachman and Payseur (2012) and Cruickshank and Hahn (2014), speciation islands need to qualify as both differentiation islands and divergence islands. The term has found broader application such as to refer to any genomic islands involved in speciation, including genomic regions under divergent selection in context of ecological speciation in the presence of gene flow. In such cases, the usefulness of the above criterion may be limited because at these timescales divergence is determined uniquely by ancestral polymorphism (Box 1) (see also Burri 2017a). A terminology making a clearer distinction may therefore be useful. In reference to their relation to primary and secondary speciation-with-gene flow (Cruickshank and Hahn 2014), I suggest the terms "primary speciation island" and "secondary speciation island."

Incidental islands

A term coined by Turner and Hahn (2010) to describe genomic islands unrelated to speciation that evolve as a consequence of linked selection following speciation.

Box 3. Empirical approaches for the inference of lineage-specific accentuated differentiation using quantitative evolutionary controls.

Amongst the first who recognized that differentiation landscapes followed a discernible pattern along chromosomes, Roesti et al. (2012) accounted for generally high differentiation in chromosome centers observed among stickleback populations pairs using residual differentiation (F_{ST}') after accounting for chromosome-wide large-scale differentiation by smoothing raw F_{ST}. Following this approach that considered single comparisons, Berner and Salzburger (2015) suggested to account for differentiation patterns resulting from BGS in low-recombination regions by adjusting observed differentiation for background differentiation observed across multiple comparisons. This comparative approach using so-called delta differentiation (ΔF_{ST}) has found application, for example, in crow speciation genomics work (Vijay et al. 2016) to formulate baseline levels of differentiation based on mean differentiation estimated from a reference set of independent comparisons, and was extended to other population genetic parameters, such as linkage and haplotype structure in research on stickleback adaptation (Roesti et al. 2015).

This approach empirically adapts baselines across the genome to take into account heterogeneity in population genetic parameters due to processes unrelated to ecologically relevant lineage-specific positive selection. The extent to which this baseline reflects common underlying processes is expected to scale with the strength of correlation among independent observations (and by the correlation of differentiation landscapes with the underlying diversity landscapes), which is not quantified by the above approach. Principle component analysis (PCA) of multiple differentiation landscapes may offer an interesting alternative in this respect: Values along the first axis (PC1), similar to mean F_{ST}, provide expected relative values of differentiation along the genome, while the proportion of variance explained by the axis provides information on the correlation among differentiation landscapes. Outlier regions can then be inferred from type II regression of observed differentiation against PC1. The pairwise nature of F_{ST} complicates this approach, and lineage-wise estimates of differentiation, such as the population branch statistic (PBS; Shriver et al. 2004; Yi et al. 2010) or population-specific F_{ST} (e.g., Buckleton et al. 2016), or nucleotide diversity (π) which is tightly bound to F_{ST}, may be better suited for this kind of analysis.

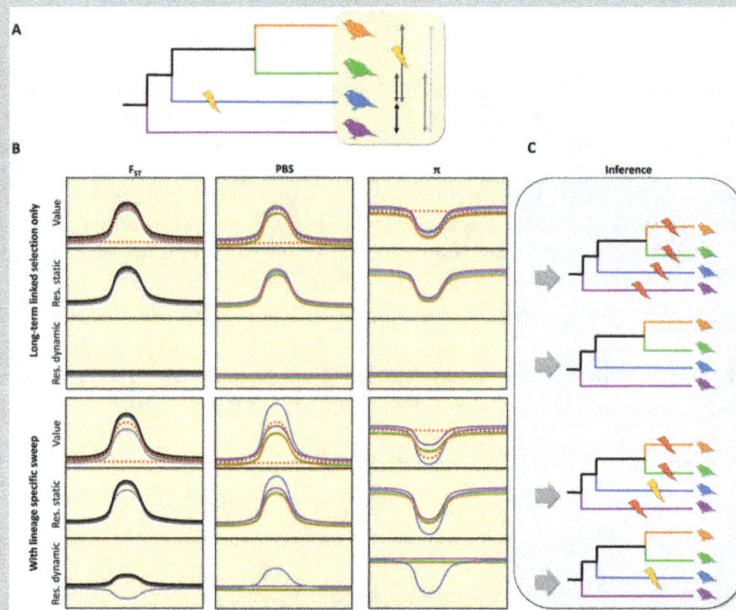

Figure Box 3. Comparative population genomics approach. (A) Phylogenetic relationships among four species. A selective sweep is indicated on the branch leading to the blue species. Pairwise distances are indicated to the right, separated according to whether (left) or not (right) they would be affected by the indicated sweep. (B) Pairwise differentiation (F_{ST}), with gray shades following the ones in (A), lineage-specific differentiation (PBS), and diversity (π) on a chromosome with low recombination in the middle, and their residuals from a static baseline (flat broken red line) and a dynamic baseline (curved broken red line). These are provided for both a scenario without (top panels) and with (bottom panels) the selective sweep depicted in (A). (C) Inference of selective sweeps based on residuals. Red flashes indicate false positive inference of selective sweeps. A static baseline leads to elevated residuals in the chromosome center and thereby false positive inference of sweeps in all instances. With a dynamic baseline only lineage-specific elevations/reductions of differentiation/diversity are reflected as positive residuals and thus inferred as footprints of a sweep. Note that with a dynamic baseline, due to its pairwise nature, F_{ST} results in a blob of negative and positive residuals in the presence of a sweep (lower left), and population-specific measures of differentiation or diversity may be better suited for this approach.

Approaches like these are expected to yield sets of genomic regions enriched for candidate regions that evolved under positive selection. To verify this prediction, the comparison of distributions of parameters more explicitly related to positive selection, such as haplotype structure (Voight et al. 2006; Sabeti et al. 2007; Tang et al. 2007; Ferrer-Admetlla et al. 2014) among candidate regions and the genomic background is helpful (e.g., Roesti et al. 2015). Furthermore, parameter distributions within genomic regions with consistently accentuated differentiation across independent comparisons may constitute more conservative baselines against which to discriminate candidate regions than the genomic background.

Outstanding questions

To identify the footprints of adaptation and speciation in the differentiation landscape, an improved understanding of the contributions of BGS, positive selection in the ancestor, and their interaction with demography toward the evolution of differentiation landscapes and their correlation is required. Previous simulations have shown that BGS in extant populations substantially affects genetic diversity and differentiation, with strongest effects close to the sites it targets and in genomic regions where recombination is infrequent, resulting in local reductions of extant N_e within the genome (Zeng and Charlesworth 2011; Zeng and Corcoran 2015) that can be mistaken as a footprint of positive selection (Huber et al. 2016). These simulations suggest that BGS acting within populations may not reduce diversity by more than two thirds relative to neutral expectations (Zeng and Charlesworth 2011) and may thus not explain the diversity reductions observed for instance in greenish warblers (see e.g., Irwin et al. 2016). Moreover, the effects of BGS in extant lineages must be stronger than in the ancestor to result in elevated differentiation (Zeng and Corcoran 2015). However, recent results suggest that the effects of BGS particularly in ancestral lineages may have been underestimated so far (Phung et al. 2016). Moreover, the effects of demography acting upon a heterogeneous diversity landscape remain largely unexplored. With regard to the model proposed here a number of questions remain open that would profit from additional simulation and empirical work:

- How strongly can BGS reduce genetic diversity in low-recombination regions relative to the remainder of the genome, if it acts over multiple subsequent speciation events over millions of years in taxa with long-term conserved recombination landscapes?
- At which timescales do correlations among differentiation landscapes build up? In particular, from which stage of differentiation onwards are effects of BGS discernible in different bins of recombination rate and functional densities, and what are the long-term contributions of BGS and recurrent hitchhiking?
- What is the effect of demographic events, such as repeated cycles of range expansions and contractions, on heterogeneous diversity landscapes produced by linked selection? Are the resulting reductions in diversity (and N_e) overproportional in regions with already reduced N_e, and may demographic events thus further amplify the heterogeneity of differentiation landscapes in addition to the effects of linked selection?
- How do recombination rate evolution, lineage-specific episodes of positive selection, and variation in effective population size among lineages interplay in the temporal dynamics of build-up and decay of the correlation among independent differentiation landscapes?
- How strong do demographic effects, such as bottlenecks, need to be to perturb the evolution of correlated differentiation landscapes?
- In how far can the evolutionary dynamics of differentiation landscapes be extrapolated among various organisms? Which roles may the mechanisms determining recombination rates have therein?

Glossary

Background selection: A process by which purifying selection against deleterious variants results in a loss of neutral genetic diversity at linked sites.

Extant lineages (extant populations, extant species): Nonancestral lineages, that is, lineages observed today including the time since the split from their ancestor.

Genomic landscapes: The distribution of parameters, such as diversity ('diversity landscape'), differentiation ('differentiation landscape') or recombination rates ('recombination landscape') along chromosomes.

Hitchhiking/recurrent hitchhiking: A process by which positive selection for a beneficial variant leads to an increase in frequency of neutral variants at linked sites, and thus loss of neutral diversity at linked sites. If hitchhiking affects the same genomic region repeatedly, the process is referred to as recurrent hitchhiking.

Linked selection: The process by which neutral genetic diversity in the genome is lost as an effect of selection at a linked site. Both background selection and hitchhiking contribute to linked selection.

AUTHOR CONTRIBUTIONS
RB conceived and wrote this paper.

ACKNOWLEDGMENTS
The author is grateful to S. J. E. Baird, A. Suh, H. Schielzeth, S. Renaut, and two anonymous reviewers for valuable comments on previous versions of the manuscript.

LITERATURE CITED

Arbeithuber, B., A. J. Betancourt, T. Ebner, and I. Tiemann-Boege. 2015. Crossovers are associated with mutation and biased gene conversion at recombination hotspots. Proc. Natl. Acad. Sci. U S A 112:2109–2114.

Bank, C., G. B. Ewing, A. Ferrer-Admettla, M. Foll, and J. D. Jensen. 2014. Thinking too positive? Revisiting current methods of population genetic selection inference. Trends Genet. 30:540–546.

Baudat, F., J. Buard, C. Grey, A. Fledel-Alon, C. Ober, M. Przeworski, et al. 2010. PRDM9 is a major determinant of meiotic recombination hotspots in humans and mice. Science 327:836–840.

Berner, D., and W. Salzburger. 2015. The genomics of organismal diversification illuminated by adaptive radiations. Trends Genet. 31:491–499.

Bhatia, G., N. Patterson, S. Sankararaman, and A. L. Price. 2013. Estimating and interpreting FST: the impact of rare variants. Genome Res. 23:1514–1521.

Buckleton, J., J. Curran, J. Goudet, D. Taylor, A. Thiery, and B. S. Weir. 2016. Population-specific FST values for forensic STR markers: a worldwide survey. Forensic Sci. Int. 23:91–100.

Burri, R. 2017a. Dissecting differentiation landscapes: a linked selection's perspective. J. Evol. Biol.: in press.

Burri, R. 2017b. Linked selection, demography, and the evolution of correlated genomic landscapes in birds and beyond. Mol. Ecol.: in press.

Burri, R., A. Nater, T. Kawakami, C. F. Mugal, P. I. Olason, L. Smeds, et al. 2015. Linked selection and recombination rate variation drive the evolution of the genomic landscape of differentiation across the speciation continuum of Ficedula flycatchers. Genome Res. 25:1656–1665.

Charlesworth, B. 2013. Background selection 20 years on. J. Hered. 104:161–171.

Charlesworth, B. 1998. Measures of divergence between populations and the effect of forces that reduce variability. Mol. Biol. Evol. 15:538–543.

Charlesworth, B., M. T. Morgan, and D. Charlesworth. 1993. The effect of deleterious mutations on neutral molecular variation. Genetics 134:1289–1303.

Comeron, J. M. 2014. Background selection as baseline for nucleotide variation across the *Drosophila* genome. PLoS Genet. 10:e1004434.

Corbett-Detig, R. B., D. L. Hartl, and T. B. Sackton. 2015. natural selection constrains neutral diversity across a wide range of species. PLoS Biol. 13:e1002112.

Cruickshank, T. E., and M. W. Hahn. 2014. Reanalysis suggests that genomic islands of speciation are due to reduced diversity, not reduced gene flow. Mol. Ecol. 23:3133–3157.

Cutter, A. D., and B. A. Payseur. 2013. Genomic signatures of selection at linked sites: unifying the disparity among species. Nat. Rev. Genet. 14:262–274.

Dutoit, L., N. Vijay, C. F. Mugal, C. M. Bossu, R. Burri, J. B. W. Wolf, et al. 2017. Covariation in levels of nucleotide diversity in homologous regions of the avian genome long after completion of lineage sorting. Proc. R. Soc. B 284, 20162756.

Ellegren, H., L. Smeds, R. Burri, P. I. Olason, N. Backström, T. Kawakami, et al. 2012. The genomic landscape of species divergence in *Ficedula* flycatchers. Nature 491:756–760.

Ellegren, H., and N. Galtier. 2016. Determinants of genetic diversity. Nat. Rev. Genet. 17:422–433.

Elyashiv, E., S. Sattath, T. T. Hu, A. Strutsovsky, G. McVicker, P. Andolfatto, et al. 2016. A genomic map of the effects of linked selection in Drosophila. PLoS Genet. 12:e1006130.

Eyre-Walker, A., and P. D. Keightley. 2007. The distribution of fitness effects of new mutations. Nat. Rev. Genet. 8:610–618.

Feder, J. L., X. Xie, J. Rull, S. Velez, A. Forbes, B. Leung, et al. 2005. Mayr, Dobzhansky, and Bush and the complexities of sympatric speciation in Rhagoletis. Proc. Natl. Acad. Sci. U S A 102:6573–6580.

Ferrer-Admetlla, A., M. Liang, T. Korneliussen, and R. Nielsen. 2014. On detecting incomplete soft or hard selective sweeps using haplotype structure. Mol. Biol. Evol. 31:1275–1291.

Feulner, P. G. D., F. J. J. Chain, M. Panchal, Y. Huang, C. Eizaguirre, M. Kalbe, et al. 2015. Genomics of divergence along a continuum of parapatric population differentiation. PLoS Genet. 11:e1004966.

Flowers, J. M., J. Molina, S. Rubinstein, P. Huang, B. A. Schaal, and M. D. Purugganan. 2012. Natural selection in gene-dense regions shapes the genomic pattern of polymorphism in wild and domesticated rice. Mol. Biol. Evol. 29:675–687.

Francioli, L. C., P. P. Polak, A. Koren, A. Menelaou, S. Chun, I. Renkens, et al. 2015. Genome-wide patterns and properties of de novo mutations in humans. Nat. Genet. 47:822–826.

Fraser, B. A., A. Künstner, D. N. Reznick, C. Dreyer, and D. Weigel. 2015. Population genomics of natural and experimental populations of guppies (Poecilia reticulata). Mol. Ecol. 24:389–408.

Haasl, R. J., and B. A. Payseur. 2016. Fifteen years of genomewide scans for selection: trends, lessons and unaddressed genetic sources of complication. Mol. Ecol. 25:5–23.

Hey, J. 1991. The structure of genealogies and the distribution of fixed differences between DNA sequence samples from natural populations. Genetics 128:831–840.

Hodgkinson, A., and A. Eyre-Walker. 2011. Variation in the mutation rate across mammalian genomes. Nat. Rev. Genet. 12:756–766.

Huber, C. D., M. DeGiorgio, I. Hellmann, and R. Nielsen. 2016. Detecting recent selective sweeps while controlling for mutation rate and background selection. Mol. Ecol. 25:142–156.

Hudson, R. R., M. Slatkin, and W. P. Maddison. 1992. Estimation of levels of gene flow from DNA sequence data. Genetics 132:583–589.

Irwin, D. E., M. Alcaide, K. E. Delmore, J. H. Irwin, and G. L. Owens. 2016. Recurrent selection explains parallel evolution of genomic regions of high relative but low absolute differentiation in a ring species. Mol. Ecol. 25:4488–4507.

Jones, F. C., M. G. Grabherr, Y. F. Chan, P. Russell, E. Mauceli, J. Johnson, et al. 2012. The genomic basis of adaptive evolution in threespine sticklebacks. Nature 484:55–61.

Kaplan, N., R. Hudson, and C. Langley. 1989. The hitchhiking effect revisited. Genetics 123:887–899.

Kawakami, T., L. Smeds, N. Backström, A. Husby, A. Qvarnström, C. F. Mugal, et al. 2014. A high-density linkage map enables a second-generation collared flycatcher genome assembly and reveals the patterns of avian recombination rate variation and chromosomal evolution. Mol. Ecol. 23:4035–4058.

Kronforst, M. R., M. E. B. Hansen, N. G. Crawford, J. R. Gallant, W. Zhang, R. J. Kulathinal, et al. 2013. Hybridization Reveals the Evolving Genomic Architecture of Speciation. Cell Rep. 5:666–677.

Lohmueller, K. E., A. Albrechtsen, Y. Li, S. Y. Kim, T. Korneliussen, N. Vinckenbosch, et al. 2011. Natural selection affects multiple aspects of genetic variation at putatively neutral sites across the human genome. PLoS Genet 7:e1002326.

Martin, S. H., K. K. Dasmahapatra, N. J. Nadeau, C. Salazar, J. R. Walters, F. Simpson, et al. 2013. Genome-wide evidence for speciation with gene flow in Heliconius butterflies. Genome Res. 23:1817–1828.

Maynard Smith, J., and J. Haigh. 1974. The hitch-hiking effect of a favourable gene. Genet Res. 23:23–35.

McVicker, G., D. Gordon, C. Davis, and P. Green. 2009. Widespread genomic signatures of natural selection in hominid evolution. PLoS Genet 5:e1000471.

Munch, K., K. Nam, M. H. Schierup, and T. Mailund. 2016. Selective sweeps across twenty millions years of primate evolution. Mol. Biol. Evol. 33:3065–3074.

Nachman, M. W., and B. A. Payseur. 2012. Recombination rate variation and speciation: theoretical predictions and empirical results from rabbits and mice. Phil. Trans. R. Soc. B 367:409–421.

Nam, K., and H. Ellegren. 2012. Recombination drives vertebrate genome contraction. PLoS Genet. 8:e1002680.

Nei, M. 1972. Genetic distance between populations. Am. Nat. 106:283–292.

Nei, M., and W.-H. Li. 1979. Mathematical model for studying genetic variation in terms of restriction endonucleases. Proc. Natl. Acad. Sci. U S A 79:5269–5273.

Noor, M. A. F., and S. M. Bennett. 2009. Islands of speciation or mirages in the desert? Examining the role of restricted recombination in maintaining species. Heredity 103:439–444.

Ohta, T. 2002. Near-neutrality in evolution of genes and gene regulation. Proc. Natl. Acad. Sci. U S A 99:16134–16137.

Oliver, P. L., L. Goodstadt, J. J. Bayes, Z. Birtle, K. C. Roach, N. Phadnis, et al. 2009. Accelerated evolution of the Prdm9 speciation gene across diverse metazoan taxa. PLoS Genet. 5:e1000753.

Payseur, B. A., and M. W. Nachman. 2002. Gene density and human nucleotide polymorphism. Mol. Biol. Evol. 19:336–340.

Pennisi, E. 2014. Disputed islands. Science 345:611–613.

Phung, T. N., C. D. Huber, and K. E. Lohmueller. 2016. Determining the effect of natural selection on linked neutral divergence across species. PLoS Genet. 12:e1006199.

Renaut, S., C. J. Grassa, S. Yeaman, B. T. Moyers, Z. Lai, N. C. Kane, et al. 2013. Genomic islands of divergence are not affected by geography of speciation in sunflowers. Nat. Commun. 4:1827.

Renaut, S., G. L. Owens, and L. H. Rieseberg. 2014. Shared selective pressure and local genomic landscape lead to repeatable patterns of genomic divergence in sunflowers. Mol. Ecol. 23:311–324.

Roesti, M., A. P. Hendry, W. Salzburger, and D. Berner. 2012. Genome divergence during evolutionary diversification as revealed in replicate lake–stream stickleback population pairs. Mol. Ecol. 21:2852–2862.

Roesti, M., D. Moser, and D. Berner. 2013. Recombination in the threespine stickleback genome—patterns and consequences. Mol. Ecol. 22:3014–3027.

Roesti, M., B. Kueng, D. Moser, and D. Berner. 2015. The genomics of ecological vicariance in threespine stickleback fish. Nat. Commun. 6:8767.

Sabeti, P. C., P. Varilly, B. Fry, J. Lohmueller, E. Hostetter, C. Cotsapas, et al. 2007. Genome-wide detection and characterization of positive selection in human populations. Nature 449:913–918.

Schlötterer, C. 2015. Genes from scratch - the evolutionary fate of *de novo* genes. Trends Genet. 31:215–219.

Shriver, M., G. Kennedy, E. Parra, H. Lawson, V. Sonpar, J. Huang, *et al.* 2004. The genomic distribution of population substructure in four populations using 8,525 autosomal SNPs. Hum. Genomics 1:274–286.

Singhal, S., E. M. Leffler, K. Sannareddy, I. Turner, O. Venn, D. M. Hooper, *et al.* 2015. Stable recombination hotspots in birds. Science 350:928–932.

Smukowski, C. S., and M. A. F. Noor. 2011. Recombination rate variation in closely related species. Heredity 107:496–508.

Soria-Carrasco, V., Z. Gompert, A. A. Comeault, T. E. Farkas, T. L. Parchman, J. S. Johnston, *et al.* 2014. Stick insect genomes reveal natural selection's role in parallel speciation. Science 344:738–742.

Stephan, W. 2010. Genetic hitchhiking versus background selection: the controversy and its implications. Phil. Trans. R. Soc. B 365:1245–1253.

Tang, K., K. R. Thornton, and M. Stoneking. 2007. A new approach for using genome scans to detect recent positive selection in the human genome. PLoS Biol. 5:e171.

Terekhanova, N. V., V. B. Seplyarskiy, R. A. Soldatov, and G. A. Bazykin. 2017. Evolution of local mutation rate and its determinants. Mol. Biol. Evol 34:1100–1109.

Turner, T. L., and M. W. Hahn. 2010. Genomic islands of speciation or genomic islands and speciation? Mol. Ecol. 19:848–850.

Turner, T. L., M. W. Hahn, and S. V. Nuzhdin. 2005. Genomic islands of speciation in *Anopheles gambiae*. PLoS Biol. 3:e285.

Van Doren, B. M., L. Campagna, B. Helm, J. C. Illera, I. J. Lovette, and M. Liedvogel. 2017. Correlated patterns of genetic diversity and differentiation across an avian family. Mol. Ecol: in press.

Vijay, N., C. M. Bossu, J. W. Poelstra, M. H. Weissensteiner, A. Suh, A. P. Kryukov, *et al.* 2016. Evolution of heterogeneous genome differentiation across multiple contact zones in a crow species complex. Nat. Commun. 7:13195.

Vijay, N., M. Weissensteiner, R. Burri, T. Kawakami, H. Ellegren, and J. B. W. Wolf. 2017. Genome-wide signatures of genetic variation within and between populations—a comparative perspective. Mol. Ecol.: in press.

Voight, B. F., S. Kudaravalli, X. Wen, and J. K. Pritchard. 2006. A map of recent positive selection in the human genome. PLoS Biol. 4:e72.

Weir, B. S., and C. C. Cockerham. 1984. Estimating F-statistics for the analysis of population-structure. Evolution 38:1358–1370.

Whitlock, M. C. 2011. G'ST and D do not replace FST. Mol. Ecol. 20:1083–1091.

Yi, X., Y. Liang, E. Huerta-Sanchez, X. Jin, Z. X. P. Cuo, J. E. Pool, *et al.* 2010. Sequencing of 50 human exomes reveals adaptation to high altitude. Science 329:75–78.

Zeng, K., and B. Charlesworth. 2011. The joint effects of background selection and genetic recombination on local gene genealogies. Genetics 189:251–266.

Zeng, K., and P. Corcoran. 2015. The effects of background and interference selection on patterns of genetic variation in subdivided populations. Genetics 201:1539–1554.

7

Heterogeneous gene duplications can be adaptive because they permanently associate overdominant alleles

Pascal Milesi,[1] Mylène Weill,[1] Thomas Lenormand,[2] and Pierrick Labbé[1,3]

[1]Institut des Sciences de l'Evolution de Montpellier (UMR 5554, CNRS-Université de Montpellier-IRD-EPHE), Campus Université de Montpellier, Place Eugène Bataillon 34095, Montpellier, CEDEX 05, France

[2]Centre d'Ecologie Fonctionnelle et Evolutive (UMR 5175, CNRS-Université de Montpellier-Université Paul-Valéry Montpellier-EPHE) 1919 route de Mende, F-34293 Montpellier, CEDEX 05, France

[3]E-mail: pierrick.labbe@umontpellier.fr

Gene duplications are widespread in genomes, but their role in contemporary adaptation is not fully understood. Although mostly deleterious, homogeneous duplications that associate identical repeats of a locus often increase the quantity of protein produced, which can be selected in certain environments. However, another type exists: heterogeneous gene duplications, which permanently associate two (or more) alleles of a single locus on the same chromosome. They are far less studied, as only few examples of contemporary heterogeneous duplications are known. Haldane proposed in 1954 that they could be adaptive in situations of heterozygote advantage, or overdominance, but this hypothesis was never tested. To assess its validity, we took advantage of the well-known model of insecticide resistance in mosquitoes. We used experimental evolution to estimate the fitnesses associated with homozygous and heterozygous genotypes in different selection regimes. It first showed that balanced antagonist selective pressures frequently induce overdominance, generating stable polymorphic equilibriums. The frequency of equilibrium moreover depends on the magnitude of two antagonistic selective pressures, the survival advantage conferred by the resistant allele versus the selective costs it induces. We then showed that heterogeneous duplications are selected over single-copy alleles in such contexts. They allow the fixation of the heterozygote phenotype, providing an alternative and stable intermediate fitness trade-off. By allowing the rapid fixation of divergent alleles, this immediate advantage could contribute to the rarity of overdominance. More importantly, it also creates new material for long-term genetic innovation, making a crucial but underestimated contribution to the evolution of new genes and gene families.

KEY WORDS: genetic polymorphism, gene duplication, genome evolution, overdominance, balancing selection, insecticide resistance.

Impact Summary

Understanding the maintenance of polymorphism in natural populations despite the erosion due to natural selection and genetic drift has been one of the challenges of the early 20th century. One of the propositions, rapidly put aside due to the lack of examples was overdominance: when the heterozygote phenotype is the fittest, none of the alleles can fix and only half of the progeny of two heterozygotes carry the best phenotype (this is called segregation burden). However, in 1954 Haldane suggested that heterogeneous duplications associating the two alleles on the same chromosome could be selected, but could also allow the fixation of the heterozygote phenotype. This hypothesis was however never tested, because both overdominance and gene duplications were considered too rare. Recent genome-wide studies showed that duplications are actually more common than substitutions. However, they usually associate identical repeats of the same allele, which can increase the protein production and be selected for (a quantitative advantage). The recent discovery of several heterogeneous duplications implicated in insecticide resistance in mosquitoes provided means to finally test Haldane's hypothesis. Insecticide-resistant individuals are advantaged in the presence of insecticide (they survive) but endure severe selective costs (their general performances are reduced) compared to susceptible individuals. Using experimental evolution and manipulating the selective pressures (insecticide dose for the advantage and the rearing conditions for the cost), we first showed that balanced antagonist pressures generate overdominance. Introducing a duplicated allele latter showed that it was indeed selected over single copies, as predicted by Haldane. Our study thus shows that heterogeneous duplications associating two already divergent alleles can be selected when the environmental conditions result in trade-offs favoring overdominance. However, they also create new and already divergent material for genetic innovation and could play a crucial role in the evolution of gene families.

Introduction

Recent genomic studies have shown that gene duplications are widespread, but mostly deleterious (Schrider and Hahn 2010). Some are nevertheless adaptive, and many examples of quantitative advantages (i.e., increased protein production) have been reported associated with identical repeats of a locus (homogeneous duplications) (reviewed in Innan and Kondrashov 2010; Katju and Bergthorsson 2013). Another kind of duplication associates two or more alleles of a single locus (i.e., divergent copies) on the same chromosome (heterogeneous duplications, Fig. 1). Their adaptive role is much harder to assess (Lenormand et al. 1998; Labbé et al. 2007a; Innan and Kondrashov 2010), as only a few examples of recent heterogeneous duplications have been described to date (Labbé et al. 2007a; Djogbénou et al. 2008; Remnant et al. 2013; Sonoda et al. 2014). In 1954, Haldane suggested that they could be selected for by enabling the permanent association of overdominant heterozygous alleles, with no segregation burden (Haldane 1954). In the early 20th century, overdominance (i.e., the situation in which the heterozygote is fitter than either of the homozygotes) was originally put forward as one of several mechanisms underlying the hybrid vigor observed in crosses between inbred strains, and as the only condition allowing a polymorphic equilibrium with constant fitnesses in population genetics equations (Fisher 1922; East 1936; Dobzhansky 1952). Sickle cell anemia is an iconic and much cited example (Lewontin 1974). However, empirical evidence for overdominance in natural populations remains scarce (Hedrick 2012; Llaurens et al. 2017), and its contribution to the selection of heterogeneous duplications has never been assessed.

In several mosquito species, including the malaria vector *Anopheles gambiae* and the West Nile virus vector *Culex pipiens*, the *ace-1* R resistance allele encodes an acetylcholinesterase protein differing from that encoded by the susceptible S allele by a single amino-acid substitution (G119S): this substitution prevents organophosphate insecticides (OPs) from binding to their target, resulting in resistance (Weill et al. 2003). However, the G119S substitution also greatly decreases the activity of the protein (Alout et al. 2008), probably accounting for the high selective cost associated with homozygous RR individuals: higher larval mortality, lower fecundity, and lower mating success for RR males than for homozygous SS males (Duron et al. 2006; Assogba et al. 2015, 2016). This situation results in a fitness trade-off, with the selection of RR individuals in the presence of insecticides, and selection against these individuals in the absence of pesticides. RS individuals have intermediate characteristics, with lower resistance than RR individuals, at a lower cost (Labbé et al. 2014; Assogba et al. 2015). This resistance/cost trade-off, as well as the migration/selection balance resulting from the alternating treated and nontreated areas, promote *ace-1* polymorphism (Lenormand et al. 1999; Labbé et al. 2007b).

Gene duplications (D alleles) bringing a susceptible S copy and a resistant R copy together in the same haplotype, on a single chromosome, have been identified in both *A. gambiae* and *C. pipiens* (Lenormand et al. 1998; Labbé et al. 2007a; Djogbénou et al. 2008). In *C. pipiens*, several independent duplications have been identified, and some may fix in natural populations (Labbé et al. 2007a; Alout et al. 2010; Osta et al. 2012). Both R and S copies are identical to single-copy S and R alleles found in the same populations, so that the D alleles probably result from unequal crossing-overs in heterozygotes (Labbé et al. 2007a). Most importantly these heterogeneous duplications confer a heterozygote-like phenotype [RS] (Labbé et al. 2014).

This model thus provides us with a unique opportunity to test Haldane's hypothesis: moderate doses of insecticide could result in balanced selective pressures, where the advantage/cost trade-off could favor the intermediate performances of heterozygotes, and thus the duplicated alleles.

We used experimental evolution to test these hypotheses. Our study showed that antagonist selective pressures could lead to overdominance, in which case the ultimate allele frequencies depended on the balance between the relative intensities of these selective pressures. It also showed that heterogeneous duplications were indeed selected in these contexts. Half a century later, our study thus provides strong experimental support for a theoretical claim made by one of the fathers of modern evolutionary theory that has long remained untested.

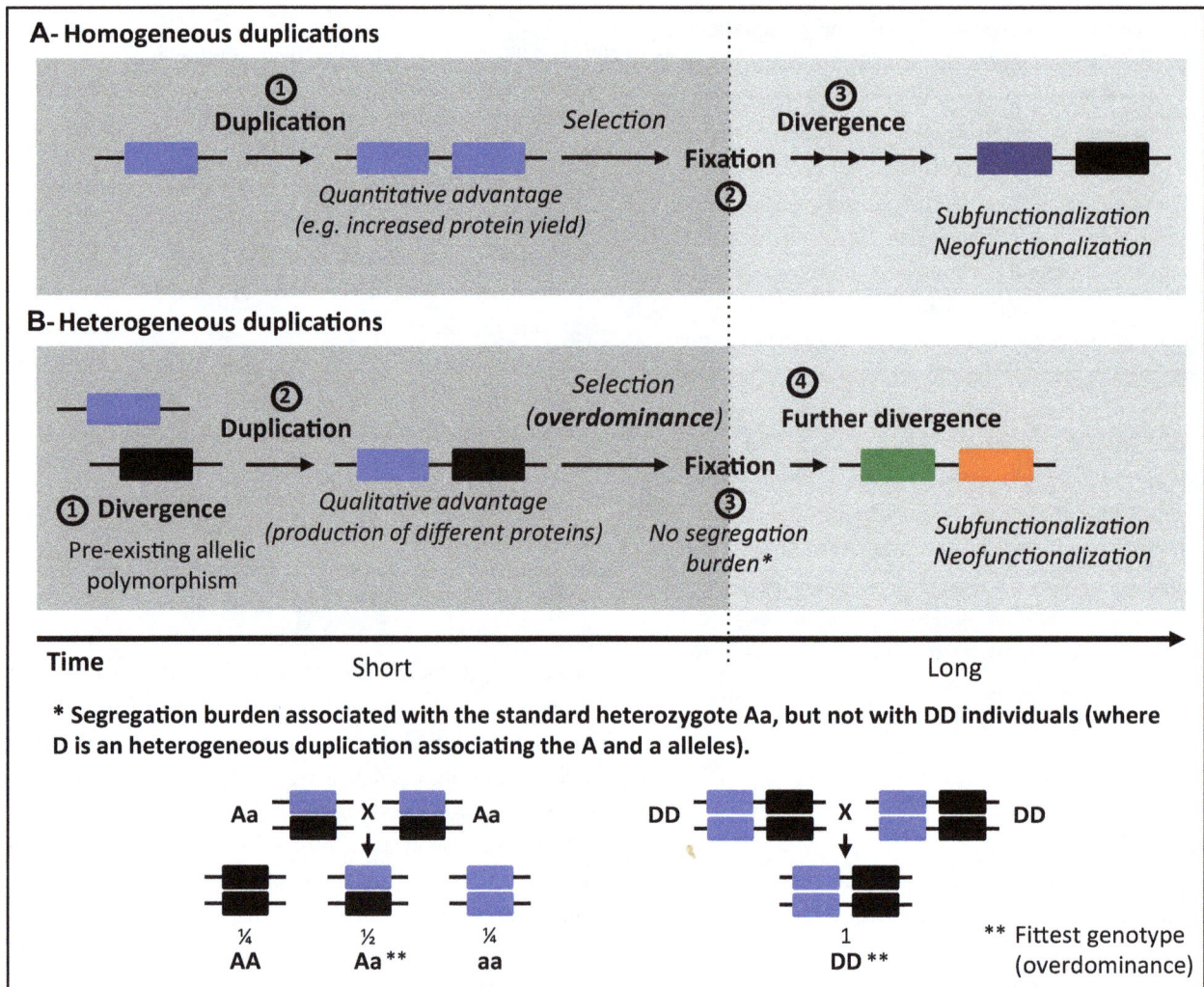

Figure 1. Origins and adaptive fates of the different types of gene duplications.
Each rectangle represents one copy of a given locus; the colors indicate that two copies/alleles differ in sequence and are associated with different functions. The timing of the various processes is indicated by the long black arrow, and the background colors differentiate the short and longer term processes.
(A) Homogeneous duplications associate two identical copies of a gene (1). They can be selected, for example, for quantitative advantages due to increased protein yield, and get fixed (2). On the long term, they can diverge and acquire new functions (3).
(B) In the case of heterogeneous duplications, the allelic divergence (1) precedes the duplication (2). They can be selected in case of overdominance because they allow the production of different proteins, and get fixed because they do not endure the segregation burden associated with standard heterozygotes (the * diagram illustrates the progenies of the two crosses and their proportions) (3). They carry different functions before fixation, but can then further diverge and acquire new functions (4).

Methods

MOSQUITO STRAINS

Three *C. pipiens* mosquito laboratory lines sharing the same genetic background (>99%) were used in this study: Slab (Georghiou et al. 1966), SR (Berticat et al. 2002), and Ducos-DFix (Labbé et al. 2014). All share the Slab genetic background, but they carry different *ace-1* alleles: the single-copy susceptible S (isolated in California), the single-copy resistance R (isolated in Southern France), and the heterogeneous resistance duplication D (D_1 allele, isolated in Martinique), respectively. D copies are

identical except for the G119S mutation, and differ from the S and R alleles by a few synonymous mutations (Labbé et al. 2007a).

EXPERIMENTAL EVOLUTION IN POPULATION CAGES

At the start of each experiment replicate, we mixed 500 second-instar larvae (L_2) of the different strains in various proportions, to control for any effect of initial conditions or genetic drift. The larvae were reared until the emergence of the adult. The adults were allowed to mate freely and the females were provided with a blood meal (chicken). Thereafter, the experiment was conducted

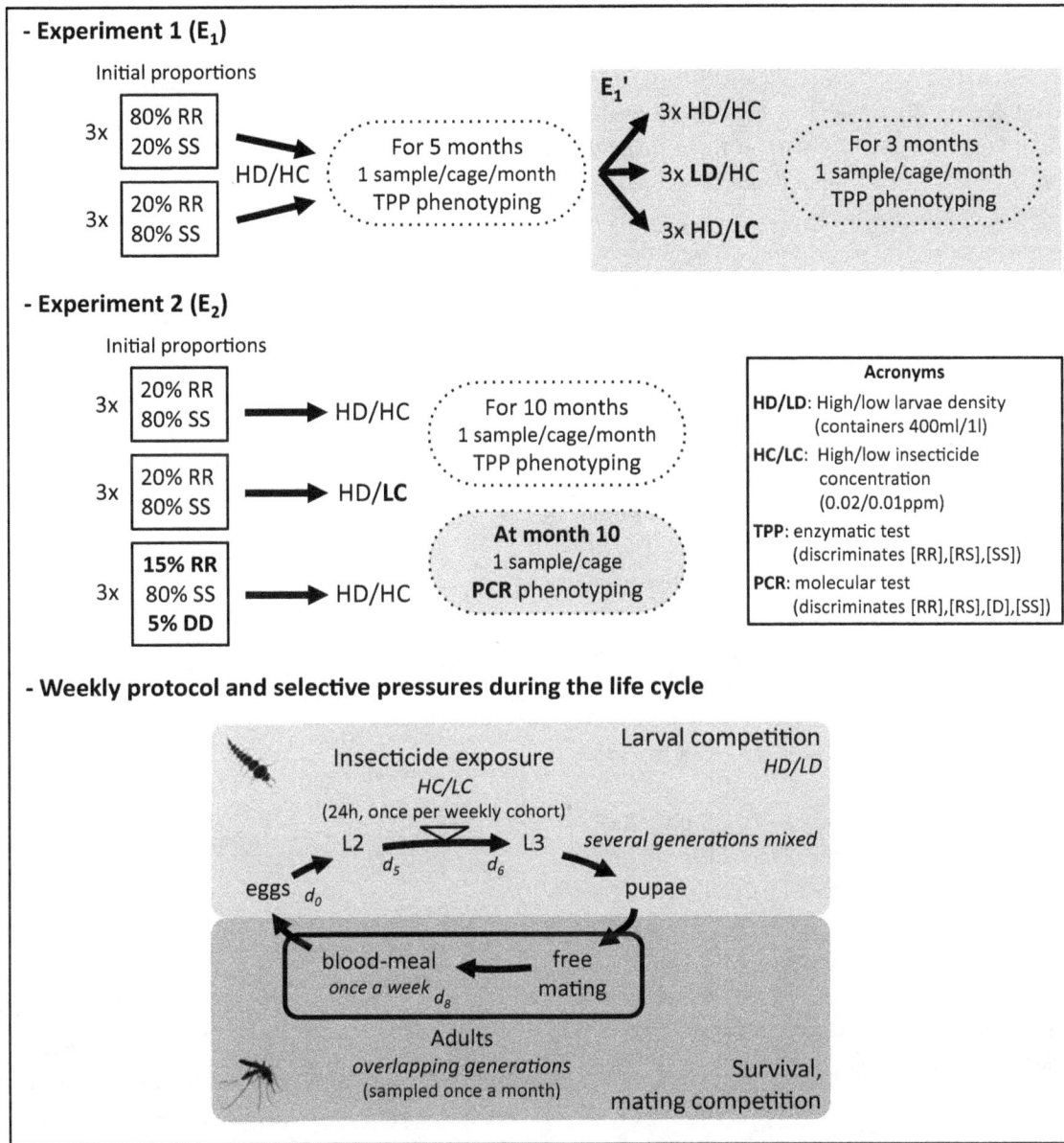

Figure 2. Experimental design.

The experimental design of the two experiments (E₁ and E₂) is presented (see text also): number of replicates, initial genotype proportions, experimental conditions, duration, sampling design, and phenotyping methods (the acronyms are defined in the box).

For experiment E₁, the replicates were split at month 4 (E₁′) and the experimental conditions were modified (reference = HD/HC) in the different subsets; the modifications are bolded in red. For experiment E₂, the conditions differing from the reference (HD/HC) and the alterations to the initial genotype proportions are bolded in red. A specific phenotyping method was used only for the 10th month (bolded in red).

The weekly protocol is also represented (see also text); the timing is indicated in days from the egg rafts' collection (d_i). The various selective pressures are indicated in red and their origins are italicized.

in weekly cycles, under standard conditions (25°C, >60% humidity, 12-h light:12-h dark). On the first day (d_0), the egg rafts were collected and placed together in a single container with ~400 ml of water, resulting in a high larval density (high density or HD, >1000 larvae/l). On day five (d_5), larvae (~L2) were exposed for 24 h to 0.02 ppm of temephos (OP insecticide, Bayer®), the

high insecticide concentration (or HC), estimated from bioassays (Labbé et al. 2014). On day six (d_6), the survivors were collected and placed together in a single container in the cage, to allow the adults to emerge; the adults mated freely. On day eight (d_8), the females were provided with a new blood meal. These weekly cycles generated overlapping generations (adults remained in the cage until their death, two to six weeks later) and ensured that all individuals were exposed to the insecticide once in their lives. The control conditions were HD/HC. These conditions were modified by reducing larval density (low density or LD) during rearing, through the use of larger containers (~1 l), or by reducing the concentration of insecticide (low concentration or LC = 0.01 ppm). The experimental design is summarized in Figure 2. Each month, we sampled an average of 48 adults at random from each cage for the estimation of phenotypic frequencies (total number of individuals analyzed = 6547; Table A1).

PHENOTYPING

For each sample, individual phenotypes were established with the TPP test (Bourguet et al. 1996) (Table A1), based on the activity of the acetylcholinesterase AChE1 in the presence and absence of insecticide (propoxur, Baygon®). This test discriminates between three phenotypes: [SS], [RR], and [RS]. The first two are always unambiguous and correspond to the homozygous SS and RR genotypes, respectively. When only the R and S alleles are present, the last phenotype corresponds to the standard heterozygote RS (i.e., the phenotypic frequencies correspond to the genotypic frequencies). However, when the D allele is present, the [RS] phenotype becomes ambiguous, as it can result from four genotypes: RS, DD, DS, and DR.

In the last month of the experiment E_2 (Fig. 2), we distinguished standard heterozygotes (RS) from individuals carrying at least one D allele (DD, DS, and DR; these three genotypes cannot be distinguished), using a PCR test specific for the susceptible copy of the D allele (DucosEx3dir – DucosEx3rev; Labbé et al. 2014), directly on second-instar larvae (no DNA extraction was required, the larvae dissolve in the buffer during the first PCR 95°C denaturation step) (Table A2).

ESTIMATION OF ALLELE FREQUENCIES

When only the R and S alleles were present, the R allele frequency $f(R)$ was calculated directly from the phenotypic frequencies for each generation of each replicate.

However, this was not possible when the D allele was present (see, section "Phenotyping"). In this case, we calculated the apparent R frequency $f^*(R) = f([RR]) + 0.5f([RS])$, that is, as if only R and S were present (D carriers then appear as heterozygotes [RS]).

For the last month of experiment E_2 (Fig. 2), we inferred the D, R, and S allele frequencies from the phenotypic frequencies (Table A2), assuming panmixia and using the maximum-likelihood approach developed by Lenormand et al. (1998): for each replicate, we calculated the log-likelihood L of observing the phenotypic data as

$$L = \sum_i n_i ln(f_i),$$

with n_i and f_i the observed number and the predicted frequency of individuals with phenotype i, respectively. L was then maximized (L_{max}) with a simulated annealing algorithm (Lenormand et al. 1998). For each allele frequency, the support limits (SL) were then calculated as the minimum and maximum values that this frequency could take without significantly decreasing the likelihood (i.e., $L_{max} -1.96$, roughly equivalent to 95% confidence intervals).

ESTIMATION OF SELECTION COEFFICIENTS

We estimated the relative fitness of the various phenotypes, using a simple deterministic (i.e., no drift) population genetics model considering infinite populations and discrete generations (two per month). This model does not completely reflect our experiments. First, drift did play a role in the allele dynamics, but it had a much smaller impact than selection, which could lead to more dispersion than expected in the observed data; this was handled statistically by controlling for overdispersion in the likelihood model (see below). Second, the generations in the experiment were overlapping, so that the actual generation number is probably below two, which should, conservatively, lead to an underestimation of fitness differences. Hence, our approach was statistically robust to these simplifications.

We used a two-step model: (1) Reproduction: the frequency f_{gi} of each genotype g among the larvae in generation i was calculated from the allele frequencies (f_R and f_S for the R and S alleles, respectively) in the gametes of the previous generation ($i − 1$), assuming panmixia ($f_{RRi} = f_{R(i-1)}^2$, $f_{SSi} = f_{S(i-1)}^2$ and $f_{RSi} = 2f_{R(i-1)}f_{S(i-1)}$); (2) Selection was taken into account between the larval stage and the adult stage, to calculate the frequency f'_{gi} of each genotype g in the adults of generation i ($f'_{gi} = (f_{gi}w_g)/\Sigma(f_{gi}w_g)$, where w_g is the fitness of the genotype g). The fitness of heterozygous individuals, [RS], $w_{RS} = 1$ was used as a reference. The relative fitnesses of the SS and RR genotypes were set as $w_{SS} = 1 + s_{SS}$ and $w_{RR} = 1 + s_{RR}$, respectively, where s_{SS} and s_{RR} are the corresponding selection coefficients. The allelic frequencies in the gametes produced by the surviving adults of generation i were then calculated from the genotypic frequencies of these individuals ($f_{Ri} = f'_{RRi} + 0.5f'_{RSi}, f_{Si} = f'_{SSi} + 0.5f'_{RSi}$).

This model was adjusted to the data (phenotypic frequencies) through a maximum likelihood approach (R software version 2.15.1 https://www.r-project.org/ package optim, method L-BFGS-B). The SL associated with the selection coefficients s_{SS} and s_{RR} were calculated from the likelihood profile ($L_{max} -1.96$) established from 10^6 simulations. For each model, we calculated the percentage of the total deviance explained as $\%TD = (D_{max} - D_{mod}) / (D_{max} - D_{min})$ (where D_{min}, D_{max}, and D_{mod}, are the minimal, maximal, and model deviances, respectively, with $D = -2 L$), and the overdispersion as $od = D_{res} / df_{res}$ (where D_{res} and df_{res} are the residual deviance and the residual degrees of freedom, respectively). We assessed the significance of differences in selection coefficients between rearing conditions (HC/HD, HC/LD, and LC/HD), by adjusting a complete model with two parameters for each set of conditions considered (sSS and sRR). A simplified model, with only one sSS or sRR parameter for all conditions, was then computed. The two models were then compared using likelihood ratio tests corrected for overdispersion (LRT_{od}): when significant, the simplified model was rejected, and the tested selection parameter was considered to differ significantly between the conditions tested.

Results and Discussion

OVERDOMINANCE RESULTS FROM BALANCED ANTAGONIST SELECTIVE PRESSURES

We first investigated the possibility that intermediate insecticide doses could result in overdominance by setting up six population evolution experiments in cages containing a mixture of *C. pipiens* RR and SS genotypes, with three replicates having an initial R frequency $f_0(R) = 0.8$ and three replicates having an initial $f_0(R) = 0.2$ (experiment E_1, Fig. 2). Each new generation was exposed to a high concentration (HC) of insecticide that killed almost all SS individuals, a few RS but no RR individuals, thereby favoring selection of the R allele. The larvae were also reared at HD, to increase competition, resulting in higher levels of larval mortality. The emerging adults were released into the same cage: they mated freely and were able to reproduce every week until their death (overlapping generations), favoring the selection of individuals with longer lifespans and higher mating success, that is, those carrying the less costly S allele. We monitored changes in R allele frequency $f(R)$ every month (about two overlapping generations), by genotyping adults randomly sampled from each of the six cages (there are only RR, SS, and RS individuals; Fig. 2 and Table A1). If these conditions result in overdominance, it is possible to predict the dynamics of the alleles: (1) $f(R)$ should reach a stable equilibrium ($\hat{f}(R)$): even when RS is the fittest genotype, when two heterozygotes mate, only half of their progeny carries the RS genotype (i.e., there is a segregation burden), the other half carrying the less fit genotypes RR or SS; the polymorphism is thus stable; (2) the frequency at equilibrium should depend on the relative fitness of each genotype in the environmental conditions considered, but should be independent of the initial frequencies of the alleles (f_0).

This is what we observed, despite variations due to drift and sampling: the frequency of R in all replicates ultimately converged around $\hat{f}(R) = 0.68 \pm 0.08$ (mean $\pm$ standard deviation), regardless of its initial frequency (Fig. 3A1). Estimates based on a population genetics model confirmed that the relative fitness of the heterozygotes (w_{RS}) was significantly higher than those of the two homozygotes (confirming overdominance) and that selec-

tion for heterozygotes was strong (E_1, Table 1): RR individuals were fitter than SS individuals (as expected from $\hat{f}(R) > 0.5$), but this resistance advantage did not compensate for the fitness cost relative to RS mosquitoes.

We then confirmed that the frequencies at equilibrium were constrained by the evolutionary trade-offs of the various genotypes by altering the environmental conditions. After four months, we kept three of the cages in the original conditions (controls: HD, HC), but the other three cages were each split in two (experiments E_1', Table 1, Fig. 2). In one half of these three cages, we reduced larval rearing density (low density, LD), but maintained the HC of insecticide treatment (Fig. 2). These conditions were expected to reduce the competition between larvae, and thus, the cost of the R allele. As expected, $f(R)$ increased (Fig. 3A2, brown stars). Accordingly, the relative fitness ranking changed significantly (LRT_{od}, $F = 10.1$; $\Delta df = 2$; $P < 0.001$): RR became the fittest genotype (E_1', Table 1), but RS remained fitter than SS. In the other half of these three cages, we reduced the insecticide concentration (LC), but maintained the high larval density (HD) (Fig. 2). These conditions were expected to reduce the selective advantage of the R allele. As expected, $f(R)$ decreased (Fig. 3A2, blue squares). Due to the shorter duration, equilibriums were not reached: the selective coefficient estimates obtained are less precise, RS remained the fittest genotype, but SS became slightly fitter than RR (LRT_{od}, $F = 32.7$; $\Delta df = 2$; $P < 0.001$; E_1', Table 1). These experiments thus provided evidence that environmental modifications affecting the selective advantage or the cost can alter the frequency equilibrium.

We assessed the robustness of these conclusions by replicating the study described above. Three cages were set up as controls (HD, HC), with an initial $f_0(R) = 0.2$ (experiment E_2, Fig. 2): after 10 months, the mean R frequency had reached an equilibrium at $\hat{f}(R) = 0.73 \pm 0.04$ (Fig. 3B1, red triangles), similar to that in E_1 ($\hat{f}(R) = 0.68 \pm 0.08$; Student's t test, $P = 0.34$). This robustness (nine cages stabilizing at about the same frequency, regardless of their initial frequency) confirmed that the allele dynamics in our experiments were mostly driven by selection. In parallel, three other cages were set up with a low insecticide concentration (HD, LC), again with an initial $f_0(R) = 0.2$ (E_2, Fig. 2). The mean R frequency had also reached equilibrium by 10 months, but at a lower value, $\hat{f}(R) = 0.24 \pm 0.14$ (Fig. 3B1, blue squares). As expected, model estimates confirmed that RS individuals were the fittest in both conditions (Table 1). However, fitness ranking changed according to the different conditions (LRT_{od}, $F = 172.7$; $\Delta df = 2$; $P < 0.001$; Table 1): in control (HD, HC) conditions, RR individuals were fitter than SS individuals (WRS > WRR > WSS), whereas SS individuals were fitter than RR individuals in LC conditions (WRS > WSS > WRR), indicating that RR resistance relative advantage cannot compensate its cost at lower insecticide concentrations. E_2 thus confirmed the conclusions of E_1:

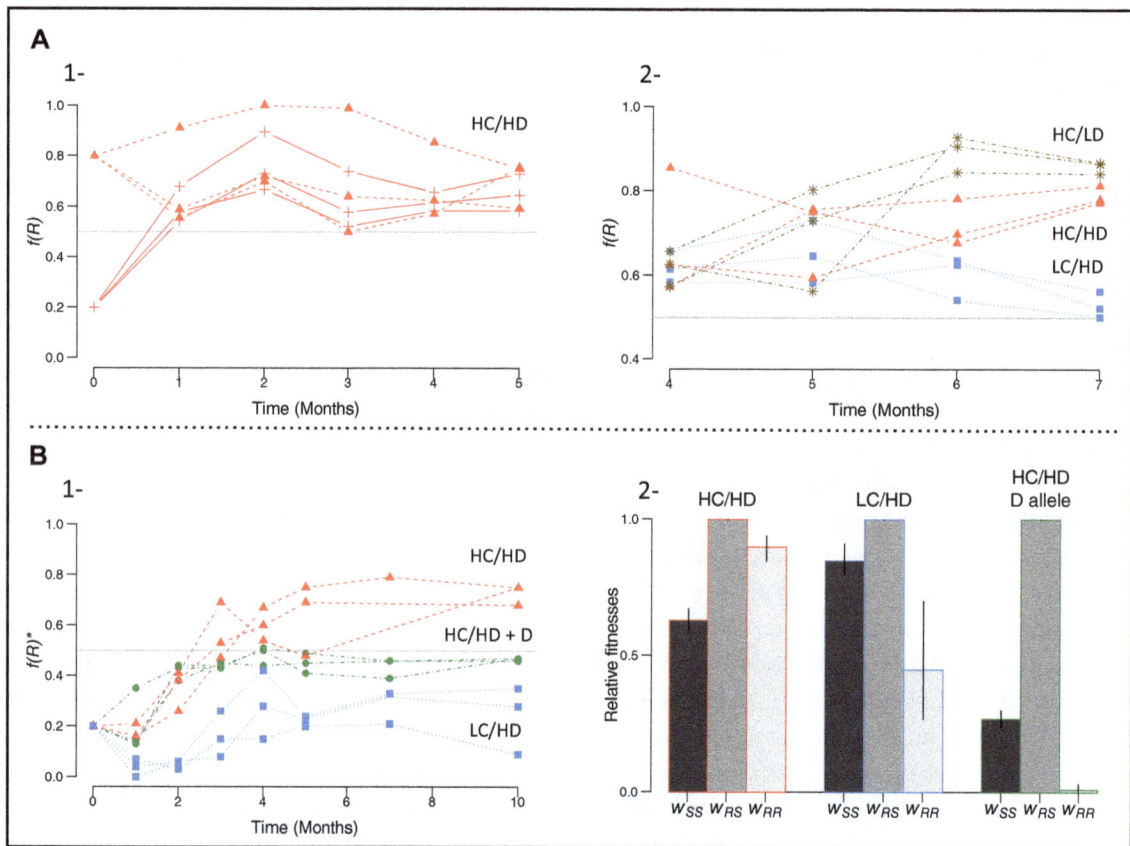

Figure 3. R allele dynamics in the various experimental evolution studies and fitness estimations

(A) Changes in R allele frequency *f(R)* in E_1 (Table 1). (A1) Original rearing conditions at high density (HD) and high insecticide concentration (HC), R and S alleles only, with $f_0(R) = 0.8$ (triangles, dashed lines) or $f_0(R) = 0.2$ (plus signs, solid lines). (A2) Altered rearing conditions after four months (E_1', Table 1): controls (HD/HC; red triangles, dashed lines), reduced density (LD/HC; brown stars, dotted-dashed lines), reduced insecticide concentration (HD/LC; blue squares, dotted lines).

(B1) Changes in *f(R)* in E_2 (Table 1): controls (HD/HC, R and S; red triangles, dashed lines), reduced insecticide concentration (HD/LC, R and S; blue squares, dotted lines), duplicated allele assay, that is, control rearing conditions (HD/HC), R, S, and D alleles, with $f_0(D) = 0.05$, $f_0(R) = 0.15$, and $f_0(S) = 0.8$ (*NB: in this set up, as all genotypes carrying D confer a heterozygote phenotype, *f(R)* is estimated as if only R and S were present, see text). (b2) The relative fitnesses of the [SS] (w_{SS}, dark gray bars), [RS] ($w_{RS} = 1$, medium gray bars), and [RR] (w_{RR}, light gray bars) phenotypes (see "Estimation of selection coefficients" in section Methods) are presented for the different conditions of the E_2 experiment (B1 plot colors are conserved).

different overdominance equilibria may exist, depending on selective pressure intensities, which can alter the relative fitness trade-offs of the different genotypes (providing a clear alternative example of overdominance to sickle cell anemia; Lewontin 1974).

OVERDOMINANCE FAVORS THE HETEROGENEOUS DUPLICATED ALLELE

We then tested whether the heterogeneous *ace-1* duplication (D) was favored when the environmental conditions result in overdominance. Previous studies suggested that D alleles could confer fitness trade-offs on their carriers similar to those of standard RS heterozygotes (Labbé et al. 2014; Assogba et al. 2015). Moreover, DD individuals should not suffer from the segregation burden associated with the RS genotype. We set up three replicates in

control conditions (HC, HD), but introduced a D allele, such that $f_0(D) = 0.05$, $f_0(R) = 0.15$, $f_0(S) = 0.80$ (E_2, Fig. 2). If D is indeed favored, then most individuals should be phenotypic heterozygotes [RS], with an apparent R frequency at equilibrium $\hat{f}(R)^*$ close to 0.5 (see, section Methods).

This is precisely what we observed: after 10 months, most of the individuals were indeed [RS] and no [RR] individuals were observed ($\hat{f}(RS) = 0.93 \pm 0.02$), with $\hat{f}(R)^* = 0.47 \pm 0.01$ (Fig. 3B1, green circles). The estimated relative fitnesses of [SS] and [RR] were both close to 0 (E_2, Table 1 and Fig. 3B2). The persistence of a few [SS] individuals while all [RR] disappeared suggests a slightly asymmetrical trade-off: as phenotyping preceded selection, it could result from a higher fitness of DS than DR individuals, thereby still generating a few new [SS]

Table 1. Experimental design and relative fitness estimations.

	Allele	Conditions	W_{SS} (SL)	W_{RR} (SL)	%TD	od
E_1	S / R	HC/HD	0.20 (0.16–0.24)	0.58 (0.54–0.62)	0.66	5.55
E_1'	S / R	HC/HD	0.10 (0.04–0.21)	0.69 (0.62–0.76)	0.85	2.8
	S / R	HC/LD	0.59 (0.35–0.90)	1.22 (1.09–1.39)	0.89	2.73
	S / R	LC/HD	0.91 (0.68–1.04)	0.78 (0.67–0.89)	0.84	2.16
E_2	S / R	HC/HD	0.63 (0.59–0.67)	0.90 (0.85–0.94)	0.76	3.04
	S / R	LC/HD	0.85 (0.80–0.91)	0.45 (0.27–0.70)	0.73	5.41
	S / R / D	HC/HD	0.27 (0.24–0.30)	0.01 (0.00–0.03)	0.89	3.54

For the various evolution experiments (E_i), the alleles in competition are indicated (single-copy susceptible S, single-copy resistant R, and heterogeneous duplication D). The conditions in which the larvae were reared are also indicated: high or low insecticide concentration (HC = 0.02 ppm and LC = 0.01 ppm temephos, respectively) and high or low larval density (HD or LD, respectively); controls are indicated in italics; conditions differing from the controls are shown in bold. For each set of conditions, a population genetics model was used to calculate the fitnesses of the single-copy susceptible (wss) and resistance (wRR) homozygotes relative to that of the heterozygote (wRS = 1). When the associated support limits (SL, in brackets) include 1, w_{SS} and/or w_{RR} are not significantly different from w_{RS}; significant differences are shown in bold. The percentage of the total deviance explained by each model (%TD) and its overdispersion (od) are also indicated.

individuals. D invasion was further confirmed with a specific molecular test applied to about 90 individuals from each replicate: the frequency of this allele increased from 0.05 to an estimated $\hat{f}(D) = 0.72 \pm 0.07$ (Table A2). All the [RS] individuals carried D (i.e., no standard RS heterozygotes were found in the cages); the [RS] phenotype fitness therefore corresponded to genotypes DS, DR, or DD. Heterogeneous duplications, by conferring the heterozygous phenotype without the associated segregation burden, can thus be fixed when selective trade-off favors overdominance, that is, when antagonist selective pressures are balanced.

OVERDOMINANCE IS PROBABLY COMMON BUT TRANSITORY

Overdominance is almost certainly more widespread than generally thought (Hedrick 2012; Llaurens et al. 2017): (i) multivariate stabilizing selection probably frequently leads to overdominance for new mutations that are beneficial in the heterozygous state (Manna et al. 2011), and (ii) recent adaptation usually involves trade-offs (Orr 2005), so balanced antagonist selective pressures would result in overdominance; some studies measuring the fitness associated to new mutations in the laboratory have confirmed these expectations (Peters et al. 2003). Our study contributes to explain the discrepancy between the prevalence of overdominance in newly arising mutations and the relative rarity of segregating overdominant alleles in the field (Manna et al. 2011): it shows that overdominance may not be robust, as limited modifications of the environment can favor one allele over the other; more importantly, we showed that, as predicted by Haldane (1954), overdominance can be rapidly abolished by the occurrence and selective spread of a heterogeneous duplication, a situation that has probably contributed to the scarcity of persistent cases of overdominance in natural populations.

HETEROGENEOUS DUPLICATIONS CAN BE IMMEDIATELY ADAPTIVE AND COULD FUEL FUTURE EVOLUTION

Overdominance (and more generally heterozygote advantage) has been proposed to explain the diversification of several multigenic families, such as MHC genes in vertebrates (Spurgin and Richardson 2010), R genes in plants (Michelmore and Meyers 1998; Panchy et al. 2016), and MAT genes in basidiomycetes (May et al. 1999): heterozygotes display a higher fitness because they can resist to more pathogens (MHC and R genes) or mate with more sexual types (MAT). However, these duplications are ancient and it is difficult to determine whether copy sequence polymorphism existed before the duplications (i.e., heterogeneous duplications associating existing alleles) or resulted from postduplication divergence (i.e., originally homogeneous duplications; Fig. 1). The difficulty to identify ancient heterogeneous duplications probably explains why their potentially crucial adaptive role has been overlooked.

Fortunately, a handful of contemporaneous heterogeneous duplications have been described that allow assessing how and why they are selected. Interestingly, they all concern insecticide target genes, probably because these duplications are recent (occurring in response to a few decades of anthropic environmental modification), associated with irreducible trade-offs, and highly scrutinized due to their impact on vector control and public health (Labbé et al. 2007a; Djogbénou et al. 2008; Remnant et al. 2013; Sonoda et al. 2014). The *ace-1* duplications in *C. pipiens* remain however the most deeply studied: so far, at least 13 duplicated alleles have been identified, most of them resulting from independent duplication events (Labbé et al. 2007a; Alout et al. 2010; Osta et al. 2012). This suggests that the conditions for overdominance are probably quite frequent: insecticide treatment practices typically result in a patchy environment, with alternating treated and non-treated areas: if the grain of the environment is smaller than the dispersal distance of the mosquito, it could result in marginal over-dominance (Lenormand et al. 1998; Labbé et al. 2007b), or even full overdominance if field conditions result in low insecticide doses (this study). Moreover, the treatment intensities typically vary in time, as they are applied usually only in some periods of the year (Lenormand et al. 1999): these fluctuations of antagonist selective pressures could also on average favor the heterozygote phenotype (i.e., marginal overdominance resulting from fluctuating selection). This frequent selection of D alleles could seriously hinder mosquito control: because they display a lower cost than R, these resistance alleles could make OP and CX insecticides virtually obsolete and threaten control strategies based on insecticide alternation. This is particularly pressing in the case of *A. gambiae*, the malaria vector, where these insecticides have been suggested as replacements for the widely used pyrethroids that face high and widespread resistance (Assogba et al. 2015, 2016).

This unique example shows that heterogeneous duplications can result in an immediate qualitative advantage in fluctuating or patchy environments, or more generally in environments with balanced antagonistic selective pressures. As predicted by Haldane in 1954, we indeed demonstrated that these duplications can be selected because they allow the permanent association of overdominant heterozygous alleles, with no segregation burden (Haldane 1954). These properties could prove useful for breeders (plant breeders in particular), as new genome-editing tools (e.g., CRISPR-cas9; Sander and Joung 2014) could be used to generate heterogeneous duplications to create stable lines displaying specific heterosis otherwise found only in hybrids (Fu et al. 2014).

However, heterogeneous duplications should also be studied in more detail in terms of their role in long-term evolution, as they probably bear witness to ancestral situations of transitory overdominance. As homogeneous duplications, they create new material for genetic innovation (Lynch and Force 2000; Osada and Innan 2008; Innan and Kondrashov 2010). However, as they result from the association of two already divergent alleles, their dissimilar copies immediately carry different functions, that is, copy functional divergence precedes fixation (Fig. 1). They are thus more likely to evolve further through subfunctionalization and neofunctionalization, and should do it more rapidly than homogeneous duplications; these can be first selected, for example, for increased protein quantity, but would diverge later (Lenormand et al. 1998; Labbé et al. 2007a; Innan and Kondrashov 2010). These heterogeneous duplications could thus play a major, albeit yet underestimated, role in the evolution of gene families.

Author Contributions
PL, PM, MW, and TL conceived and designed the study; PL, PM, and MW performed the experiments; PL, PM, MW, and TL analyzed and interpreted the data; PL and PM drafted the manuscript and all authors contributed to later versions of the manuscript.

ACKNOWLEDGMENTS
We would like to thank N. Pasteur, H. Alout, M. Sicard, P. H. Gouyon and M. Raymond for helpful comments on the manuscript, and P. Makoundou and S. Unal for technical support. This work was funded by French ANR program (project "SilentAdapt", BIOADAPT 2013–2015, Grant number ANR-13-ADAP-0016). This study is the contribution 2017–150 of the *Institut des Sciences de l'Evolution de Montpellier* (UMR 5554, CNRS-UM-IRD-EPHE).

LITERATURE CITED
Alout, H., L. Djogbénou, C. Berticat, F. Chandre, and M. Weill. 2008. Comparison of *Anopheles gambiae* and *Culex pipiens* acetycholinesterase 1 biochemical properties. Comp. Biochem. Physiol. B Biochem. Mol. Biol. 150:271–277.

Alout, H., P. Labbé, N. Pasteur, and M. Weill. 2010. High incidence of *ace-1* duplicated haplotypes in resistant *Culex pipiens* mosquitoes from Algeria. Insect Biochem. Mol. Biol. 41:29–35.

Assogba, B. S., L. S. Djogbénou, P. Milesi, A. Berthomieu, J. Perez, D. Ayala, F. Chandre, M. Makoutodé, P. Labbé, and M. Weill. 2015. An *ace-1* gene duplication resorbs the fitness cost associated with resistance in *Anopheles gambiae*, the main malaria mosquito. Sci. Rep. 5:1–12.

Assogba, B. S., P. Milesi, L. S. Djogbénou, A. Berthomieu, P. Makoundou, L. S. Baba-Moussa, A.-S. Fiston-Lavier, K. Belkhir, P. Labbé, and M. Weill. 2016. The *ace-1* locus is amplified in all resistant *Anopheles gambiae* mosquitoes: fitness consequences of homogeneous and heterogeneous duplications. PLoS Biol. 14:e2000618.

Berticat, C., G. Boquien, M. Raymond, and C. Chevillon. 2002. Insecticide resistance genes induce a mating competition cost in *Culex pipiens* mosquitoes. Genet. Res. 79:41–47.

Bourguet, D., N. Pasteur, J. A. Bisset, and M. Raymond. 1996. Determination of Ace.1 genotypes in single mosquitoes: toward an ecumenical biochemical test. Pestic. Biochem. Physiol. 55:122–128.

Djogbénou, L., F. Chandre, A. Berthomieu, R. K. Dabiré, A. Koffi, H. Alout, and M. Weill. 2008. Evidence of introgression of the *ace-1R* mutation and of the *ace-1* duplication in west African *Anopheles gambiae s. s.* PLoS One 3:e2172:1–7.

Dobzhansky, T. 1952. Nature and origin of heterosis. Pp. 218–223 in J. W. Gowan, ed. Heterosis. Iowa State College Press, Ames, IA.

Duron, O., P. Labbé, C. Berticat, F. Rousset, S. Guillot, M. Raymond, and M. Weill. 2006. High *Wolbachia* density correlates with cost of infection for insecticide resistant *Culex pipiens* mosquitoes. Evolution 60:303–314.

East, E. M. 1936. Heterosis. Genetics 21:375–397.

Fisher, R. A. 1922. On the dominance ratio. Proc. R. Soc. London B 42:321–341.

Fu, D., M. Xiao, A. Hayward, Y. Fu, G. Liu, G. Jiang, and H. Zhang. 2014. Utilization of crop heterosis: a review. Euphytica 197:161–173.

Georghiou, G. P., R. L. Metcalf, and F. E. Gidden. 1966. Carbamate-resistance in mosquitoes: selection of *Culex pipiens fatigans* Wied (= *Culex quinquefasciatus*) for resistance to Baygon. Bull. World Heal. Organ. 35:691–708.

Haldane, J. B. S. 1954. The biochemistry of genetics. George Allen and Unwin, Ltd., Lond.

Hedrick, P. W. 2012. What is the evidence for heterozygote advantage selection? Trends Ecol. Evol. 27:698–704.

Innan, H., and F. Kondrashov. 2010. The evolution of gene duplications: classifying and distinguishing between models. Nat. Rev. Genet. 11:97–108.

Katju, V., and U. Bergthorsson. 2013. Copy-number changes in evolution: rates, fitness effects and adaptive significance. Front. Genet. 4:1–12.

Labbé, P., A. Berthomieu, C. Berticat, H. Alout, M. Raymond, T. Lenormand, and M. Weill. 2007a. Independent duplications of the acetylcholinesterase gene conferring insecticide resistance in the mosquito *Culex pipiens*. Mol. Biol. Evol. 24:1056–1067.

Labbé, P., C. Berticat, A. Berthomieu, S. Unal, C. Bernard, M. Weill, and T. Lenormand. 2007b. Forty years of erratic insecticide resistance evolution in the mosquito *Culex pipiens*. PLoS Genet. 3:e205.

Labbé, P., P. Milesi, A. Yébakima, N. Pasteur, M. Weill, and T. Lenormand. 2014. Gene-dosage effects on fitness in recent adaptive duplications: *ace-1* in the mosquito *Culex pipiens*. Evolution 68:2092–2101.

Lenormand, T., T. Guillemaud, D. Bourguet, and M. Raymond. 1998. Appearance and sweep of a gene duplication: adaptive response and potential for new functions in the mosquito *Culex pipiens*. Evolution 52:1705–1712.

Lenormand, T., D. Bourguet, T. Guillemaud, and M. Raymond. 1999. Tracking the evolution of insecticide resistance in the mosquito *Culex pipiens*. Nature 400:861–864.

Lewontin, R. C. 1974. The genetic basis of evolutionary change. Columbia Univ. Press, New York.

Llaurens, V., A. Whibley, and M. Joron. 2017. Genetic architecture and balancing selection: the life and death of differentiated variants. Mol. Ecol. 26:2430–2448.

Lynch, M., and A. Force. 2000. The probability of duplicate gene preservation by subfunctionalization. Genetics, 154, 459–473.

Manna, F., G. Martin, and T. Lenormand. 2011. Fitness landscapes: an alternative theory for the dominance of mutation. Genetics 189:923–937.

May, G., F. Shaw, H. Badrane, and X. Vekemans. 1999. The signature of balancing selection: fungal mating compatibility gene evolution. Proc. Natl. Acad. Sci. U S A 96:9172–9177.

Michelmore, R. W., and B. C. Meyers. 1998. Clusters of resistance genes in plants evolve by divergent selection and a birth-and-death process. Genome Res. 8:1113–1130.

Orr, H. A. 2005. The genetic theory of adaptation: a brief history. Nat. Rev. Genet. 6:119–127.

Osada, N., and H. Innan. 2008. Duplication and gene conversion in the *Drosophila melanogaster* genome. PLoS Genet. 4:e1000305.

Osta, M., Z. Rizk, P. Labbé, M. Weill, and K. Knio. 2012. Insecticide resistance to organophosphates in *Culex pipiens* complex from Lebanon. Parasit. Vectors 5:1–6.

Panchy, N., M. Lehti-shiu, and S. Shiu. 2016. Evolution of gene duplication in plants. Plant Physiol. 171:2294–2316.

Peters, A. D., D. L. Halligan, M. C. Whitlock, and P. D. Keightley. 2003. Dominance and overdominance of mildly deleterious induced mutations for fitness traits in *Caenorhabditis elegans*. Genetics 165:589–599.

Remnant, E. J., R. T. Good, J. M. Schmidt, C. Lumb, C. Robin, P. J. Daborn, and P. Batterham. 2013. Gene duplication in the major insecticide target site, *Rdl*, in *Drosophila melanogaster*. Proc. Natl. Acad. Sci. U. S. A. 110:14705–14710.

Sander, J. D., and J. K. Joung. 2014. CRISPR-Cas systems for editing, regulating and targeting genomes. Nat. Biotechnol. 32:347–355.

Schrider, D. R., and M. W. Hahn. 2010. Gene copy-number polymorphism in nature. Proc. Biol. Sci. 277:3213–3221.

Sonoda, S., X. Shi, D. Song, P. Liang, X. Gao, Y. Zhang, J. Li, Y. Liu, M. Li, M. Matsumura, et al. 2014. Duplication of acetylcholinesterase gene in diamondback moth strains with different sensitivities to acephate. Insect Biochem. Mol. Biol. 48:83–90.

Spurgin, L. G., and D. S. Richardson. 2010. How pathogens drive genetic diversity: MHC, mechanisms and misunderstandings. Proc. R. Soc. B Biol. Sci. 277:979–988.

Weill, M., G. Lutfalla, K. Mogensen, F. Chandre, A. Berthomieu, C. Berticat, N. Pasteur, A. Philips, P. Fort, and M. Raymond. 2003. Insecticide resistance in mosquito vectors. Nature 423:423–426.

Appendix

Table A1. Population cages phenotypic data.

	Alleles and conditions	Month	Replicate 1				Replicate 2				Replicate 3			
			[SS]	[RS]	[RR]	Tot	[SS]	[RS]	[RR]	Tot	[SS]	[RS]	[RR]	Tot
E$_1$	S / R HC / HD	1	0	41	4	45	5	17	20	42	4	30	11	45
		3	1	24	23	48	0	10	38	48	1	30	17	48
		3	5	28	12	45	1	21	22	44	6	31	8	45
		4	5	27	16	48	2	29	17	48	3	34	11	48
		5	6	22	20	48	0	26	22	48	4	32	12	48

		Month	Replicate 4				Replicate 5				Replicate 6			
			[SS]	[RS]	[RR]	Tot	[SS]	[RS]	[RR]	Tot	[SS]	[RS]	[RR]	Tot
		1	1	38	6	45	3	31	11	45	0	8	37	45
		3	4	21	23	48	3	21	24	48	0	0	48	48
		3	3	19	3	25	3	25	15	43	0	1	43	44
		4	9	23	16	48	0	36	12	48	1	12	35	48
		5	1	19	23	43	0	39	9	48	2	12	18	32

	Alleles and conditions	Month	Replicate 1				Replicate 2				Replicate 3			
			[SS]	[RS]	[RR]	Tot	[SS]	[RS]	[RR]	Tot	[SS]	[RS]	[RR]	Tot
E$_1$′	S / R HC / HD	4	9	23	16	48	1	12	35	48	0	36	12	48
		5	1	19	23	43	2	12	18	32	0	39	9	48
		6	0	21	27	48	1	29	18	48	1	27	20	48
		7	0	18	30	48	0	5	6	11	0	19	24	43
	S / R HC / LD	4	9	23	16	48	2	29	17	48	0	36	12	48
		5	5	16	27	48	0	19	29	48	8	26	14	48
		6	1	13	34	48	0	9	39	48	0	7	41	48
		7	0	15	32	47	0	13	35	48	0	12	33	45
	S / R LC / HD	4	5	27	16	48	2	29	17	48	3	34	11	48
		5	6	22	20	48	0	26	22	48	4	32	12	48
		6	4	14	6	24	8	19	21	48	6	24	18	48
		7	13	22	13	48	8	28	10	46	8	26	14	48

	Alleles and conditions	Month	Replicate 1				Replicate 2				Replicate 3			
			[SS]	[RS]	[RR]	Tot	[SS]	[RS]	[RR]	Tot	[SS]	[RS]	[RR]	Tot
E$_2$	S / R HC / HD	1	39	19	0	58	39	19	0	58	28	20	0	48
		2	12	36	0	48	26	19	3	48	11	35	2	48
		3	6	33	9	48	11	29	8	48	3	24	21	48
		4	1	36	11	48	1	26	16	43	4	36	8	48
		5	3	24	21	48	1	22	25	48	5	11	4	20
		6	–	–	–	–	–	–	–	–	–	–	–	–
		7	–	–	–	–	0	14	20	34	–	–	–	–
		8	–	–	–	–	–	–	–	–	–	–	–	–
		9	–	–	–	–	–	–	–	–	–	–	–	–
		10	2	57	37	96	2	44	50	96	0	48	48	96

(Continued)

Table A1. Continued.

Alleles and conditions	Month	Replicate 1				Replicate 2				Replicate 3			
		[SS]	[RS]	[RR]	Tot	[SS]	[RS]	[RR]	Tot	[SS]	[RS]	[RR]	Tot
S / R HC / LD	1	48	2	1	51	47	0	0	47	41	7	0	48
	2	42	6	0	48	42	6	0	48	45	3	0	48
	3	27	17	4	48	40	8	0	48	35	12	1	48
	4	16	24	8	48	21	27	0	48	34	14	0	48
	5	25	23	0	48	15	13	0	28	28	19	0	47
	6	–	–	–	–	–	–	–	–	–	–	–	–
	7	9	14	1	24	17	31	0	48	27	19	0	46
	8	–	–	–	–	–	–	–	–	–	–	–	–
	9	–	–	–	–	–	–	–	–	–	–	–	–
	10	30	64	2	96	42	52	0	96	79	17	0	96
S / R / D* HD / HC	1	35	13	0	48	15	34	0	49	36	12	0	48
	2	12	36	0	48	7	41	0	48	5	26	1	32
	3	5	43	0	48	7	41	0	48	6	42	0	48
	4	3	23	0	26	0	57	1	58	1	46	1	48
	5	6	40	1	47	1	47	0	48	9	39	0	48
	6	–	–	–	–	–	–	–	–	–	–	–	–
	7	–	–	–	—	4	44	0	48	11	37	0	48
	8	–	–	–	–	–	–	–	–	–	–	–	–
	9	–	–	–	–	–	–	–	–	–	–	–	–
	10	9	130	0	139	12	128	0	139	7	129	0	136

The number of individuals for each phenotype ([RR], [RS], [SS], TPP test) is indicated for each replicate and each month. The total number of individuals tested (Tot) was usually 48, but varied due to technical issues.

*Due to the presence of the D allele, the [RS] phenotype corresponds to RS, DS, DR, or DD genotypes in these replicates.

Table A2. Phenotypic data used to estimate D frequency at month 10 in E_2.

Phenotype*	Replicate 1	Replicate 2	Replicate 3
[RR]	0	0	0
[RS]	0	0	0
[D]	83	80	86
[SS]	7	11	4
Tot	90	91	90

The number of individuals for each phenotype ([RR], [RS], [D], [SS]) and the total number of tested individuals (Tot) are indicated for each replicate.

*All phenotypes correspond to only one genotype, except [D] that corresponds to DS, DR, or DD genotypes.

8

No evidence that kin selection increases the honesty of begging signals in birds

Kat Bebbington[1,2] (iD) and Sjouke A. Kingma[2,3]

[1] School of Biological Sciences, University of East Anglia, Norwich Research Park, Norwich NR4 7TJ, UK

[2] Behavioural & Physiological Ecology, GELIFES, University of Groningen, 9700CC Groningen, The Netherlands

[3] E-mail: sjoukeannekingma@gmail.com

Providing plausible mechanisms to explain variation in the honesty of information communicated through offspring begging signals is fundamental to our understanding of parent–offspring conflict and the evolution of family life. A recently published research article used comparative analyses to investigate two long-standing hypotheses that may explain the evolution of begging behavior. The results suggested that direct competition between offspring for parental resources decreases begging honesty, whereas indirect, kin-selected benefits gained through saving parental resources for the production of future siblings increase begging honesty. However, we feel that evidence for a role of kin selection in this context is still missing. We present a combination of arguments and empirical tests to outline alternative sources of interspecific variation in offspring begging levels and discuss avenues for further research that can bring us closer to a complete understanding of the evolution of offspring signaling.

KEY WORDS: Comparative studies, competition, kin selection, signaling.

A Short Introduction to Offspring Begging Signals

Across a diverse range of taxa, offspring direct behaviorally complex begging displays toward caregiving parents. The function and evolution of such behavior has intrigued biologists for decades, spawning a myriad of different explanatory hypotheses that make diverse assumptions about the balance of power between parents and their offspring (Royle et al. 2002a), the reliability of information that begging signals convey to parents (Kilner and Johnstone 1997) and the roles of kin selection and competition among offspring (Trivers 1972).

If parents make active choices about how to partition resources between their offspring, there are two scenarios where we can expect offspring begging to be an honest signal about the state of the offspring. First, if the cost of expressing a begging signal outweighs the marginal fitness gained from successfully securing parental resources, begging signals should honestly reflect the offspring's marginal fitness gain per unit of additional parental investment (Godfray 1995). Second, if there is a high risk that not all offspring survive to adulthood, begging signals should honestly

reflect quality because individuals are selected to boast their own quality and/or parents are selected to preferentially invest in the most valuable offspring (Grafen 1990). However, the degree to which begging behaviors honestly reflect any such information seems to vary greatly between species (Mock et al. 2011). Although many studies support predictions of honest begging (e.g., Redondo and Castro 1992; Andrews and Smiseth 2013), others suggest that begging is a form of scramble competition for resources passively allocated by parents to the most conspicuous display (e.g., Smith and Montgomerie 1991; Parker et al. 2002). One possible source of this variation may be interspecific differences in the degree of evolutionary conflict within the family over the allocation of parental resources. Specifically, where high relatedness between family members means that their evolutionary interests in terms of resource allocation are more aligned (Trivers 1974), honesty should prevail. In contrast, where evolutionary interests are less aligned, for example when the direct fitness benefits of acquiring resources outweigh the benefits of sharing them with relatives, scramble competition and dishonesty should be more prevalent (Briskie et al. 1994).

Figure 1. Modified from Caro et al. (2016). Kin selection predicts that offspring should be honest about their need when parents are likely to produce full siblings in future (left-hand panel). If this is the case, the death of one parent (middle panel) should promote offspring dishonesty because of reduced relatedness to future offspring (relatedness = 1 × 0.25). However, we argue that divorce (right-hand panel) does not promote dishonesty in this way because both parents will continue breeding and hence produce two sets of half-siblings, which together have equal or even higher value than one set of full siblings (total relatedness ≥ 2 × 0.25 = 0.5).

Drivers of Honest Begging Signals: A Recent Case Study

In a recent comparative analysis across avian taxa, Caro et al. (2016) aimed to explain interspecific variation in honesty of begging signals in relation to variation in conflict between family members over the allocation of parental care. Caro et al. (2016) first tested the hypothesis that begging honesty decreases with increasing competition for parental resources (Mock and Parker 1997). They showed convincing evidence that the correlation between begging and some measured component of offspring "need" (such as hunger levels) becomes weaker with the presence and increasing number of siblings in both current and future broods. These interspecific patterns provide important validation for the hypothesis that offspring competition for limited resources selects for exaggerated, and thus dishonest, begging signals (Royle et al. 2002a). Moving onto a second hypothesis, Caro et al. (2016) aimed to test whether begging is more honest when relatedness to future offspring, and hence the inclusive fitness benefit of sharing parental resources, is higher (Trivers 1974). According to Caro et al.'s (2016) interpretations, the results they present appear to support this second hypothesis; in doing so, they may provide the first empirical evidence that relatedness between competitors can effectively reduce parent–offspring conflict and offer one explanation for variation in begging honesty, one of the most widely debated phenomena in behavioral ecology. Below, we explain why it is premature to embrace the conclusions of Caro et al. (2016) as evidence for a role for kin selection in this context, and that multiple other processes may, alternatively or additionally, explain their findings.

Estimating the Inclusive Fitness Value of Future Siblings

When an individual's parents can produce more offspring in the future, inclusive fitness benefits (i.e., the transfer of shared genes to future generations) may favor offspring who adopt strategies that facilitate the production of future siblings. Producing honest signals about current nutritional state to preserve parental resources (i.e., energy or food) for future broods (Trivers 1974) is one potential strategy. How then should we calculate expected inclusive fitness benefits from the perspective of current offspring? In their comparative analysis, Caro et al. (2016) suggested that relatively low inclusive fitness benefits arise when parents do not breed together to produce future broods, as is the case when (1) one or both of the parents die or (2) parents divorce. By combining these two measures, Caro et al. (2016) showed that offspring begging signals are less honest when parents have a lower likelihood to breed together in the future, which they interpreted as evidence that kin selection drives honesty of begging signals.

Although we agree that the death of one parent indeed reduces future indirect benefits, it is incorrect to assume the same for divorce, and we therefore question whether the conclusion that kin selection underlies begging honesty is correct. As demonstrated in Figure 1, divorced parents will both produce half-siblings with a total inclusive fitness value equal to that produced when they were together. In fact, the inclusive fitness benefits gained from offspring produced from divorced parents are (on average) greater than those from parents who remain together. In another recent comparative study, Culina et al. (2015) showed that divorce generally improves parents' subsequent reproductive success, suggesting that offspring in species with high divorce rates should, on average, be under greater selection to beg honestly. Thus, combining divorce and mortality rates to generate a proxy for kin-selected benefits may lead to an erroneous conclusion about whether kin selection may underlie variation in begging honesty.

Having established that parental divorce may not necessarily reduce the kin-selected incentives of current offspring to beg honestly, we retested the hypothesis that high inclusive fitness benefits of future offspring select for honest signaling, using an identical set of species and the same sample size as Caro et al. (2016). To provide a more accurate calculation of inclusive fitness benefits in

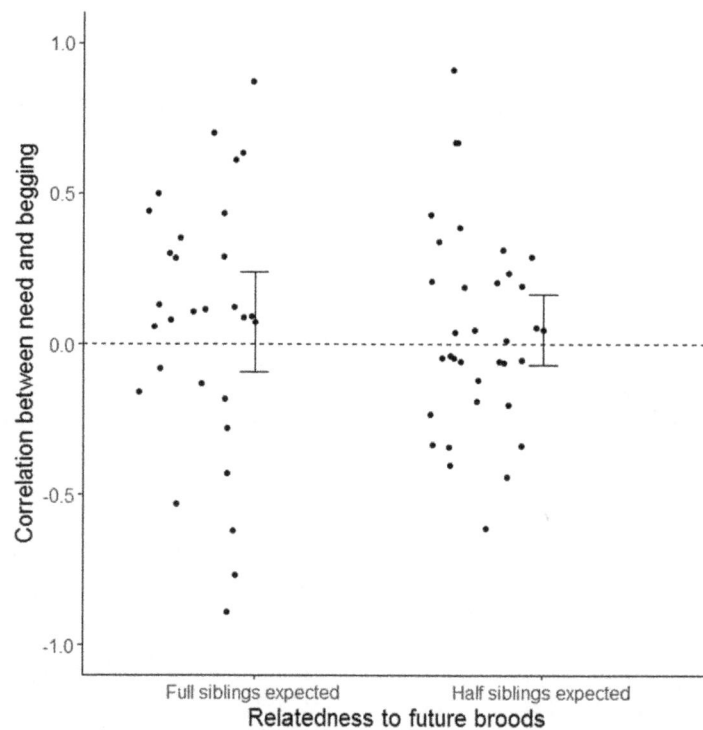

Figure 2. Relationship between begging honesty (measured as the correlation between begging intensity and need) and relatedness to future broods in 63 bird species. Full siblings are expected when there is <50% chance of at least one parent death before next year (34 species) and half siblings are expected when there is >50% chance of at least one parent death before next year (29 species). Raw data were plotted and error bars represent 95% CIs.

terms of the likelihood of full siblings being produced in future, we disregarded divorce rates and only used the likelihood of both parents surviving to reproduce next season. Data were obtained from Caro et al. (2016), and phylogenetic generalized least square (PGLS) analyses were implemented in the Caper package (Orme et al. 2013) in R 3.3.0 (R Development Core Team 2016). We accounted for phylogenetic uncertainty by applying the models to a set of 100 phylogenetic trees (using the Hackett backbone with all species) obtained from http://www.birdtree.org (Jetz et al. 2012). In contrast to the results reported by Caro et al. (2016), we found no difference in the correlation between begging and need (i.e., begging honesty) according to whether the chance of both parents surviving was greater than 50% (PGLS: $\beta \pm SE = -0.026 \pm 0.098$, $t_{61} = -0.271$, $P = 0.787$, Fig. 2). We also were unable to support the conclusion of Caro et al. (2016) if, instead of using this classification, we tested the effect of the absolute probability that both parents survive (range = 2–88%) (PGLS: $\beta \pm SE = 0.073 \pm 0.2071$, $t_{61} = 0.351$ $P = 0.727$).

As we outline above and in Figure 1, variation in divorce rates is unlikely to reduce future inclusive fitness. Because divorce accounted for on average ($\pm SE$) $49 \pm 4\%$ of Caro et al.'s measure of "likelihood that pairs did not breed together the following year" (44 species, range = 0–99%), it is perhaps not surprising that when we omit divorce rates from the equation we cannot

support the conclusion that kin selection plays a role in honest begging. However, an important question remains: how then can we explain the intriguing relationship found by Caro et al. (2016) that begging is more honest if pairs have a low likelihood to breed again together the following year? In the final section of this article, we outline a series of arguments that provide both potential explanations for the results presented in Caro et al. (2016) and exciting avenues for further research on the evolution of offspring begging signals more generally.

Beyond Kin Selection: Explaining Variation in Offspring Begging Honesty

As outlined above, scramble competition for limited parental resources may be an important mechanism that decreases signal honesty (Royle et al. 2002a); in line with this, Caro et al. (2016) show that begging signals are less honest in the face of competition with coexisting offspring. We propose that the relationship between begging honesty and pair bond duration (or the "likelihood that parents reproduce together in future") can also be explained in terms of offspring competition. Below, we present three potential mechanisms by which offspring competition could drive this relationship, one of which we were able to test using

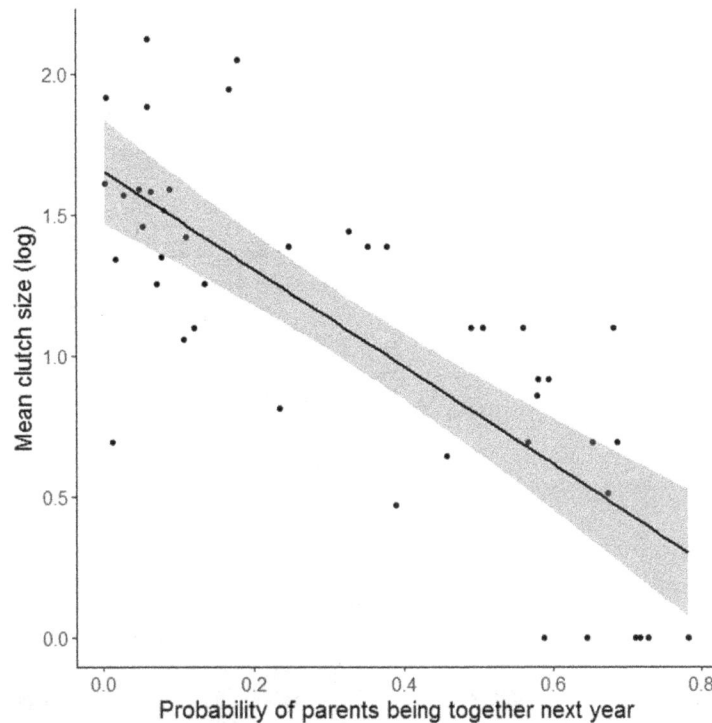

Figure 3. Relationship between mean clutch size (log transformed) and the probability of parents reproducing together in the next year across 44 bird species. Untransformed raw data were plotted (with the regression line through the raw data) and shaded areas represent 95% CIs.

Caro et al.'s (2016) dataset. Although we explicitly consider variation in offspring begging in the context of the relationship between honesty and pair bond duration, it is important to note that these alternative hypotheses are entirely speculative at this point. Nonetheless, we believe that these and other mechanisms are interesting to consider in relation to offspring begging signals more generally and may stimulate future work.

(1) Pair bond duration is associated with clutch size and offspring competition

If offspring begging honesty is related to competition for limited resources, species where sibling competition is more intense are expected to be less honest. Because individuals with relatively short lifespans and hence short pair bonds produce a large number of offspring in each reproductive attempt (Charnov and Krebs 1974; Martin 2002), offspring competition might be higher within broods of species with short pair bonds. Using the dataset from Caro et al. (2016), we used PGLS analyses (as described above) to test for a correlation between the likelihood that parents produce together in future and levels of current offspring competition in terms of clutch size. Although the amount of parental resources per offspring would give a truer measure of the degree of offspring competition than the absolute number of competitors (Mock et al. 2009), measuring clutch size at least captures some of the variation between offspring raised

on their own (who by definition experience no direct competition) and those raised with siblings (where there is at least potential for competition). We found that species where parents have a higher probability of breeding together in the following year (calculated as: [(survival probability)^2 × (one-divorce rate)] produce broods of smaller size (log-transformed; PGLS: $\beta \pm$ SE $= -0.684 \pm 0.122$, $t_{42} = -5.608$, $P < 0.001$, Fig. 3), a pattern that is partly driven by the long pair bond duration in species that have only one offspring per brood (Fig. 3). This result suggests that the association between parental pair bond duration and begging honesty can, at least partly, be explained by the fact that competition in current broods of species with short pair bonds is higher. Further exploration of this pattern with a more accurate representation of offspring competition, such as the proportion of offspring that recruit per brood, the degree of asymmetry in offspring size, or the amount of parental provisioning, would be very useful to confirm this relationship.

(2) Parental divorce is linked to social mate competition

Among bird species, parental divorce rates are linked to extra-pair paternity (Cezilly and Nager 1995) and mutual ornamentation (Kraaijeveld 2003; Botero and Rubenstein 2012); both these factors reflect the level of social competition for mates. If species with high divorce rates are characterized by high levels of mate competition, we can expect strong selection for competition-related behavioral traits in such species. Because traits that increase reproductive success in adulthood (in this case, traits that increase competitive ability) will be present in the offspring of successful adults, we can predict that offspring in species with high divorce rates are more competitive and less likely to beg honestly. The idea that offspring behaviors may be influenced by selection for adult behaviors is not new (Kölliker et al. 2012), but perhaps revisiting models of offspring begging behavior in the light of parent–offspring coadaptation may reveal intriguing new patterns.

(3)   Pair bond duration is associated with increased sexual conflict over care

A key principle of life-history theory is that parents trade-off investment in current offspring with investment in future offspring (Stearns 1992). This trade-off gives rise to sexual conflict over the distribution of parental investment costs (Trivers 1972), which in turn reduces the overall parental investment each offspring receives (Royle et al. 2002b, 2010; Lessells and McNamara 2012). In species that form relatively short-term pair bonds, breeding partners have little interest in the long-term reproductive potential of their partner and sexual conflict should be more intense (Lessells and McNamara 2012; Bebbington and Hatchwell 2016); perhaps one reason why begging is less honest in divorce-prone species is that offspring have to compete more for relatively little parental investment. The interplay between sexual conflict over parental care, parent–offspring conflict, and sibling rivalry is being increasingly recognized as an important source of information about the evolution of family life (Kölliker et al. 2012; Royle et al. 2014, 2016); considering the role of social evolution in the light of multiple levels of conflict will hopefully inspire future studies of offspring begging.

Concluding Remarks

Interspecific variation in honesty of begging signals is an important source of information to make inferences about how selection acts according to social and ecological circumstances. The frequently hypothesized role of kin selection in mediating parent–offspring conflict (Trivers 1974; Mock and Parker 1997), and thus begging honesty, is intriguing and certainly merits further investigation. However, based on the current evidence, we argue that we lack any firm empirical evidence that kin selection is important in this context. In conclusion, we propose that the results of Caro et al. (2016) demonstrate convincing evidence that scramble competition for limited resources is the main driver of interspecific variation in the honesty of begging signals in birds. Although kin selection is likely to play an important role in the evolution and stability of family life (Emlen 1995), it is crucial that we ac-

count for all sources of variation in inclusive fitness to determine the mechanisms by which it acts. We suggest that considering species-specific ecology and conflict on different family levels may lead to a more balanced insight into the forces that select for begging honesty, including kin selection.

AUTHOR CONTRIBUTIONS
KB and SAK conceived the idea and developed the hypotheses, SAK performed all analyses, KB drafted the manuscript.

ACKNOWLEDGMENTS
K.B. was funded by a NERC PhD studentship, and S.A.K. by a VENI Fellowship (863.13.017) awarded by the Netherlands Organisation for Scientific Research (NWO). We thank Douglas Mock, Andy Gardner, and two anonymous reviewers for thoughtful comments on a previous version of this manuscript.

LITERATURE CITED
Andrews, C. P., and P. T. Smiseth. 2013. Differentiating among alternative models for the resolution of parent–offspring conflict. Behav. Ecol. 24:1185–1191.

Bebbington, K., and B. J. Hatchwell. 2016. Coordinated parental provisioning is related to feeding rate and reproductive success in a songbird. Behav. Ecol. 27:652–659.

Botero, C. A., and D. R. Rubenstein. 2012. Fluctuating environments, sexual selection and the evolution of flexible mate choice in birds. PLoS ONE 7:e32311.

Briskie, J. V., C. T. Naugler, and S. M. Leech. 1994. Begging intensity of nestling birds varies with sibling relatedness. Proc. R. Soc. B 258:73–78.

Caro, S. M., S. A. West, and A. S. Griffin. 2016. Sibling conflict and dishonest signaling in birds. Proc. Natl. Acad. Sci. USA 113:13803–13808.

Cezilly, F., and R. G. Nager. 1995. Comparative evidence for a positive association between divorce and extra-pair paternity in birds. Proc. R. Soc. B 262:7–12.

Charnov, E. L., and J. R. Krebs. 1974. On clutch-size and fitness. IBIS 116:217–219.

Culina, A., E. Radersma, and B. C. Sheldon. 2015. Trading up: the fitness consequences of divorce in monogamous birds. Biol. Rev. 90:1015–1034.

Emlen, S. T. 1995. An evolutionary theory of the family. Proc. Natl. Acad. Sci. USA 92:2092–8099.

Godfray, H. C. J. 1995. Signaling of need between parents and offspring: parent–offspring conflict and sibling rivalry. Am. Nat. 146:1–24.

Grafen, A. 1990. Biological signals as handicaps. J. Theor. Biol. 144:517–546.

Jetz, W., G. H. Thomas, J. B. Joy, K. Hartmann, and A. O. Mooers. 2012. The global diversity of birds in space and time. Nature 491:444–448.

Kilner, R., and R. A. Johnstone. 1997. Begging the questions: are offspring solicitation behaviours honest? Trends Ecol. Evol. 12:11–15.

Kölliker, M., N. J. Royle, and P. T. Smiseth. 2012. Parent–offspring coadaptation. Pp. 285–303 in N. J. Royle, P. T. Smiseth, and M. Kölliker, eds. The evolution of parental care. Oxford Univ. Press, Oxford U.K.

Kraaijeveld, K. 2003. Degree of mutual ornamentation in birds is related to divorce rate. Proc. R. Soc. B 270:1785–1791.

Lessells, C. M., and J. M. McNamara. 2012. Sexual conflict over parental investment in repeated bouts: negotiation reduces overall care. Proc. R. Soc. B 279:1506–1514.

Martin, T. E. 2002. A new view of avian life-history evolution tested on an incubation paradox. Proc. R. Soc. B 269:309–316.

Mock, D. W., and G. A. Parker. 1997. The evolution of sibling rivalry. Oxford Univ. Press, Oxford U.K.

Mock, D. W., P. L. Schwagmeyer, and M. B. Dugas 2009. Parental provisioning and nestling mortality in house sparrows. Anim. Behav. 78:677–684.

Mock, D. W., M. B. Dugas, and S. A. Strickler. 2011. Honest begging: expanding from signal of need. Behav. Ecol. 22:909–917.

Orme, D., R. Freckleton, G. Thomas, T. Petzoldt, S. Fritz, N. Isaac, and W. Pearse. 2013. Caper: comparative analyses of phylogenetics and evolution in R. R package version 0.5.2. Available via https://CRAN.R-project.org/package=caper.

Parker, G. A., N. J. Royle, and I. R. Hartley. 2002. Begging scrambles with unequal chicks: interactions between need and competitive ability. Ecol. Lett. 5:206–215.

R Development Core Team. 2016. R: a language and environment for statistical computing. R foundation for Statistical computing, Vienna, Austria. Available via http://www.R-project.org.

Redondo, T., and F. Castro. 1992. Signalling of nutritional need by magpie nestlings. Ethology 92:193–204.

Royle, N. J., I. R. Hartley, and G. A. Parker. 2002a. Begging for control: when are offspring solicitation behaviours honest? Trends Ecol. Evol. 17:434–440.

———. 2002b. Sexual conflict reduces offspring fitness in zebra finches. Nature 416:733–736.

Royle, N. J., S. Wiebke, and S. R. Dall. 2010. Behavioral consistency and the resolution of sexual conflict over parental investment. Behav. Ecol. 21:1125–1130.

Royle, N. J., A. F. Russell, and A. J. Wilson. 2014. The evolution of flexible parenting. Science 345:776–781.

Royle, N. J., S. A. Alonzo, and A. J. Moore. 2016. Co-evolution, conflict and complexity: what have we learned about the evolution of parental care behaviours? Curr. Opin. Behav. Sci. 12:30–36.

Smith, H. G., and R. Montgomerie. 1991 Nestling American robins compete with siblings by begging. Behav. Ecol. Sociobiol. 29:307–312.

Stearns, S. C. 1992. The evolution of life histories. Oxford Univ. Press, Oxford U.K.

Trivers, R. L. 1972. Parental investment and sexual selection. Pp. 136–208 in B. Campbell, ed. Sexual selection and the descent of man 1871–1971. Aldine Transaction, Chicago.

———. 1974. Parent–offspring conflict. Am. Zool. 14:249–264.

9

Friendly foes: The evolution of host protection by a parasite

Ben Ashby[1,2,3] (iD) and Kayla C. King[4]

[1] Department of Mathematical Sciences, University of Bath, Bath BA2 7AY, United Kingdom

[2] Department of Integrative Biology, University of California Berkeley, Berkeley 94720, California

[3] E-mail: benashbyevo@gmail.com

[4] Department of Zoology, University of Oxford, Oxford OX1 3PS, United Kingdom

Hosts are often infected by multiple parasite species, yet the ecological and evolutionary implications of the interactions between hosts and coinfecting parasites are largely unknown. Most theoretical models of evolution among coinfecting parasites focus on the evolution of virulence, but parasites may also evolve to protect their hosts by reducing susceptibility (i.e., conferring resistance) to other parasites or reducing the virulence of coinfecting parasites (i.e., conferring tolerance). Here, we analyze the eco-evolutionary dynamics of parasite-conferred resistance and tolerance using coinfection models. We show that both parasite-conferred resistance and tolerance can evolve for a wide range of underlying trade-offs. The shape and strength of the trade-off qualitatively affects the outcome causing shifts between the minimisation or maximization of protection, intermediate stable strategies, evolutionary branching, and bistability. Furthermore, we find that a protected dimorphism can readily evolve for parasite-conferred resistance, but find no evidence of evolutionary branching for parasite-conferred tolerance, in general agreement with previous work on host evolution. These results provide novel insights into the evolution of parasite-conferred resistance and tolerance, and suggest clues to the underlying trade-offs in recent experimental work on microbe-mediated protection. More generally, our results highlight the context dependence of host-parasite relationships in complex communities.

KEY WORDS: defensive mutualism, host protection, parasite evolution, parasitism mutualism, resistance, tolerance.

Impact Summary

Hosts are often infected with multiple species of parasites with a variety of evolutionary implications. Do coinfecting parasites evolve to become more or less deadly? Can some parasites evolve to protect their hosts from others, thereby providing a net benefit? Existing theory has largely focused on the first question, but relatively little is known about the evolution of host protection. Empirical evidence indicates that host protection is in fact common; various forms of defense have been observed among fungi, bacteria, protozoa, and viruses (bacteriophages) that colonize hosts. Furthermore, recent experiments have shown that a mildly virulent species of bacteria can evolve to protect animal hosts from a more virulent infection, transitioning along the parasitism-mutualism continuum. Despite this growing body of empirical research, there are few theoretical predictions for the evolution of host protection. Here, we use mathematical modeling to explore the evolution of two forms of host protection: parasite-conferred resistance and tolerance. Parasites that confer resistance reduce the likelihood that a second parasite species will be able to infect, whereas parasites that confer tolerance reduce the virulence of coinfecting parasites. We show that both forms of host protection can evolve for a wide range of evolutionary trade-offs, although there are notable differences between the two and the nature of the trade-off qualitatively changes the outcome. For example, the generation and maintenance of high and low levels of defense is possible for resistance, but does not appear to be possible for tolerance, consistent with existing theory on host evolution. Our results provide useful insights into the evolution of host protection and make several general predictions (e.g., the coexistence of high and low levels of resistance is more likely when hosts are long-lived). This study highlights the context-dependent nature of host–parasite interactions and lays the foundations for future theoretical research on the parasitism–mutualism continuum.

In nature, hosts are typically susceptible to a wide range of parasites, including many species of bacteria and fungi, protozoa, and viruses. Coinfections consisting of multiple strains or species of parasites are therefore likely to be common (Petney & Andrews 1998; Cox 2001; Telfer et al. 2010). Crucially, the dynamics of coinfections can be very different to single infections, both in terms of disease (Griffiths et al. 2011) and evolutionary outcomes (Alizon et al. 2013). For example, infection with *Mycobacterium tuberculosis* (TB) increases the risk of mortality in patients already infected by the human immunodeficiency virus (HIV) (Aaron et al. 2004), but this also decreases the infectious period, which theory predicts may select for increased virulence (Bremermann & Pickering 1983). It is clear that understanding how coinfecting parasites interact with each other and their hosts has important implications not only for infectious disease control (Brown et al. 2009; Balmer & Tanner 2011; Griffiths et al. 2011), but also for understanding the ecological and evolutionary outcomes of the community (Read & Taylor 2001; Brown et al. 2002; Alizon 2013; Johnson et al. 2015).

The literature on coinfections has predominantly focused on the evolution of virulence (reviewed in Alizon et al. 2013). In general, theory predicts that low (high) relatedness during coinfections selects for higher (lower) virulence (Hamilton 1972; Bremermann & Pickering 1983; Sasaki & Iwasa 1991; Frank 1992, 1994, 1996; van Baalen & Sabelis 1995). The core assumption of these models is that parasites interact indirectly through exploitative competition (one parasite indirectly harms the prospects of another by consuming a shared resource), but parasites can interact through many other mechanisms. For example, phenotypic plasticity and impaired host immunity select for lower virulence (Choisy & de Roode 2010), and if cooperation among kin increases growth rates then high relatedness may increase virulence (Chao et al. 2000; Brown et al. 2002; West & Buckling 2003). Alternatively, parasites may modulate the virulence of coinfecting species to prolong the life of the host, or may secrete antimicrobial toxins that actively harm competitors through interference competition (spite). For instance, *Streptococcus pneumoniae* produces hydrogen peroxide, which induces lysogenic bacteriophage in *Staphylococcus aureus* to lyse their hosts (Selva et al. 2009). Interference competition has received much less attention than exploitative competition, but is predicted to play a crucial role in parasite evolution (Gardner et al. 2004). For example, spite selects for greater virulence when relatedness is at an extreme and

lower virulence when relatedness is intermediate (Gardner et al. 2004; Massey et al. 2004; Inglis et al. 2009). The ability of parasites to protect their host from additional, perhaps more virulent, infections may therefore evolve as a by-product of interference competition.

Host protection has been found across plant and animal species (Ford & King 2016). Although protective microbes can also be parasitic and therefore costly, they may provide a net benefit to their hosts if they compete with more virulent parasites—"the enemy of my enemy is my friend" (Martinez et al. 2015). Protective microbes can form a significant component of host defense. For example, the survival of monarch butterfly larvae (*Danaus plexippus*) is higher when coinfected with a virulent protozoan parasite (*Ophryocystis elektroscirrha*) and a lethal parasitoid fly (*Lespesia archippivora*), than when only infected by the latter (Sternberg et al. 2011). Some vertically transmitted bacteria in insects, such as *Hamiltonella* (Vorburger & Gouskov 2011; Polin et al. 2014) and *Wolbachia* (Hughes et al. 2011; Blagrove et al. 2012), are costly but provide hosts with protection against other parasite species. Other known examples of parasite-conferred defense include the transfer of resistance genes by lysogenic phages (van Baalen & Jansen 2001) and protection against a virulent fungus by less virulent fungi (Michalakis et al. 1992). Recently, it was discovered that within-host antagonistic interactions between microbial parasite species drove the rapid de novo evolution of protective properties in a worm–bacteria system (King et al. 2016). The boundary between parasitism and mutualism is often blurred, with many bacteria providing context-dependent defense and retaining mild pathogenicity (Polin et al. 2014; Martinez et al. 2015). Together, these empirical observations suggest that evolutionary transitions between parasitism and mutualism are likely to be common. Moreover, this work highlights the potential for host protection to impact infectious disease ecology and evolution.

Few theoretical predictions exist to support this growing body of empirical research on the evolution of host protection (Michalakis et al. 1992; van Baalen & Jansen 2001; Jones et al. 2011). Here, we show that host protection can readily evolve, but the precise outcome depends on the shape and strength of any underlying trade-offs.

Methods

We study the evolution of host protection using two coinfection models (Choisy & de Roode 2010; Alizon 2013). First, we assume that coinfections only occur between parasites of different species (model A), as this greatly simplifies the analysis. Hence if a mutant strain arises in a given host, we assume that it is either immediately cleared or replaces the resident strain. We relax this assumption in the Supporting Information (model B), allowing coinfections to occur between strains of the same species.

MODEL DESCRIPTION

In our primary model (model A), the host population is divided into four classes according to its infection status: susceptible to both parasite species (S); infected by parasite 1 but susceptible to parasite 2 (I_1); infected by parasite 2 but susceptible to parasite 1 (I_2); and infected by both parasites (I_{12}). Hosts have a natural mortality rate of b and reproduce at a maximum per-capita rate of a subject to density-dependent competition (qN with $N = S + I_1 + I_2 + I_{12}$) giving a birth rate of $v(N) = (a - qN)N$. The maximum pairwise transmission rate for parasite j is $\tilde{\beta}_j$ and recovery occurs at rate γ_j; there is no immunity following recovery. Hosts experiencing a single infection by parasite j suffer an additional baseline mortality rate (virulence) of $\tilde{\alpha}_j$, while coinfections lead to an additional mortality rate of α_{12}.

We study the evolution of two forms of host protection by parasite 1: (i) resistance, $\beta_2(y) = \tilde{\beta}_2(1 - \delta y)$; and (ii) tolerance, $\alpha_{12}(y) = \alpha_1(y) + \tilde{\alpha}_2[1 - (1 - \delta)y]$, with $\delta = 0$ or $\delta = 1$. The strength of host protection is denoted by $0 \leq y \leq 1$, with $y = 0$ corresponding to no protection and $y = 1$ to maximum protection. Hence, infection by parasite 1 may either reduce susceptibility to subsequent infection by parasite 2 (resistance, $\delta = 1$), or reduce the virulence of parasite 2 in mixed infections (tolerance, $\delta = 0$). For example, parasite 2 may struggle to establish itself in hosts that are already infected by parasite 1, or parasite 1 may actively harm parasite 2 through physiological defenses (resistance). Alternatively, parasite 1 may produce antitoxins that limit virulence factors produced by parasite 2 (tolerance). Parasites that protect their hosts incur a fitness cost, $c(y)$, which leads to either a reduction in transmission, $\beta_1(y) = \tilde{\beta}_1[1 - c(y)]$, or an increase in virulence, $\alpha_1(y) = \tilde{\alpha}_1[1 + c(y)]$, where

$$c(y) = \frac{c_1(1 - e^{c_2 y})}{1 - e^{c_2}} \quad (1)$$

The parameter $c_1 > 0$ determines the maximum strength of the cost and $c_2 \in \mathbb{R}_{\neq 0}$ determines the rate at which costs increase (accelerating: $c_2 > 0$, decelerating: $c_2 < 0$). Costs associated with host protection may arise due to changes in either the allocation or consumption of host resources. For example, the protective parasite may divert resources from making transmission stages to producing antimicrobials or antivirulence compounds (transmission cost). Alternatively, a parasite may cause additional damage to the host by consuming more resources so that it can maintain its transmission rate and defend against another parasite (virulence cost). It is possible that both transmission and virulence will vary with host protection, but the results are likely to be similar to the single-cost scenarios (e.g., if virulence increases/decreases in addition to a transmission rate cost then the overall cost is slightly stronger/weaker compared to when virulence is fixed). We therefore only consider single costs.

The epidemiological dynamics of monomorphic parasites in well-mixed populations are:

$$\frac{dS}{dt} = v(N) - \left[b + \lambda_1(y) + \lambda_{2,S}\right]S + \gamma_1 I_1 + \gamma_2 I_2 \quad (2a)$$

$$\frac{dI_1}{dt} = \lambda_1(y)S - \left[\Gamma_1(y) + \lambda_2(y)\right]I_1 + \gamma_2 I_{12} \quad (2b)$$

$$\frac{dI_2}{dt} = \lambda_{2,S}S - \left[\Gamma_2 + \lambda_1(y)\right]I_2 + \gamma_1 I_{12} \quad (2c)$$

$$\frac{dI_{12}}{dt} = \lambda_1(y)I_2 + \lambda_2(y)I_1 - \Gamma_{12}(y)I_{12} \quad (2d)$$

where $\Gamma_1(y) = b + \alpha_1(y) + \gamma_1$, $\Gamma_2 = b + \tilde{\alpha}_2 + \gamma_2$, and $\Gamma_{12}(y) = b + \alpha_{12}(y) + \gamma_1 + \gamma_2$ are the inverse of the infectious periods, and $\lambda_{2,S} = \tilde{\beta}_2(I_2 + I_{12})$ and $\lambda_j(y) = \beta_j(y)(I_j + I_{12})$ are the forces of infection ($j = 1, 2$). The initial dynamics of a rare mutant, y_m, when the resident is at equilibrium ($N^* = S^* + I_1^* + I_2^* + I_{12}^*$) are:

$$\frac{dI_m}{dt} = \lambda_1(y_m)S^* - \left[\Gamma_1(y_m) + \lambda_2^*(y_m)\right]I_m + \gamma_2 I_{m2} \quad (3a)$$

$$\frac{dI_{m2}}{dt} = \lambda_1(y_m)I_2^* + \lambda_2^*(y_m)I_m - \Gamma_{12}(y_m)I_{m2} \quad (3b)$$

where I_m is hosts infected with the mutant and I_{m2} is hosts coinfected with the mutant and parasite 2.

ANALYSIS

We use a combination of numerical analysis and simulations to explore the evolution of host protection. Using evolutionary invasion analysis (Metz et al. 1992; Dieckmann & Law 1996; Geritz et al. 1998), we first derive the fitness of a rare mutant, $w(y_m)$—assumed to be phenotypically similar to the resident for parasite 1—when the resident population is at equilibrium. Since there is no analytic solution for the multiparasite endemic equilibrium, we solve the system of equations over a sufficiently long time period to ensure that the system is close to a stable state (verified numerically). The population will evolve in the direction of the selection gradient, $s(y) = \frac{dw}{dy_m}\big|_{y_m=y}$, until a singular strategy, y^*, is reached at $s(y^*) = 0$. The singular strategy is locally "evolutionarily stable" (ES) if $\frac{ds}{dy}\big|_{y=y^*} < 0$ and is "convergence stable" (CS) if $s(y) < 0$ for $y = y^* + \epsilon$ and $s(y) > 0$ for $y = y^* - \epsilon$ for sufficiently small $\epsilon > 0$. ES implies that a singular strategy is a local fitness maximum and CS implies that the strategy is locally attracting (i.e., it can be reached by recurrent small mutations). We evaluate whether y^* is ES and CS, in which case it is a "continuously stable strategy" (CSS). If y^* is CS but not ES, then the singular strategy is a branching point (BR), which indicates that disruptive selection will occur leading to a protected dimorphism. If y^* is neither CS nor ES, then the singular strategy is a repeller

(RE), which may lead to bistability (i.e., the outcome depends on the initial conditions). If a repeller is the only singular strategy, then $y = 0$ and $y = 1$ are both locally attracting. Global minimisation (MN) occurs when $s(y) < 0$ for all $y > 0$, and global maximization (MX) occurs when $s(y) > 0$ for all $y < 1$. Finally, the singular strategy is referred to as a "Garden of Eden" when y^* is ES but is not CS (the singular strategy is evolutionarily stable but is unattainable through small mutations).

The above method assumes a separation of ecological and evolutionary timescales (mutations are rare) and that selection is weak (mutations have a small effect). We relax these assumptions in our simulations, which allow mutations to occur when the system is not close to its dynamical attractor (simulation code in the online Supporting Information). Starting with a single resident trait, y_r, we solve the ODE system for a given time period $[0, T]$ ($T = 100$), then introduce a mutant, $y_m = y_r \pm \epsilon_1$ (mutation size $\epsilon_1 = 0.02$), at low frequency. We then rerun the ODE solver over the period $[T, 2T]$ and remove any strains that have fallen below a frequency of $\epsilon_2 = 10^{-3}$. If more than one trait is still present in the population, then the next mutant is chosen based on a weighted probability of the trait frequencies. The process is repeated for $n = 2000$ iterations.

Results

IMPACT OF HOST PROTECTION ON THE ECOLOGICAL DYNAMICS

We begin by examining how host protection affects the ecological dynamics by analyzing the basic reproductive ratios, $R_0(i, j)$, which give the average number of secondary infections for parasite j when rare given that parasite i is already at equilibrium (Choisy & de Roode 2010). The equations for $R_0(i, j)$ are (see Supporting Information):

$$R_0(2, 1) = \frac{\beta_1(y)\left(\beta_2(y) I_2^*\left(S^* + I_2^*\right) + I_2^*\left(\Gamma_1(y) + \gamma_2\right) + \Gamma_{12}(y) S^*\right)}{\beta_2(y) I_2^*\left(\Gamma_{12}(y) - \gamma_2\right) + \Gamma_1(y) \Gamma_{12}(y)}$$

(4a)

$$R_0(1, 2) = \frac{\tilde{\beta}_2 S^*\left(\beta_1(y) I_1^* + 1\right) + \beta_2(y) I_1^*\left(\beta_1(y) I_1^* + \Gamma_2 + \gamma_1\right)}{\Gamma_{12}(y)\left(\beta_1(y) I_1^*\left(\Gamma_{12}(y) - \gamma_1\right) + \Gamma_2 \Gamma_{12}(y)\right)}$$

(4b)

When the other parasite is not present equations 4A–B reduce to $R_0(1) = \frac{\beta_1(y)S^*}{\Gamma_1(y)}$ and $R_0(2) = \frac{\tilde{\beta}_2 S^*}{\Gamma_2}$, respectively. The parasites coexist at a stable endemic equilibrium provided both $R_0(i, j) > 1$, but if $R_0(i, j) < 1$ for one parasite then it will be excluded. In general, tolerance increases $R_0(1, 2)$ and the prevalence of parasite 2 (Fig. 1A and B), as is the case with single parasite systems (Boots et al. 2009). From the perspective of the host, the benefits of parasite-conferred tolerance are likely to be rather limited, as increased survival at the individual level leads to

increased disease prevalence at the population level; the net effect may therefore be negative for the host (Fig. 1C). For parasite-conferred resistance, both $R_0(1, 2)$ and the prevalence of parasite 2 initially decline as host protection increases, but if host protection is costly then the prevalence of parasite 1 will eventually fall, causing a resurgence for parasite 2 (Fig. 1A and B). This means that stronger resistance can increase the prevalence of parasite 2, although such a situation is unlikely to be evolutionarily stable. Parasite-conferred resistance can be extremely beneficial for the host, leading to a marked increase in host density at equilibrium (Fig. 1C).

PARASITE FITNESS AND SELECTION GRADIENT

Using the next-generation method (Hurford et al. 2010), we derive the following expression which is sign equivalent to the invasion fitness of a rare mutant, y_m (see Supporting Information):

$$w(y_m) = \frac{\beta_1(y_m) A(y_m)}{B(y_m)} - 1$$

(5)

where $A(y_m) = S^*[\Gamma_{12}(y_m) + \lambda_2^*(y_m)] + I_2^*[\Gamma_1(y_m) + \gamma_2 + \lambda_2^*(y_m)]$ and $B(y_m) = \Gamma_{12}(y_m)[\Gamma_1(y_m) + \lambda_2^*(y_m)] - \gamma_2 \lambda_2^*(y_m)$. The selection gradient, $s(y) = \frac{dw}{dy_m}\big|_{y_m=y}$, is:

$$s(y) = \frac{1}{B(y)}\left\{A(y)\left.\frac{d\beta_1}{dy_m}\right|_{y_m=y} + \beta_1(y)\left.\frac{dA}{dy_m}\right|_{y_m=y}\right.$$
$$\left. - \frac{\beta_1(y) A(y)}{B(y)}\left.\frac{dB}{dy_m}\right|_{y_m=y}\right\}$$

(6)

We solve the selection gradient and its derivative numerically to determine whether each singular strategy is ES and/or CS. We primarily consider the effects of the strength and shape of the trade-off (eq. (1)), along with the effects of host lifespan ($1/b$) and the virulence of parasite 1 ($\tilde{\alpha}_1$). We focus on transmission rate costs in the main text and virulence costs in the Supporting Information. The Supporting Information also contains the results for model B, which are broadly consistent with those presented here.

EVOLUTION OF PARASITE-CONFERRED RESISTANCE

Assuming parasite 1 initially confers no protection to the host, resistance can only evolve by small mutations when the trade-off accelerates ($c_2 > 0$), or when the trade-off decelerates and the cost is small ($c_1 \ll 1, c_2 < 0$). The qualitative outcome is most sensitive to the shape of the trade-off (c_2), and there are five regions of the trade-off space that are common (Figs. 2 and 3). First, the parasite may always experience selection against host protection (minimisation). This occurs for moderate to high costs over a fairly broad range of intermediate trade-off shapes. Second, the parasite may evolve to an intermediate level of host protection (CSS) when costs accelerate ($c_2 > 0$). Third, a repeller may cause

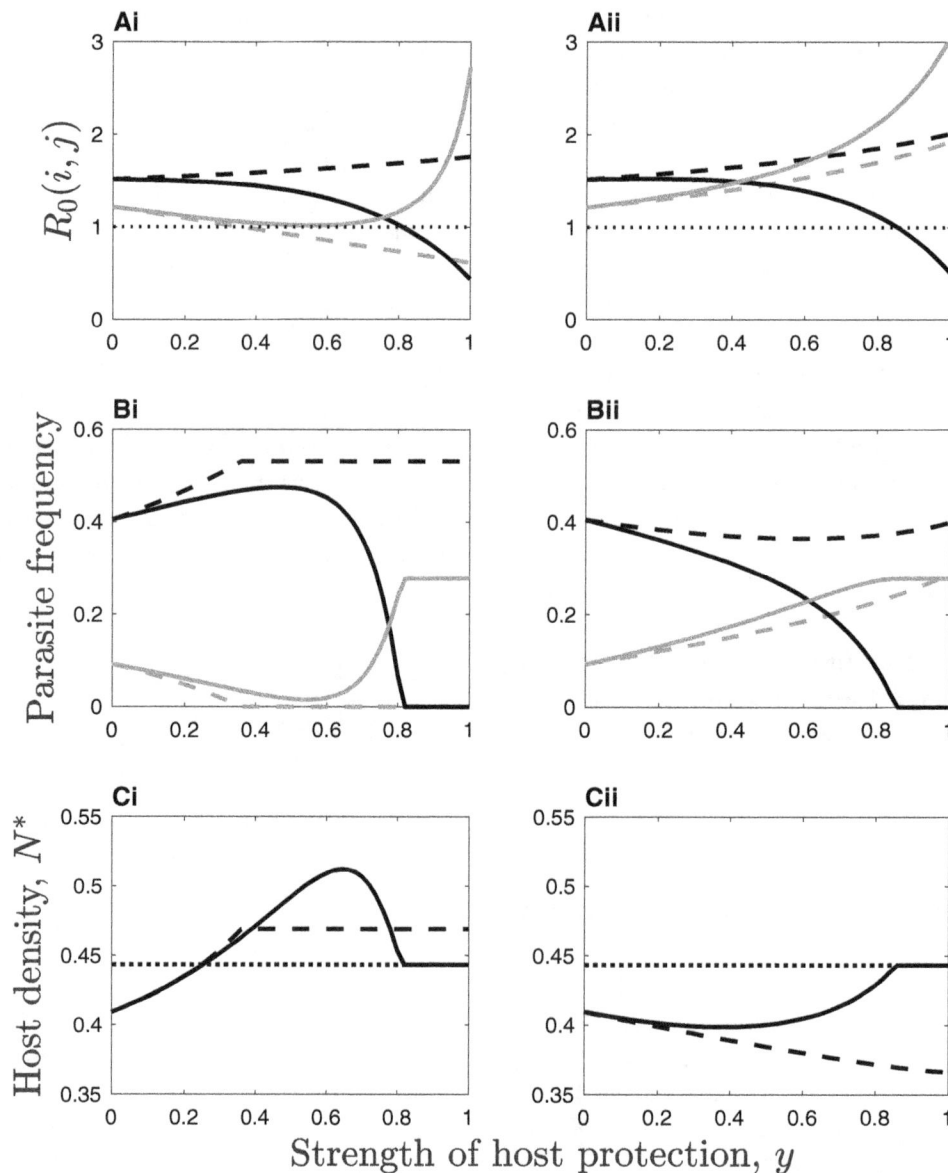

Figure 1. Impact of parasite-conferred resistance (left) and tolerance (right) on the ecological dynamics. Dashed lines correspond to $c_1 = 0$ (no costs) and solid lines to accelerating transmission rate costs ($c_1 = 0.75$, $c_2 = 3$). (Ai–ii) The basic reproductive ratio, $R_0(i, j)$, of parasite $j = 1, 2$ (black and gray, respectively) when parasite $i = 2, 1$ is at equilibrium (the dotted line shows the exclusion threshold). (Bi–ii) Parasite frequency at equilibrium. (Ci–ii) Host density at equilibrium (the protective parasite is a net mutualist when host density is above the dotted line). Parameters: $a = 1$, $b = 0.5$, $q = 0.5$, $\tilde{\alpha}_1 = 0.5$, $\tilde{\alpha}_2 = 1$, $\tilde{\beta}_1 = 5$, $\tilde{\beta}_2 = 5$, $\gamma_1 = 0.1$, $\gamma_2 = 0.1$.

bistablility so that the parasite evolves to either minimize or maximize host protection depending on the initial conditions ($y = 0$ and $y = 1$ are locally attracting). This outcome generally occurs when costs decelerate ($c_2 < 0$) and are relatively large ($c_1 \gg 0$). Fourth, the parasite may branch into two strategies through disruptive selection, eventually leading to a stable dimorphism with $y_1^* = 0$ and $y_2^* = 1$. Branching occurs when costs decelerate and are relatively low in magnitude ($c_1 \ll 1$). Finally, there may be two singular strategies: a repeller and a branching point. In all cases we found that the repeller was located below the branching point. Hence, the parasite may either minimize y ($y = 0$ is a

local attractor) or branch into two diverging strategies depending on the initial conditions. This outcome occurs for intermediate decelerating costs.

Increasing the lifespan of the host (decreasing b) and reducing the virulence of parasite 1 (decreasing $\tilde{\alpha}_1$) generally increases the size of the branching regions and makes minimisation and bistability less likely. However, for sufficiently low b and $\tilde{\alpha}_1$ we found more complex outcomes for intermediate costs that are weakly accelerating, consisting of a repeller and either one or two CSSs, or a CSS and a branching point (Fig. 2A). These regions are mostly similar to the RE and RE + BR regions described

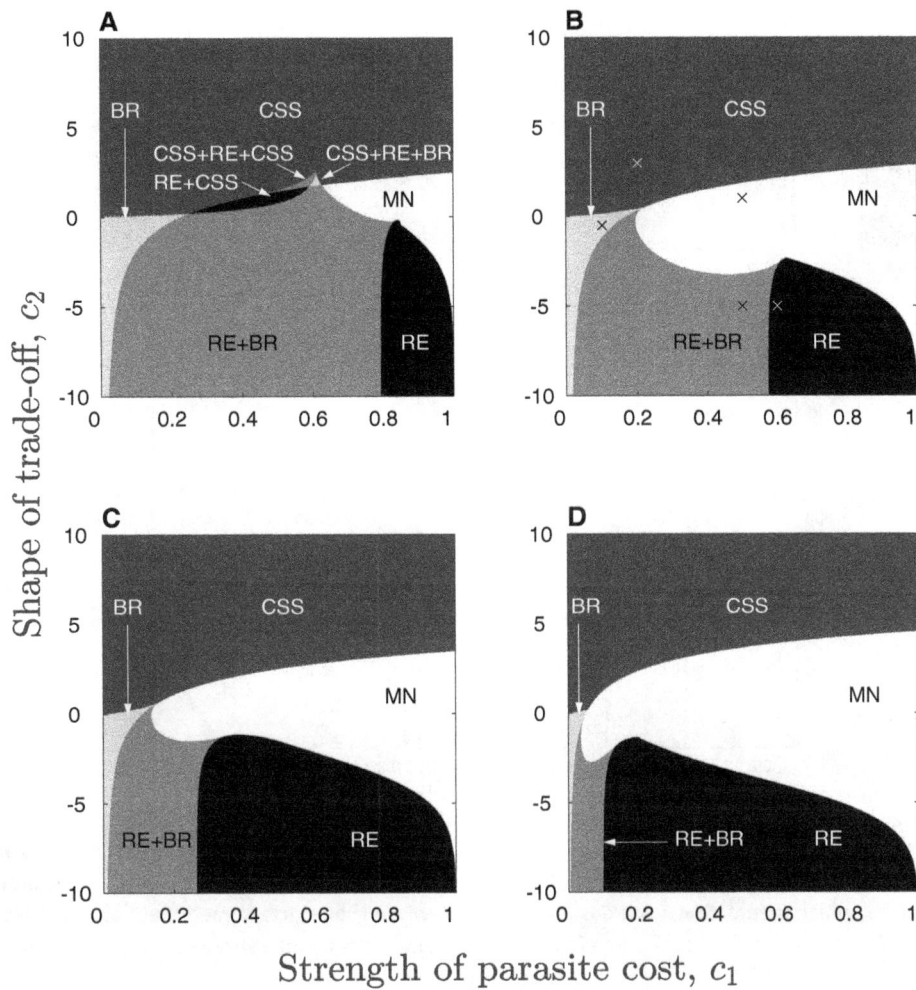

Figure 2. Evolution of parasite-conferred resistance when there is a transmission rate cost. Higher values of c_1 correspond to greater costs, and higher (lower) values of c_2 correspond to more strongly accelerating (decelerating) costs (eq. (1)). Qualitative outcomes: minimisation (MN); intermediate continuously stable strategy (CSS); repeller (RE); and evolutionary branching (BR). The natural mortality rate, b, increases from 0.05 (left column) to 0.5 (right column). The virulence of parasite 1, $\tilde{\alpha}_1$, increases from 0.1 (top row) to 1 (bottom row). Crosses in panel B correspond to Fig. 3. Remaining parameters as described in Fig. 1.

above, with the exception that $y = 0$ and $y = 1$ are no longer local attractors. We verified the numerical analysis of the model with simulations and found them to closely match the numerical results (Fig. 3).

EVOLUTION OF PARASITE-CONFERRED TOLERANCE

As with resistance, the qualitative outcome for tolerance is most sensitive to the shape of the trade-off (Fig. 4), and tolerance can only evolve by small mutations when the trade-off accelerates ($c_2 > 0$), or when the trade-off decelerates and the cost of protection is small ($c_1 \ll 1, c_2 < 0$). However, there are some notable differences between the two scenarios. When parasite 1 confers tolerance there are four main regions of the cost space describing different evolutionary outcomes (Fig. 4). First, the parasite always experiences selection against host protection (minimisation) when costs are moderate to high (over a broad range of interme-

diate trade-off shapes). Second, selection always favors greater host protection (maximization) when costs are low to moderate in magnitude, regardless of whether the trade-off accelerates or decelerates. Third, the parasite may evolve an intermediate level of host protection (CSS) for moderate to high accelerating costs. Fourth, the system may exhibit bistability due to a repeller. Bistability usually occurs for intermediate decelerating costs, although the region of bistability shrinks as the shape of the trade-off tends toward being linear ($c_2 \rightarrow 0$). A small region of the cost space exists near the intersection of these main regions corresponding to a Garden of Eden scenario (with or without a CSS). This means that the singular strategy is evolutionarily stable but is unattainable through small mutations, and so in reality it is likely to behave as a repeller (Fig. 5).

These general relationships are consistent as host lifespan and the virulence of parasite 1 are varied, although maximization

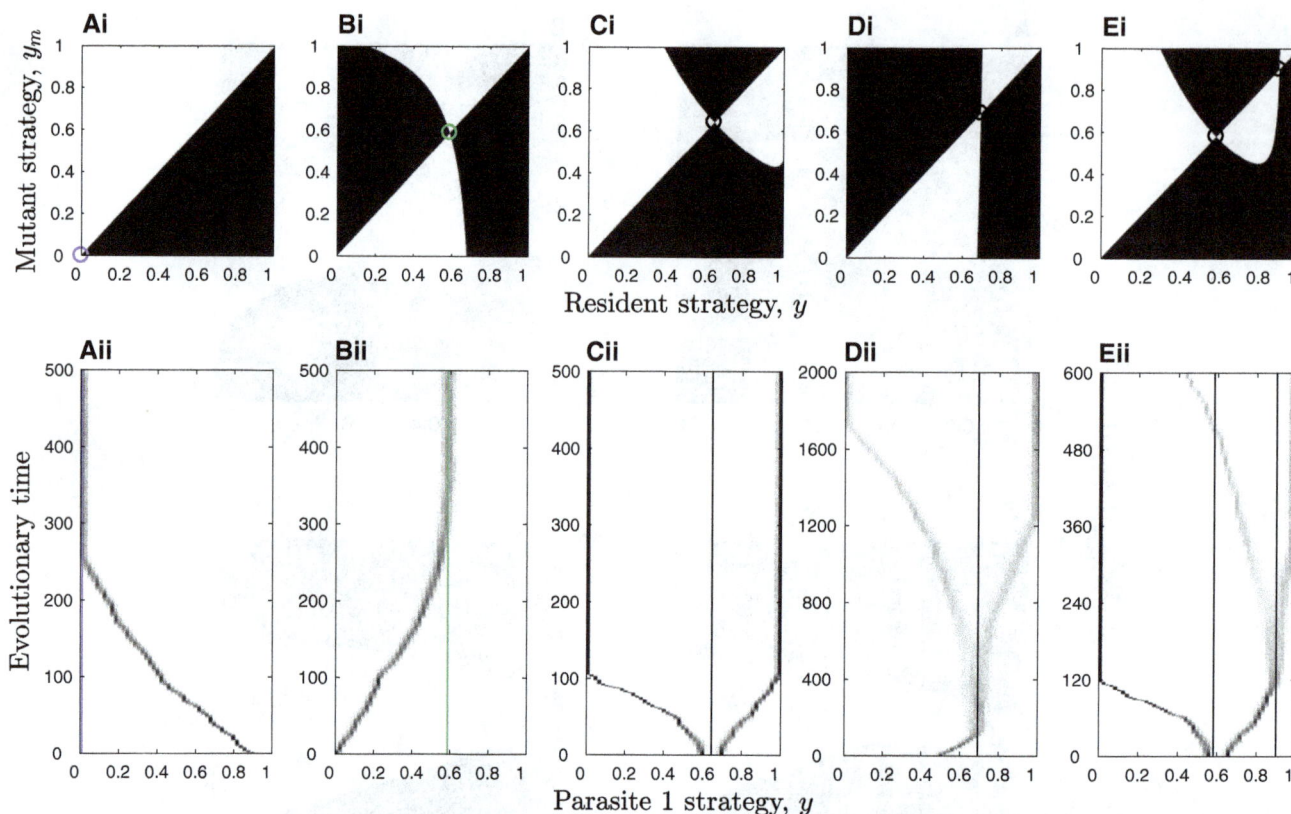

Figure 3. Pairwise invasion plots (PIPs; top) and simulations (bottom) for the points in Fig. 2B: (A) minimisation (purple); (B) CSS (green); (C) repeller (red); (D) evolutionary branching (blue); (E) repeller (red) and evolutionary branching (blue). The mutant can only invade in the black regions of the PIPs, which means that y increases (decreases) when the region immediately above (below) the line $y = y_m$ is black and the region immediately below (above) this line is white. Note that plots C and E show two separate simulations with different initial conditions either side of the repeller. Same parameters as Fig. 1, with $\tilde{\alpha}_1 = 0.1$.

tends to become more likely as b and $\tilde{\alpha}_1$ decrease. We did not find any evidence of evolutionary branching when the parasite confers tolerance, which by contrast is relatively common in the case of resistance. Again, simulations were found to closely match the numerical results (similar to Fig. 3, omitted for brevity).

Discussion

Interactions between coinfecting parasites are likely to play a crucial role in shaping the ecological and evolutionary dynamics of infectious diseases (Read & Taylor 2001; Brown et al. 2002; Alizon 2013; Johnson et al. 2015). A large body of theory has primarily focused on how competitive or cooperative strategies to exploit host resources affect the evolution of virulence in mixed infections (Bremermann & Pickering 1983; Sasaki & Iwasa 1991; Frank 1992, 1996; van Baalen & Sabelis 1995; Alizon & van Baalen 2008; Choisy & de Roode 2010; Alizon & Lion 2011). The aim of our study was to understand the extent to which parasite inter- and intraspecies interactions drive the evolution of host protection, a widely observed phenomenon (Michalakis et al. 1992; van Baalen & Jansen 2001; Ford & King 2016). Our study

was therefore more closely related to theoretical models of spite (Gardner et al. 2004) and an existing model of host protection by vertically transmitted parasites (Jones et al. 2011).

We explored how host protection evolves subject to a wide range of trade-offs. Our study has two key results. First, host protection can evolve for many types of trade-off, but the qualitative outcome depends on the mechanism of protection and the precise nature of the trade-off. For example, evolutionary branching—leading to a stable dimorphism—only appears to occur for parasite-conferred resistance, not for tolerance. This is likely due to the positive frequency dependence that is typically associated with tolerance mechanisms, which leads to an increase in the prevalence of the targeted parasite and tends to prevent branching (Roy & Kirchner 2000; Boots et al. 2009). In general, host protection is most likely to evolve if the trade-off accelerates, or if the trade-off decelerates and the cost of protection is relatively low. The qualitative outcome is more sensitive to the shape of the trade-off rather than the magnitude of the cost, with accelerating trade-offs generally selecting for a CSS, whereas decelerating trade-offs tend to produce evolutionarily unstable strategies, leading to either bistability or branching.

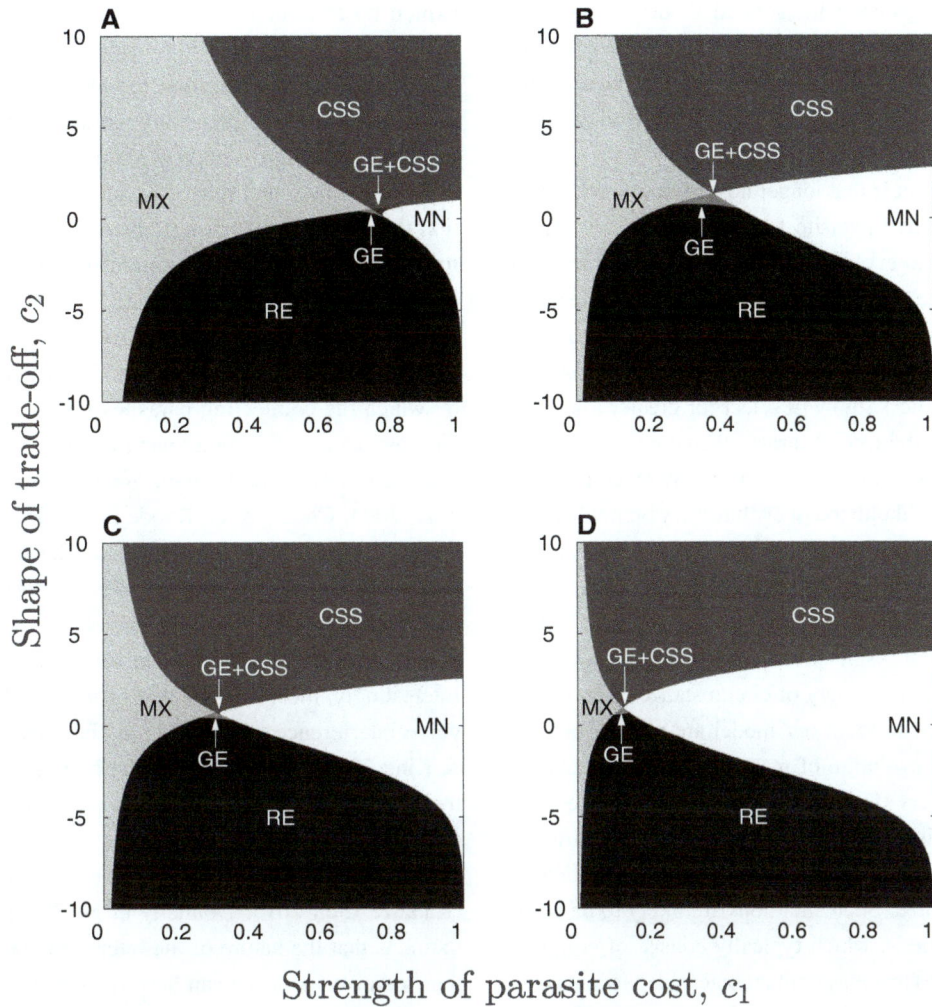

Figure 4. Evolution of parasite-conferred tolerance when host protection is associated with a transmission rate cost. In addition to most of the singular strategies described in Fig. 2 for parasite-conferred resistance, we also find: maximization (MX) for nonzero costs and Garden of Eden (GE) with or without a CSS. The natural mortality rate, b, increases from 0.05 in the plots on the left to 0.5 on the right. The virulence of parasite 1, $\tilde{\alpha}_1$, increases from 0.1 in the top row to 1 in the bottom row. The cost function and the remaining labels and parameters are described in Fig. 2.

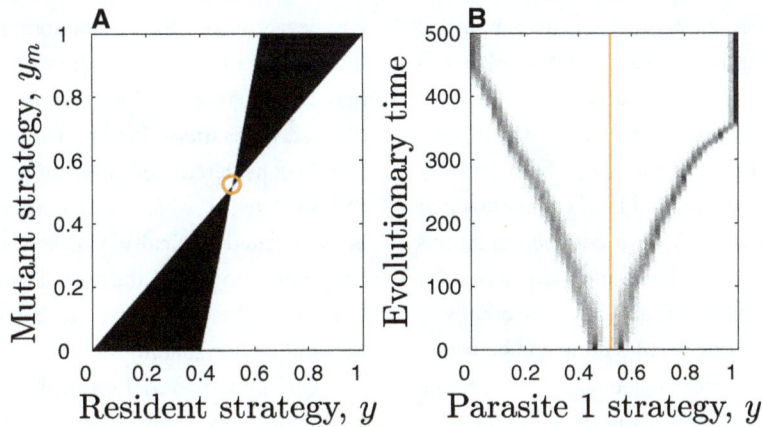

Figure 5. PIP (A) and simulation (B) for the Garden of Eden (GE; orange) outcome. The GE is evolutionarily stable but is not convergence stable, and hence it cannot be approached by small mutations (the mutant can only invade the resident in the black regions of the PIP). In reality, the GE will generally behave as a repeller, as shown in the two evolutionary trajectories in panel B. Parameters as described in Fig. 1, except $c_1 = 0.37$, $c_2 = 1$, $\tilde{\alpha}_1 = 0.1$, and $\delta = 0$.

These patterns are consistent with general theory in adaptive dynamics, which shows that strongly accelerating trade-offs produce CSSs and strongly decelerating trade-offs produce evolutionary repellers (Mazancourt & Dieckmann 2004; Bowers et al. 2005).

Our second key result is that longer host lifespans and lower virulence of the protective parasite tend to increase the range of conditions that lead to evolutionary branching (resistance) or maximization (tolerance). In both cases, there is an increase in the average infectious period and hence in the likelihood of coinfections. It is easy to see that reducing the background mortality (b) or virulence from parasite 1 ($\tilde{\alpha}_1$) will select for greater tolerance because the virulence of the second parasite then dominates the infectious period for coinfections. It is less clear why reducing these parameters increases the likelihood of evolutionary branching, but this pattern is consistent with a previous study of host–parasite range coevolution which showed that branching is more common as host lifespan (and hence the infectious period) increases (Best et al. 2010). Together, our results predict that host protection can readily evolve under a wide variety of circumstances. Moreover, the broad patterns we observe in our model are consistent with previous theory on the evolution of resistance and tolerance by the host (Boots & Bowers 1999; Boots et al. 2009). The key difference here, however, is that defense is conferred by the parasite, and thus is obtained from the environment dynamically rather than being genetically inherited. Such situations are likely to be common in natural populations, which typically consist of complex communities of parasites that may confer context-dependent costs and benefits to their hosts (Michalakis et al. 1992; van Baalen & Jansen 2001; Betts et al. 2016; Ford & King 2016).

Our study builds on previous models of coinfecting parasites, in particular the work of Choisy and de Roode (2010) (model A) and Alizon (2013) (model B). A crucial difference between the two models is that different strains of the same species are able to coinfect the same host in model B. Still, we found that our results were remarkably similar across the two frameworks (Fig. S1). In model B, we assumed that defense is specific to parasite 2, but if defense is more general (e.g., a priority effect), then other strains of parasite 1 are also likely to be negatively impacted. We also assumed that the overall level of resistance or tolerance was equal to the mean of the two coinfecting strains of parasite 1, but it is possible that the results may differ for other functional forms. A more realistic (but much more complex) approach would be to use a nested model of within- and between-host dynamics to fully account for the dynamics of coinfecting strains (Mideo et al. 2008). Future theory should examine whether the evolution of parasite-conferred resistance and tolerance is affected by within-host dynamics.

The biological arguments underlying our results are fairly intuitive. Parasites should not only evolve optimal strategies to exploit host resources, but should also evolve strategies to cope with mixed infections. While many studies of parasite evolution have considered coinfections, the motivation of our study is different to most of the preceding work, which has focused almost exclusively on the evolution of virulence. In our model, virulence does not evolve, and thus it is not the degree of host exploitation that is under selection. Rather, it is the degree to which the focal parasite defends a common resource and the mechanism by which the resource is protected that is evolvable. In spite of this key difference, there are some conceptual similarities with the evolution of virulence theory. In particular, the mechanism by which the coinfecting parasites interact with each other and the host is crucial (Bremermann & Pickering 1983; van Baalen & Sabelis 1995; Frank 1996; West & Buckling 2003; Gardner et al. 2004; Choisy & de Roode 2010). Here, host protection is likely to have a negative impact on the nonprotective parasite if the mechanism in question leads to interference competition (e.g., resistance), but conversely may be beneficial if host protection extends the longevity of mixed infections (e.g., tolerance). Interestingly, most documented examples of host protection involve interference competition as the mechanism at play (Ford & King 2016). The context of the interaction between coinfecting parasites is clearly crucial for predicting the ecological and evolutionary outcomes in both cases. Our study is also related to recent work on the impact of superinfections on host evolution (Kada & Lion 2015; Donnelly et al. 2017). Again, a common theme is that the nature of the interaction between parasites and their relative virulence can have important consequences for the evolution of defense, regardless of whether this is intrinsic to the host or conferred by another species.

We are only aware of one other theoretical model of the evolution of host protection by another species, which concerned the resistance conferred by vertically transmitted symbionts against horizontally transmitted parasites (Jones et al. 2011). The studies are clearly linked by the common theme of host protection, although there are notable differences (e.g., in our model defense may take the form of either resistance or tolerance and the protective parasite is transmitted horizontally). In particular, Jones et al. (2011) considered the impact of parasitic castration on the level of host defense, which is crucial because the defensive parasite is transmitted vertically, and hence its reproduction is intrinsically linked to that of the host. The impact of parasitic castration is likely to be much lower in our model, as both parasites are transmitted horizontally.

Our study is closely linked to recent empirical work showing the de novo evolution of microbe-mediated protection during experimental evolution of a novel, tripartite interaction between a host, and two parasites (King et al. 2016). This work showed that mildly parasitic bacteria (*Enterococcus faecalis*) living in nematodes rapidly evolve to defend their animal hosts against infection by a more virulent pathogen (*S. aureus*). Driven by frequent antagonistic interactions with coinfecting *S. aureus*, *E. faecalis* evolve to increase production of superoxides. These act as antimicrobials, which actively suppress the virulence and within-host fitness of *S. aureus*. The evolved microbes also stay mildly parasitic during single infections, demonstrating the context-dependent nature of their beneficial effects.

The theory established in the present study adds to our general understanding of the complex ecoevolutionary relationships between hosts and coinfecting parasites, and specifically, to our understanding of evolution along the mutualism–parasitism continuum (Michalakis et al. 1992; van Baalen & Jansen 2001). For simplicity, we considered the evolution of either parasite-conferred resistance ($\delta = 1$) or tolerance ($\delta = 0$). However, some parasites may confer mixed modes of protection to their hosts ($0 < \delta < 1$), in which case it is likely that the level of investment in each mode of defense may evolve. An interesting extension of our work would therefore be to allow both the strength and level of investment in each mode of host protection to coevolve. We have addressed the question of how different mechanisms of host protection evolve when hosts and nonprotective parasites are evolutionarily static, but such a constraint will need to be relaxed in future theory to understand the coevolutionary dynamics of all parties. For example, selection for mechanisms that reduce virulence in mixed infections may simply lead to selection for greater virulence among coinfecting parasites. Similarly, hosts may invest less in their own defenses and may promote the growth of less virulent parasites that offer protection against more virulent parasites, thus accelerating the transition from parasitism to mutualism. However, the host will not promote the growth of a defensive parasite unless it provides a net benefit; in our model this is most likely when host protection occurs through resistance rather than tolerance due to ecological feedbacks that decrease (resistance) or increase (tolerance) the prevalence of another parasite (Fig. 1C). While the above scenarios seem plausible, the mathematical details will need to be worked out in future studies that account for coevolutionary interactions. Indeed, a greater theoretical understanding of mixed infections beyond the realm of virulence evolution is needed to support a growing body of empirical research, especially on microbe-mediated protection in animal and plant hosts.

AUTHOR CONTRIBUTIONS

Both authors conceived the study. BA designed and analyzed the model, and drafted the initial version of the manuscript. Both authors contributed to later versions of the manuscript.

ACKNOWLEDGMENTS

We thank R. Iritani for helpful comments on the manuscript. This work was supported by the Natural Environment Research Council (grant number NE/N014979/1). KCK acknowledges funding from the Leverhulme Trust (RPG-2015-165).

LITERATURE CITED

Aaron, L., D. Saadoun, I. Calatroni, O. Launay, N. Mémain, V. Vincent, G. Marchal, B. Dupont, O. Bouchaud, D. Valeyre, *et al.* 2004. Tuberculosis in HIV-infected patients: a comprehensive review. Clin. Microbiol. Infect. 10:388–398.

Alizon, S. 2013. Co-infection and super-infection models in evolutionary epidemiology. Interface Focus 3:20130031.

Alizon, S., and M. van Baalen. 2008. Multiple infections, immune dynamics, and the evolution of virulence. Am. Nat. 172:E150–E168.

Alizon, S., and S. Lion. 2011. Within-host parasite cooperation and the evolution of virulence. Proc. R. Soc. B Biol. Sci. 278:3738–3747.

Alizon, S., J. C. de Roode, and Y. Michalakis. 2013. Multiple infections and the evolution of virulence. Ecol. Lett. 16:556–567.

van Baalen, M., and V. A. A. Jansen. 2001. Dangerous liasons: the ecology of private interest and common good. Oikos 95:211–224.

van Baalen, M., and M. W. Sabelis. 1995. The dynamics of multiple infection and the evolution of virulence. Am. Nat. 146:881–910.

Balmer, O., and M. Tanner. 2011. Prevalence and implications of multiple-strain infections. Lancet Infect. Dis. 11:868–878.

Best, A., A. White, E. Kisdi, J. Antonovics, M. A. Brockhurst, and M. Boots. 2010. The evolution of host-parasite range. Am. Nat. 176:63–71.

Betts, A., C. Rafaluk, and K. C. King. 2016. Host and parasite evolution in a tangled bank. Trends Parasitol. 32:863–873.

Blagrove, M. S. C., C. Arias-Goeta, A.-B. Failloux, and S. P. Sinkins. 2012. Wolbachia strain wMel induces cytoplasmic incompatibility and blocks dengue transmission in Aedes albopictus. Proc. Natl. Acad. Sci. U. S. A. 109:255–260.

Boots, M., and R. G. Bowers. 1999. Three mechanisms of host resistance to microparasites—avoidance, recovery and tolerance—show different evolutionary dynamics. J. Theor. Biol. 201:13–23.

Boots, M., A. Best, M. R. Miller, and A. White. 2009. The role of ecological feedbacks in the evolution of host defence: what does theory tell us? Philos. Trans. R. Soc. Lond. B. Biol. Sci. 364:27–36.

Bowers, R. G., A. Hoyle, A. White, and M. Boots. 2005. The geometric theory of adaptive evolution: trade-off and invasion plots. J. Theor. Biol. 233:363–377.

Bremermann, H. J., and J. Pickering. 1983. A game-theoretical model of parasite virulence. J. Theor. Biol. 100:411–426.

Brown, S. P., M. E. Hochberg, and B. T. Grenfell. 2002. Does multiple infection select for raised virulence? Trends Microbiol. 10:401–405.

Brown, S. P., S. A. West, S. P. Diggle, and A. S. Griffin. 2009. Social evolution in micro-organisms and a Trojan horse approach to medical intervention strategies. Philos. Trans. R. Soc. London B. 364:3157–3168.

Chao, L., K. A. Hanley, C. L. Burch, C. Dahlberg, and P. E. Turner. 2000. Kin selection and parasite evolution: higher and lower virulence with hard and soft selection. Q. Rev. Biol. 75:261–275.

Choisy, M., and J. C. de Roode 2010. Mixed infections and the evolution of virulence: effects of resource competition, parasite plasticity, and impaired host immunity. Am. Nat. 175:E105–E118.

Cox, F. E. 2001. Concomitant infections, parasites and immune responses. Parasitology 122 Suppl:S23–S38.

Dieckmann, U., and R. Law. 1996. The dynamical theory of coevolution: a derivation from stochastic ecological processes. J. Math. Biol. 34:579–612.

Donnelly, R., A. White, and M. Boots. 2017. Host lifespan and the evolution of resistance to multiple parasites. J. Evol. Biol. 30:561–570.

Ford, S. A., and K. C. King. 2016. Harnessing the power of defensive microbes: evolutionary implications in nature and disease control. PLoS Pathog. 12:e1005465.

Frank, S. A. 1992. A kin selection model for the evolution of virulence. Proc. Biol. Sci. 250:195–197.

———. 1994. Kin selection and virulence in the evolution of protocells and parasites. Proc. R. Soc. London B 258:153–161.

———. 1996. Models of parasite virulence. Q. Rev. Biol. 71:37–78.

Gardner, A., S. A. West, and A. Buckling. 2004. Bacteriocins, spite and virulence. Proc. R. Soc. B. 271:1529–1535.

Geritz, S. A. H., E. Kisdi, G. Meszena, and J. A. J. Metz. 1998. Evolutionarily singular strategies and the adaptive growth and branching of the evolutionary tree. Evol. Ecol. 12:35–37.

Griffiths, E. C., A. B. Pedersen, A. Fenton, and O. L. Petchey. 2011. The nature and consequences of coinfection in humans. J. Infect. 63:200–206.

Hamilton, W. D. 1972. Altruism and related phenomena. Annu. Rev. Ecol. Syst. 3:193–232.

Hughes, G. L., R. Koga, P. Xue, T. Fukatsu, and J. L. Rasgon 2011. *Wolbachia* infections are virulent and inhibit the human malaria parasite *Plasmodium falciparum* in *Anopheles gambiae*. PLoS Pathog. 7:e1002043.

Hurford, A., D. Cownden, and T. Day 2010. Next-generation tools for evolutionary invasion analyses. J. R. Soc. Interface 7:561–571.

Inglis, R. F., A. Gardner, P. Cornelis, and A. Buckling 2009. Spite and virulence in the bacterium. Pseudomonas aeruginosa Proc. Natl. Acad. Sci. U. S. A. 106:5703–5707.

Johnson, P. T. J., J. C. de Roode, and A. Fenton. 2015. Why infectious disease research needs community ecology. Science 349:1259504.

Jones, E. O., A. White, and M. Boots. 2011. The evolution of host protection by vertically transmitted parasites. Proc. Biol. Sci. 278:863–870.

Kada, S., and S. Lion. 2015. Superinfection and the coevolution of parasite virulence and host recovery. J. Evol. Biol. 28:2285–2299.

King, K. C., M. A. Brockhurst, O. Vasieva, S. Paterson, A. Betts, S. A. Ford, C. L. Frost, M. J. Horsburgh, S. Haldenby, and G. D. Hurst. 2016. Rapid evolution of microbe-mediated protection against pathogens in a worm host. ISME J. 10:1915–1924.

Lively, C. M., K. Clay, M. J. Wade, and C. Fuqua. 2005. Competitive coexistence of vertically and horizontally transmitted parasites. Evol. Ecol. Res. 7:1183–1190.

Martinez, J., S. Ok, S. Smith, K. Snoeck, J. P. Day, and F. M. Jiggins. 2015. Should symbionts be nice or selfish? Antiviral effects of Wolbachia are costly but reproductive parasitism is not. PLoS Pathog. 11:1–20.

Massey, R. C., A. Buckling, and R. ffrench-Constant. 2004. Interference competition and parasite virulence. Proc. R. Soc. B. 271:785–788.

de Mazancourt, C., and U. Dieckmann. 2004. Trade-off geometries and frequency-dependent selection. Am. Nat. 164:765–778.

Metz, J. A., R. M. Nisbet, and S. A. Geritz 1992. How should we define "fitness" for general ecological scenarios? Trends Ecol. Evol. 7:198–202.

Michalakis, Y., I. Olivieri, F. Renaud, and M. Raymond 1992. Pleiotropic action of parasites: how to be good for the host. Trends Ecol. Evol. 7:59–62.

Mideo, N., S. Alizon, and T. Day 2008. Linking within- and between-host dynamics in the evolutionary epidemiology of infectious diseases. Trends Ecol. Evol. 23:511–517.

Petney, T. N., and R. H. Andrews 1998. Multiparasite communities in animals and humans: frequency, structure and pathogenic significance. Int. J. Parasitol. 28:377–393.

Polin, S., J. C. Simon, and Y. Outreman 2014. An ecological cost associated with protective symbionts of aphids. Ecol. Evol. 4:826–830.

Read, A. F., and L. H. Taylor 2001. The ecology of genetically diverse infections. Science 292:1099–1102.

Roy, B. A., and J. W. Kirchner 2000. Evolutionary dynamics of pathogen resistance and tolerance. Evolution 54:51–63.

Sasaki, A., and Y. Iwasa 1991. Optimal growth schedule of pathogens within a host: switching between lytic and latent cycles. Theor. Popul. Biol. 39:201–239.

Selva, L., D. Viana, G. Regev-Yochay, K. Trzcinski, J. M. Corpa, I. Lasa, R. P. Novick, and J. R. Penadés. 2009. Killing niche competitors by remote-control bacteriophage induction. Proc. Natl. Acad. Sci. U. S. A. 106:1234–1238.

Sternberg, E. D., T. Lefèvre, A. H. Rawstern, and J. C. de Roode 2011. A virulent parasite can provide protection against a lethal parasitoid. Infect. Genet. Evol. 11:399–406.

Telfer, S., X. Lambin, R. Birtles, P. Beldomenico, S. Burthe, S. Paterson, and M. Begon. 2010. Species interactions in a parasite community drive infection risk in a wildlife population. Science 330:243–246.

Vorburger, C., and A. Gouskov 2011. Only helpful when required: a longevity cost of harbouring defensive symbionts. J. Evol. Biol. 24:1611–1617.

West, S. A., and A. Buckling 2003. Cooperation, virulence and siderophore production in bacterial parasites. Proc. R. Soc. B 270:37–44.

10

Fire ant social chromosomes: Differences in number, sequence and expression of odorant binding proteins

Rodrigo Pracana,[1,*] ![iD] Ilya Levantis,[1,*] ![iD] Carlos Martínez-Ruiz,[1] Eckart Stolle,[1] ![iD] Anurag Priyam,[1] ![iD] and Yannick Wurm[1,2] ![iD]

[1] School of Biological and Chemical Sciences, Queen Mary University of London, E1 4NS London, United Kingdom

[2] E-mail: y.wurm@qmul.ac.uk

Variation in social behavior is common yet our knowledge of the mechanisms underpinning its evolution is limited. The fire ant *Solenopsis invicta* provides a textbook example of a Mendelian element controlling social organization: alternate alleles of a genetic element first identified as encoding an odorant binding protein (OBP) named *Gp-9* determine whether a colony accepts one or multiple queens. The potential roles of such a protein in perceiving olfactory cues and evidence of positive selection on its amino acid sequence made it an appealing candidate gene. However, we recently showed that recombination is suppressed between *Gp-9* and hundreds of other genes as part of a >19 Mb supergene-like region carried by a pair of social chromosomes. This finding raises the need to reassess the potential role of *Gp-9*. We identify 23 OBPs in the fire ant genome assembly, including nine located in the region of suppressed recombination with *Gp-9*. For six of these, the alleles carried by the two variants of the supergene-like region differ in protein-coding sequence and thus likely in function, with *Gp-9* showing the strongest evidence of positive selection. We identify an additional OBP specific to the Sb variant of the region. Finally, we find that 14 OBPs are differentially expressed between single- and multiple-queen colonies. These results are consistent with multiple OBPs playing a role in determining social structure.

KEY WORDS: Green beard, olfaction, pheromone, social behavior, social chromosome, *Solenopsis invicta*, supergene.

Impact Summary

The invasive red fire ant provides a unique opportunity to investigate how changes in social behavior can evolve. In this species, two distinct forms of social organization coexist: colonies either have strictly one queen or up to dozens of reproductive queens.

Pioneering work in the early 2000s demonstrated that this social dimorphism has a genetic basis: one of the two alleles (versions) of the gene *Gp-9* is always and exclusively present in multiple-queen colonies but never in single-queen colonies. *Gp-9* encodes an odorant binding protein (OBP), a type of protein that can be involved in the production and in the perception of pheromones. The two alleles of *Gp-9* differ in amino acid sequence, and thus likely also in function. Furthermore, workers with the *b* allele of *Gp-9* behave spitefully toward queens that lack the *b* allele—in this sense the *b* allele appears to be a selfish gene with a so-called "green beard" effect.

However, it was recently shown that the two alleles of *Gp-9* are part of a pair of "social chromosomes." Like a pair of sex chromosomes, the two social chromosomes differ from each other, so that hundreds of genes in addition to *Gp-9* have alleles present exclusively in multiple-queen colonies. This challenges the idea that *Gp-9* is pivotal in determining social organization.

Here, to determine whether there is still reason to believe that OBPs play roles in this system, we characterize all fire ant OBPs. We find that fire ants have 24 OBPs, 10 of which are in the social chromosome. Two of the social chromosome

*These authors are the joint first authors.

OBPs are exclusive to alternate social chromosome variants, and four OBPs in addition to *Gp-9* have differences in amino acid sequence between the variants. We also find differences in the activity levels of 14 OBP genes between single- and multiple-queen colonies.

In sum, our study provides evidence that multiple OBPs may be responsible for differences between single- and multiple-queen colonies. These results represent a significant step toward understanding the mechanisms by which the social chromosomes function.

Introduction

Variation in social behavior is common yet our knowledge of the mechanisms underpinning its evolution is limited (Robinson et al. 2005; Johnson and Linksvayer 2010). The fire ant *Solenopsis invicta* provides a rare, textbook example of variation in a fundamental social trait: some colonies have one queen, whereas others have up to dozens of queens. Queens that will form their own single-queen colony typically disperse over greater distances and can effectively colonize newly available habitats. In contrast, multiple-queen colonies can outcompete single-queen colonies in saturated habitats and harsh environments, and can split by fission (Bourke and Heinze 1994; Ross and Keller 1995; Tschinkel 2006). Multiple additional traits differ between the two social forms, including in queen fecundity, colony size, worker size distribution, and worker aggressiveness (Ross and Keller 1995; DeHeer et al. 1999; Keller and Ross 1999; Goodisman et al. 2000; DeHeer 2002; Buechel et al. 2014; Huang et al. 2014).

A series of landmark studies (Ross 1997; Keller and Ross 1998; Ross and Keller 1998) demonstrated that the two social forms are under the control of a Mendelian element. This element was first identified in a screen of electrophoretic markers as a polymorphic protein coding gene, *Gp-9*, with two alleles: *Gp-9B* and *Gp-9b* (Ross 1997). If a colony includes only *Gp-9 BB* workers, they will accept a single *Gp-9 BB* queen and execute any additional queens. In contrast, if more than ~20% of the workers in a colony are *Gp-9 Bb* heterozygotes, they will execute reproductively active *Gp-9 BB* queens but accept dozens of *Gp-9 Bb* queens (Ross 1997; Keller and Ross 1998, 1999; Ross and Keller 1998, 2002; DeHeer et al. 1999; Gotzek & Ross 2007). In contrast, *Gp-9 bb* queens die before becoming reproductively active (Ross 1997; DeHeer et al. 1999; Keller and Ross 1999; Gotzek and Ross 2007; Trible and Ross 2016). The workers discriminate between queens of alternate genotypes based on olfactory cues (Keller and Ross 1998; Ross and Keller 1998, 2002), such as differences in the queens' cuticular hydrocarbon profiles (Eliyahu et al. 2011; Trible and Ross 2016). Because workers carrying the *Gp-9b* allele recognize whether queens also carry this allele

and execute those that do not, this system represents a rare example of a "green beard gene" (Keller and Ross 1998), named after a theoretical model of a behavioral selfish genetic element (West and Gardner 2010).

In another landmark study, Krieger and Ross (2002) demonstrated that *Gp-9* encodes an odorant binding protein (OBP). OBPs are essential components of insect communication systems: they bind and transport pheromones and other semiochemicals, generally mediating their perception and sometimes their secretion (Pelosi et al. 2006, 2014; Leal 2013). Furthermore, tests of historical selection on *Gp-9* reveal a significant excess of nonsynonymous (amino acid replacing) substitutions relative to synonymous (silent) substitutions between the lineage of *Gp-9* b-like alleles and *Gp-9* B-like alleles in the fire ant and its relatives. This implies that directional or diversifying selection has driven the molecular evolution of *Gp-9*, and is associated with differentiation between the two forms of social organization in these ants (Krieger and Ross 2002, 2005). Several models lay out the potential function of *Gp-9*, generally involving differential production or perception of pheromones in queens as well as workers of alternate genotypes (Krieger 2004; Gotzek and Ross 2007, 2009).

However, recent genome-wide analyses of the social dimorphism revealed that the association between genotype and form of social organization is not limited to *Gp-9* (Wang et al. 2013). Instead, genetic maps obtained using Restriction site Associated DNA (RAD) markers from crosses in seven families showed that this association extends over a large chromosomal region of suppressed recombination. The two variants of this region, respectively, marked by the *Gp-9B* and *Gp-9b* alleles are carried by a pair of "social chromosomes" named SB and Sb. The region is genetically differentiated over 10.8 Mb (55%) of the mapped assembly of the social chromosomes, although its total length could be 19.4–31.5 Mb given the estimated size of the nonassembled portion of the genome (Pracana et al. 2017). Based on the current NCBI gene set, this region contains at least 443 protein coding genes, including *Gp-9*. The two chromosomes differ by at least one large inversion affecting a large portion of the region and an additional small (48 kb) inversion. The region of suppressed recombination can be described as a supergene, a locus containing multiple genes with tightly linked allelic combinations that control a complex polymorphic phenotype (Linksvayer et al. 2013; Schwander et al. 2014; Thompson and Jiggins 2014).

A study of general patterns of divergence and diversity showed that Sb has two orders of magnitude lower diversity than SB and than the rest of the genome, and that there is high ratio of nonsynonymous to synonymous substitutions between SB and Sb (Pracana et al. 2017). These results suggest that the evolution of Sb has been shaped by Hill–Robertson effects (the effects of selection on linked loci) due to the rarity of recombination in Sb (Wang et al. 2013; Pracana et al. 2017). However, little work has been done to characterize the genes present in the supergene region and to identify the mechanisms by which SB and Sb control the phenotypic differences between single- and multiple-queen colonies. Studies using cDNA microarrays representing 3673 genes demonstrated that the supergene region is enriched for genes that are differentially expressed between queens (Nipitwattanaphon et al. 2014) and workers (Wang et al. 2008, 2013) of the two colony types. This suggests that genes other than *Gp-9* could be responsible for the social dimorphism. Given that the determination of queen number requires the differential production and perception of semiochemicals by individuals of each genotype, it remains likely that OBPs play a part in determining the dimorphism.

Here, we determine to which extent OBPs have potentially functional divergence between social forms. For this, we identify all OBPs in the fire ant reference genome and map them to their genomic locations. Subsequently, we use population-sequencing data to identify allelic differences between OBPs found on alternate variants of the social chromosome supergene. We also sequence an outgroup species, *Solenopsis geminata*, which allows us to determine which supergene variant carries the derived allele for each substitution. Finally, we compare gene expression profiles of all OBPs and gene coexpression modules between social forms. We show that there are nucleotide and amino acid sequence level differences between SB and Sb in the supergene OBPs, and that OBPs inside and outside the supergene are differentially expressed between single- and multiple-queen colonies.

Methods

OBP DISCOVERY AND MANUAL GENE MODEL CURATION

The sequences of 18 fire ant OBP genes were previously reported, based on searches of Sanger-sequenced Expressed Sequence Tag (EST) libraries (Table S1; Wang et al. 2007; Xu et al. 2009; Gotzek et al. 2011; Wurm et al. 2011). We used a curation approach similar to those previously used on other genes (Ingram et al. 2012; Corona et al. 2013; Kulmuni et al. 2013; Privman et al. 2013) to find the position of these OBP genes in the fire ant genome assembly (Wurm et al. 2011) and to discover previously unreported OBP genes. Our curation pipeline is described in detail in Supporting Information Methods. Briefly, we iteratively performed blastn and blastp (Camacho et al. 2009; Priyam et al. 2015) searches of the fire ant genome assembly (Wurm et al. 2011) using as queries the previously known fire ant OBP sequences as well as UniProt sequences that are part of the Pfam family "PBP_GOBP" (Finn et al. 2014; UniProt Consortium 2015). We manually curated the results of these searches by inspecting alignments of transcriptomic and genomic reads, which allowed us to infer intron–exon boundaries and coding sequences of these OBPs. We

labeled the curated gene predictions that correspond to the previously known OBP genes (*SiOBP1–17*) according to the notation used by Gotzek et al. (2011) and we labeled newly discovered loci *SiOBPZ1–Z7*. We used a genetic map (Pracana et al. 2017) to assign OBPs to linkage groups. We generated a codon-level alignment of the *S. invicta* OBPs using MAFFT-linsi (version 6.903b; Katoh and Toh 2008) and PRANK (version 120626; Löytynoja and Goldman 2005), and built a phylogenetic tree using RaxML (version 8.2.9; Stamatakis 2006).

IDENTIFYING ALLELIC DIFFERENCES FOR OBPs CARRIED BY ALTERNATE VARIANTS OF THE SOCIAL CHROMOSOME

We used whole-genome sequences from one *SB* and one *Sb* male from each of seven colonies that had been sequenced at low coverage (Illumina 2^*100 bp paired-end genome shotgun sequences; ~6×–8× coverage) in 2012 (NCBI SRP017317; Wang et al. 2013). Each of these samples is a haploid male (ants have a haplodiploid sex determination system). We filtered the reads, aligned them to the reference genome using bowtie2 (version 2.1.0; Langmead and Salzberg 2012), and used samtools and bcftools (version 1.3.1 for both; Li et al. 2009) to call variants among the individuals (Supporting Information Methods).

We produced whole-genome sequencing reads of the outgroup species *S. geminata*. We sequenced a pool of 10 workers (sampled in Thailand by Dr. Adam Devenish, University College London, United Kingdom) using Illumina HiSeq 4000 (×11 coverage; Supporting Information Methods). We called variants between the sample and the reference assembly (using freebayes version 1.0.2-33-gdbb6160; Garrison and Marth 2012) within the coding sequence of each OBP using freebayes (Supporting Information Methods). We classed the alleles in each SB-Sb substitution as ancestral or derived based on the allele carried in the outgroup species. We estimated the rate of synonymous and nonsynonymous divergence (dS and dN, respectively) between SB and Sb using seqinR (version 3.0-7; Charif and Lobry 2007).

DETECTION OF COPY NUMBER AND STRUCTURAL VARIATION IN OBPs

We visually inspected the alignments of the seven *SB* and the seven *Sb* haploid male samples against each OBP region. Deletions were identified as regions with no coverage and duplications were identified as regions where the coverage was higher than the background (Supporting Information Methods). Using the *de novo* assembler MIRA (version 4.0.2; Chevreux et al. 1999), we produced the sequence of the duplicate copy of *SiOBP12*, which we named *SiOBPZ5* (approach detailed in Supporting Information Methods).

GENE EXPRESSION OF *S. INVICTA* OBPs IN PUBLICALLY AVAILABLE RNA SEQUENCING DATASETS

We analyzed all available RNA sequencing (RNA-Seq) data from the NCBI SRA database for *S. invicta* (data from Wurm et al. 2011; Morandin et al. 2016 and PRJNA266847; details in Table S2). We determined the expression levels of *S. invicta* transcripts using the Kallisto count mode (version 0.43.0; Bray et al. 2016). Each sample was independently normalized using the DESeq2 method (version 1.14.1; Love et al. 2014). Additionally, we performed genome-wide analysis of differential expression of data from Morandin et al. (2016), comparing three pools of queens from multiple-queen colonies with two pools from single-queen colonies, as well as two pools of workers from multiple-queen colonies with three pools from single-queen colonies. The pools of workers from multiple-queen colonies contain a mix of individuals of both genotypes, whereas the pool of queens from multiple-queen colonies has only *SB/SB* queens. We used a standard DESeq2 approach to identify expression differences between single- and multiple-queen samples in queens and in workers. Additional details regarding these analyses are in Supporting Information Methods.

DIFFERENTIAL EXPRESSION OF GENE COEXPRESSION MODULES ACROSS SOCIAL FORMS

We created gene coexpression modules from two cDNA microarray datasets (Platform GPL6930, with 25,344 probes representing 3673 genes; Supporting Information Methods and Table S3; Wang et al. 2007), one with queen samples (GSE42062; Nipitwattanaphon et al. 2013), the other with worker samples (E-GEOD-11694; Wang et al. 2008). Both datasets included *SB/SB* and *SB/Sb* samples. We created modules for each set using weighted gene coexpression network analysis (WGCNA) (version 1.49; Langfelder and Horvath 2008). We used *t*-tests to determine whether any module eigengene is correlated with genotype or social form. In queens, we compared *SB/SB* to *SB/Sb* samples because all samples originate in multiple-queen colonies. In workers, we separated the effect of genotype from the effect of social form following the approach in Wang et al. (2008): we compared genotypes (*SB/SB* vs *SB/Sb*) using samples from multiple-queen colonies, and we compared across social forms (single queen vs multiple queen) using *SB/SB* samples only.

EVIDENCE FOR SELECTION BASED ON NUCLEOTIDE DIVERSITY

Genomic regions that underwent recent selective sweeps are characterized by low nucleotide diversity (π) (Smith and Haigh 1974; Nei 1987; Nachman 2001). We used measurements of π along a sliding window of the genome, originally produced by Pracana et al. (2017), to identify selection pressure acting on *S. invicta*

OBPs. Measurements of π were taken from nonoverlapping 10 kb windows (Supporting Information Methods).

Results

THE FIRE ANT REFERENCE GENOME ASSEMBLY CONTAINS 23 PUTATIVE OBPs

We combined automatic and manual curation approaches incorporating genomic and gene expression data to identify the sequence, exon structure, and location of 23 putative OBP genes in the *S. invicta* reference genome. Seventeen of these matched fire ant OBP gene sequences that had been previously reported, although with differences in sequence or in their inferred location in linkage groups (Table S1 and Supporting Information Methods). The remaining seven putative OBP genes are novel to *S. invicta* (Table S4). Interestingly, the coverage depth of *SiOBPZ6* is fourfold higher (95% confidence interval [3.66–4.78]; *t*-test $t_{df=6} = 14.0$, $P < 10^{-5}$) than that of 1000 randomly selected genes, suggesting that there are four copies of this gene. There is little genetic variation among reads mapping to this gene across the 14 individuals in our dataset (4.2 Single Nucleotide Polymorphisms [SNPs] per 1000 bp). The alignment of whole-genome sequencing reads of the outgroup species *S. geminata* to the *S. invicta* reference assembly shows that all OBPs are covered in this outgroup species. The coverage depth of *SiOBPZ6* is threefold higher in *S. geminata* (95% confidence interval [2.78–3.16]; *t*-test $t_{df=999} = 20.7$, $P < 10^{-15}$), suggesting that this species also carries multiple copies of this gene.

Nine of the 23 OBPs in the genome are adjacent to unrelated genes, the remainder are organized into gene clusters. There are three locations in the genome each containing a cluster of four OBPs (two in linkage group 16, one in linkage group 3) and one containing a cluster of two OBPs (in linkage group 6). Intriguingly, none of these clusters appear to be completely monophyletic (Fig. 1). For previously known OBPs, the topology of our phylogenetic tree agrees with previously published trees (Gotzek et al. 2011; Zhang et al. 2016), with the exception of the position of *SiOBP15* (low bootstrap values in all trees) and *SiOBP5*.

NONSYNONYMOUS DIFFERENTIATION BETWEEN SB And Sb IN OBPs

Eight of the OBPs are located in scaffolds of the SB fire ant genome assembly that map to the supergene region, with two clusters of four OBPs (Fig. 2). One of the clusters includes *Gp-9* (which was named *SiOBP3* in Gotzek et al. 2011). A ninth gene, *SiOBP9*, is located in an unmapped scaffold that likely also belongs to the supergene region based on high levels of SB-Sb differentiation (Fig. 2). To determine whether the supergene OBPs have allelic differences between SB and Sb, we used whole-genome sequence data from seven *SB* males and seven *Sb* males.

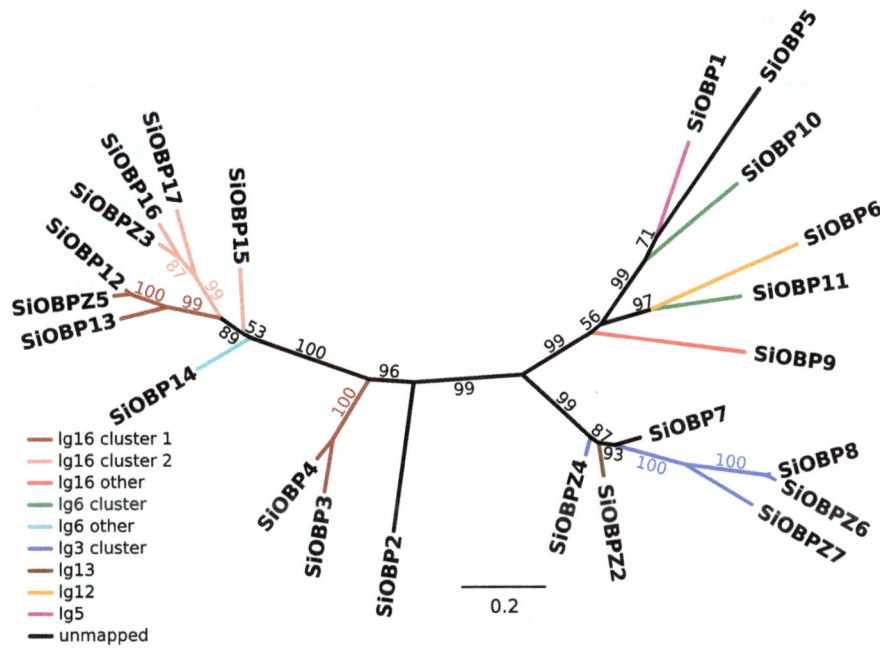

Figure 1. Phylogenetic tree based on a codon-level alignment of revised gene predictions for previously described OBPs (*SiOBP1–17*) and novel OBPs (*SiOBPZ2–Z6*). Branches are colored by gene cluster and linkage group (lg). *SiOBPZ1* was removed from this analysis because the high divergence of its sequence led to unreliable alignments and positioning in the phylogeny. All OBPs on linkage group 16 (lg16) are within the supergene-like region of the social chromosomes (Fig. 2).

These data confirmed the previous finding that *Gp-9/SiOBP3* has eight nonsynonymous and one synonymous fixed single nucleotide substitutions between SB and Sb in the North American study population (Krieger and Ross 2002). Of the other OBPs in the supergene region, *SiOBP4* has three nonsynonymous and two synonymous substitutions. Two additional supergene OBPs have one fixed nonsynonymous substitution between SB and Sb (Table 1). Performing an analysis of the ratio of nonsynonymous to synonymous substitutions between alleles (dN/dS) was only possible for the two genes with the most divergent alleles: *Gp-9/SiOBP3* had the highest ratio of nonsynonymous to synonymous substitutions (dN/dS = 1.48), followed by *SiOBP4* (dN/dS = 0.74).

We analyzed the OBP sequences from an outgroup species, *S. geminata*, estimated to have diverged from *S. invicta* 3–3.5 million years ago (Moreau and Bell 2013; Ward et al. 2015), that is, before the divergence between SB and Sb in *S. invicta* (estimated 0.35–0.42 million years ago; Wang et al. 2013). These sequences allowed us to determine the ancestral allele in each substitution. Sb carried the derived allele in most of the positions with nonsynonymous substitutions between SB and Sb (seven out of eight in *Gp-9/SiOBP3* and all in *SiOBP4* and *SiOBPZ3*; we could not derive the two *SiOBP13* substitutions, as *S. geminata* read coverage was too low for this gene). This pattern is consistent with most nonsynonymous substitutions between SB and Sb having arisen in the lineage leading to Sb.

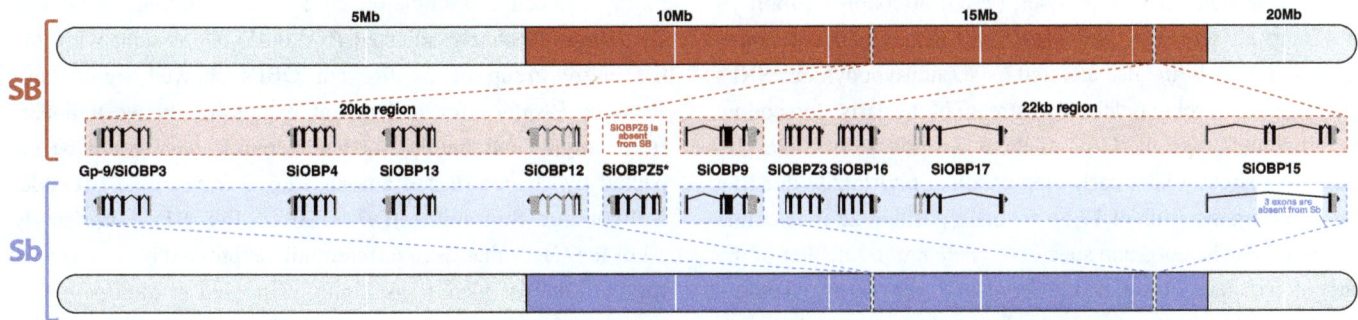

Figure 2. Relative positions on the social chromosome (i.e., linkage group 16) of 10 OBP loci, highlighting intron–exon structures and differences between the supergene region of Sb (blue) and SB (red). *SiOBPZ5* is specific to Sb but we do not know its exact location; *SiOBP15* is missing a 3-exon region in Sb; *SiOBP9* is in an unmapped scaffold that likely belongs to the supergene region based on high levels of SB-Sb differentiation (Pracana et al. 2017).

Table 1. OBP differentiation between SB and Sb: the number of sequence-level differences between SB and Sb and differential OBP gene expression between multiple- and single-queen colonies.

S. invicta OBP locus	Nonsynonymous differences	Synonymous differences	Total differences	Significant differential expression between colonies types	
				In queens	In workers
SiOBP3 (*Gp-9*)	8	1	9	Yes	No
SiOBP4	3	2	5	Yes	Yes
SiOBP13	1	1	2	Yes	Yes
SiOBP12	Frameshift insertion in SB and duplication in Sb			Yes	No
SiOBPZ5	Present exclusively in Sb				
SiOBPZ3	1	0	1	No	No
SiOBP9	0	1	1	No	No
SiOBP16	0	0	0	Yes	No
SiOBP17	0	0	0	Yes	Yes
SiOBP15	~2600 bp deletion in Sb			No	No

All differentially expressed genes between social forms were overexpressed in multiple-queen colonies.

COPY NUMBER AND STRUCTURAL DIFFERENTIATION BETWEEN SB AND Sb IN OBPs

We also found structural differences between SB and Sb affecting two OBPs. For the first, *SiOBP15*, we detected a ~2600 bp deletion unique to *Sb* individuals (Fig. 2, Table 1). This deletion is derived (i.e., it is not present in the outgroup species, *S. geminata*) and causes the loss of three out of five coding exons (89 out of 139 amino acids), although it does not cause a frameshift. The second OBP with a major structural difference is *SiOBP12*. In Sb individuals, this gene is duplicated, forming the Sb-specific *SiOBPZ5* (Fig. 2, Table 1). This gene increases the total OBP count of *S. invicta* to 24. There are 18 fixed amino acid differences between *SiOBPZ5* and the SB allele of *SiOBP12* sequence (one deleted codon, 21 nonsynonymous and four synonymous nucleotide-level fixed differences; four codons each contain two single-nucleotide fixed differences; dN/dS = 2.67). Intriguingly, *SiOBP12* has an early stop codon (TAG) at codon position 16 of 176 in all seven *SB* individuals and the reference genome. These individuals are also affected by six nonsynonymous SNPs and two polymorphic indels downstream of the early stop codon. *Sb* individuals have the CAG allele at position 16 of *SiOBP12*, but have a slightly later early stop codon at position 37 due to a frameshifting insertion of 17 bp at codon position 25 (nucleotide position 74). The outgroup species *S. geminata* has neither of the early stop codons. However, the very low *S. geminata* read coverage observed in the two terminal exons of this gene (median < 3; $t_{df = 999} = -11.29$, $P < 10^{-27}$) could indicate a deletion in this species. *SiOBP12* is thus nonfunctional in *Sb* and *SB* individuals, and putatively nonfunctional in the outgroup species. The Sb-specific gene *SiOBPZ5* appears to be functional as it has no

early stop codons. None of the other OBPs showed differences in structure or in copy number between SB or Sb.

FOURTEEN OBPs ARE DIFFERENTIALLY EXPRESSED BETWEEN SOCIAL FORMS

We compared expression levels between single- and multiple-queen colonies in workers and in queens (Fig. 3; Table 1) using RNA-Seq data from Morandin et al. (2016). General expression patterns showed an enrichment in differentially expressed genes in the supergene region in queens (expected proportion = 0.022, observed proportion = 0.059, $Chi^2_{df = 1} = 32.84$, $P = 10^{-8}$) but not in workers (expected proportion = 0.021, observed proportion = 0.024, $Chi^2_{df = 1} = 0.05$, $P = 0.82$).

In queens, fourteen OBPs, including seven in the supergene region, were significantly differentially expressed between multiple-queen and single-queen colonies (DESeq2 Wald test; Benjamini–Hochberg adjusted $P < 0.05$). Consistent with this, the entire group of 24 fire ant OBPs showed significantly stronger *P*-values for differential expression between queens from single- and multiple-queen colonies than would be expected by chance (tested among 12,693 transcripts, two-sided Kolmogorov–Smirnov test; $P < 10^{-11}$; Fig. S1). Surprisingly, all of the OBPs that were differentially expressed between social forms in queens were more highly expressed in multiple-queen colonies than in single-queen colonies (14 significant OBPs in queens, binomial test 14 out of 14; null probability = 0.5; $P < 10^{-4}$). In workers, only four OBPs (all in the supergene region) were significantly differentially expressed between social forms (DESeq2's Wald test Benjamini–Hochberg adjusted

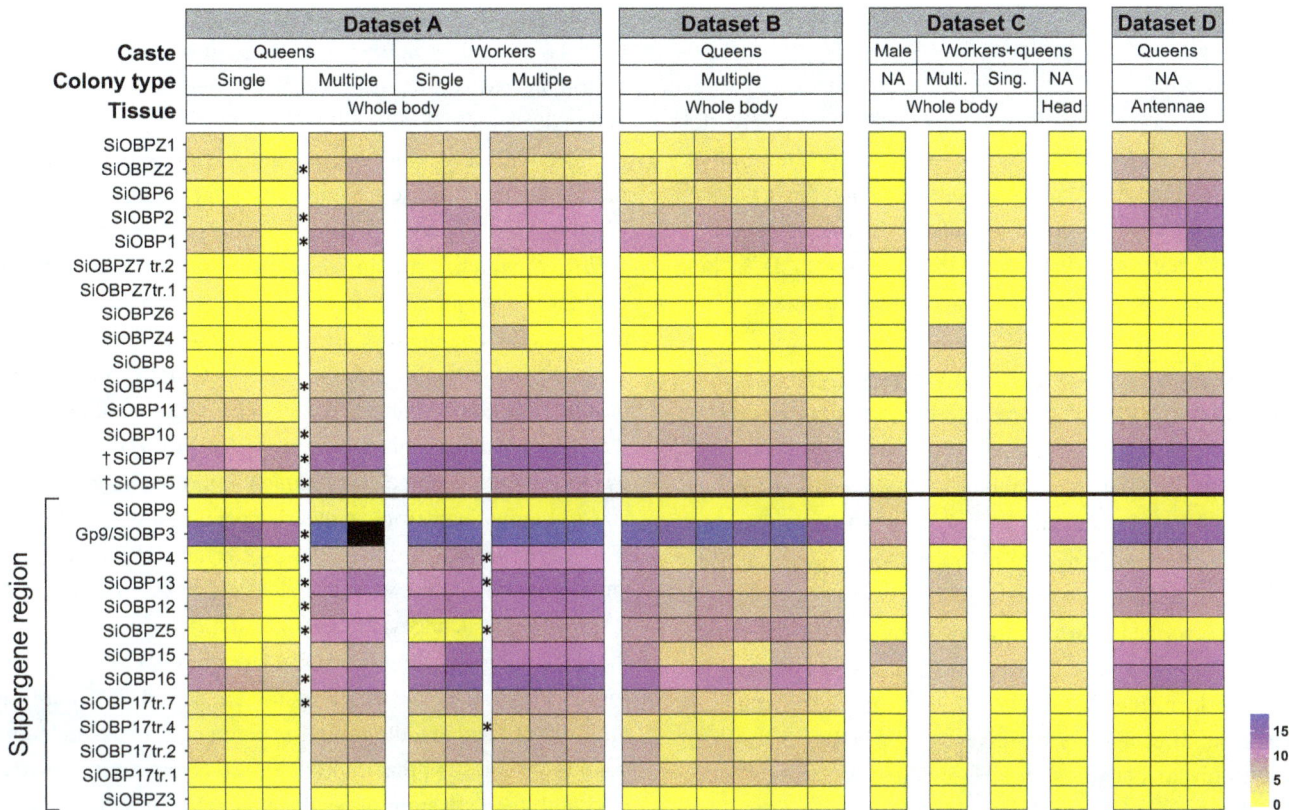

Figure 3. Expression patterns for all analyzed RNAseq datasets. Each tile represents the logarithm base 2 of DESeq normalized transcript counts. The rows with asterisks (*) correspond to those OBPs with significant differential expression between social forms within castes in dataset A (Morandin et al. 2016). Information about each dataset is available in Table S2 (A: PRJDB4088, B and C: PRJNA49629, D: PRJNA266847). † The exons of *SiOBP5* and *SiOBP7* are split across three unmapped scaffolds; we do not know whether these genes are within or outside the supergene region.

$P < 0.05$). All the differentially expressed OBPs in workers were also differentially expressed in queens. For one of these OBPs (*SiOBP17*), a different splice form was differentially expressed between colony types in queens than in workers (Fig. 3).

We additionally obtained qualitative gene expression profiles of all OBPs across 18 additional samples, in total representing seven different conditions of body part, social form, and caste (Table S2). We find generally consistent expression patterns for OBPs across all independent samples (Fig. 3). For instance, in every sample, *Gp-9/SiOBP3* was the most highly expressed of all OBPs, whereas *SiOBPZ3* was only residually expressed (0.26 or fewer transcripts per million reads). Six OBPs had only residual expression in queen antennae and in heads, although most of these showed at least some expression in whole-body samples. The expression of one of these genes, *SiOBP9*, appears to be limited to males.

GENE COEXPRESSION MODULES CORRELATED WITH SOCIAL FORM

We used WGCNA (Langfelder and Horvath 2008) to produce modules of coexpressed genes from a set of worker samples

(Wang et al. 2008) and a set of queen samples (Nipitwattanaphon et al. 2013). Both datasets compare *SB/SB* and *SB/Sb* samples. The queen and worker datasets, respectively, clustered into 30 and 37 coexpression modules (Table S5). Most modules in one dataset share a significant number of probes with a module in the other dataset (30 out of 31 in queens and 35 out of 37 in workers; Fisher's exact test for the overlap of the pairs of modules across the datasets, Bonferroni corrected $P > 0.05$; Fig. S2). However, in most cases there was no one-to-one correspondence between datasets (17 out of 31 modules in queens and 19 out of 37 in workers have significant overlaps with more than one module). Eight of the OBPs discovered in the present study are represented in the microarray (Table S5). In the worker dataset, the module "worker_D" includes four of the OBPs (*SiOBP3, SiOBP12, SiOBP13,* and *SiOBP16*), accounting 25% of the 16 genes in the module. OBPs were present in nine other modules, although in all nine cases the OBP represented a very small proportion of the genes in the module (Table S5).

We tested whether there were gene coexpression modules with differential eigengene expression between genotypes or social forms. In queens, four modules had differential expression between genotypes (Table S7). In workers, one module had differential eigengene expression between genotypes, and one module had differential gene expression between social forms (Table S7). One of the modules that had differential expression between genotypes in queens ("queen_X") corresponded with the module with differential expression between genotypes in workers ("worker_Z"). Only one of the modules with differential eigengene expression includes an OBP (*SiOBP15* in "queen_D"). None of these modules were enriched for any GO term.

THREE OBPs ARE IN A REGION OF THE GENOME WITH CHARACTERISTICS OF A RECENT SELECTIVE SWEEP

We used measurements of π among *SB* individuals in nonoverlapping 10 kb windows from Pracana et al. (2017) to determine whether any OBPs are in regions of low π, characteristic of recent selective sweeps. Among windows overlapping OBPs, two neighboring windows had π within the lower quartile of the whole-genome distribution ($\pi < 0.0004$; Fig. S3). These two windows overlap the loci *SiOBPZ4*, *SiOBPZ7*, and *SiOBPZ6*, which are within 19 kb of each other on linkage group 3. We did not perform an equivalent analysis on *Sb* individuals because the entire region of suppressed recombination has the signature of a recent sweep in Sb (Pracana et al. 2017).

Discussion
THE PUTATIVE ROLE OF OBPs IN DETERMINING SOCIAL DIMORPHISM

The description of *Gp-9* as a green beard gene (Keller and Ross 1998) and its subsequent characterization as an OBP (Krieger and Ross 2002) led to the proposal of different models of how this single gene can control the dimorphism in social organization (reviewed by Gotzek and Ross 2007). At their most basic level, these models propose that *Gp-9* controls the production of a green-beard odor in queens and the differential perception of this odor by workers of alternate genotypes. However, it was also proposed that *Gp-9* additionally controls differential odor production in workers (Gotzek and Ross 2007), as well as a number of physiological and morphological traits in queens (Keller and Ross 1995; DeHeer et al. 1999; DeHeer 2002) and males (Lawson et al. 2012). The discovery that *Gp-9* is tightly linked to hundreds of other genes (Wang et al. 2013; Pracana et al. 2017)—including the nine additional OBPs we report here—suggests that the roles previously attributed to *Gp-9* could be split between multiple genes.

The key roles of OBPs in semiochemical perception (Leal 2013) and secretion (Li et al. 2008; Iovinella et al. 2011; Sun et al. 2012) lead to the prediction that such proteins are involved in determining the two colony types. Our results support this

hypothesis, as we find divergence in protein coding sequence between SB and Sb in the OBPs in the supergene region, as well as differences in the regulation of OBP expression between single- and multiple-queen colonies.

The differences in protein coding sequence affect seven of the ten OBPs in the supergene region, including *Gp-9/SiOBP3*. The biggest differences are in *SiOBPZ5*, absent in SB, and in *SiOBP15*, which is missing three exons in Sb. Such differences could have a major effect on semiochemical communication. Additionally, among the four intact OBPs with nonsynonymous divergence between SB and Sb, both *Gp-9/SiOBP3* and *SiOBP4* have dN/dS ratios indicative of adaptive differentiation between the alleles of these genes (Krieger and Ross 2002). This interpretation comes with some caution due to our relatively low sample size (14 individuals from an invasive population).

Additionally, 14 out of the 24 fire ant OBPs were differentially expressed between social forms in queens or in workers. Our analysis uncovers three potentially important aspects of the differential regulation of OBP expression in the two social forms. First, all of the differentially expressed OBPs are more highly expressed in multiple-queen colonies than in single-queen colonies, suggesting that multiple-queen colony traits are associated with the activation of semiochemical communication pathways. Second, this activation seems to be stronger in queens, as more OBPs were differentially expressed between social forms in queens (14 OBPs) than in workers (four OBPs). This result reflects the more general pattern that the supergene region was enriched for differentially expressed genes between colony types in queens, but not in workers. The pools of workers from multiple-queen colonies contain a mix of individuals of both genotypes (36% *SB/SB* and 64% *SB/Sb* workers expected; Buechel et al. 2014), which could mask differences between *SB/SB* workers from single-queen colonies and *SB/Sb* workers from multiple-queen colonies. Indeed, previous studies using cDNA microarray data and a different gene set suggest that the supergene region is enriched for differentially expressed genes in both queens (Nipitwattanaphon et al. 2013) and workers (Wang et al. 2013). Third, several of the queen-specific differentially expressed OBPs are located outside the supergene, implying that they are regulated in *trans* by elements in the supergene. It is important to note that all three patterns could be affected by our use of samples from whole bodies, which is known to introduce several types of biases if the differences in expression are tissue specific (Johnson et al. 2013; Montgomery and Mank 2016). A particular issue is differences in allometry (i.e., relative body-size proportions) between the individuals of different groups, for instance the larger gaster of queens in single-queen colonies relative to queens in multiple-queen colonies (Tschinkel 2006). These biases cannot be resolved by standard normalization methods, which are designed to normalize by entire library size rather than by the relative abundance of different transcripts (Dillies et al. 2013). Tissue-specific gene expression profiling (Bastian et al. 2008; Robinson et al. 2013; Jasper et al. 2015) would be needed to control for such allometric differences.

Our results also support the idea that along with OBPs, other genes are likely involved in defining the social polymorphism of *S. invicta*. For instance, only one of the coexpression modules with significantly different eigengene expression contained an OBP. Furthermore, other genes inside and outside the supergene region were differentially expressed between social forms. Thus, a venue of further investigation would be to examine the potential roles of other genes, including genes from families known to be involved in communication, including chemosensory proteins (Kulmuni et al. 2013), desaturases (Helmkampf et al. 2015), fatty-acid reductases (Lassance et al. 2010; Niehuis et al. 2013), and olfactory (Wurm et al. 2011), gustatory (Robertson et al. 2003; Zhou et al. 2012), and ionotropic receptors (Benton et al. 2009; Zhou et al. 2012). It is important to note that additional experimental work would be necessary to demonstrate whether OBPs or any of these proteins have a functional role. An interesting approach would be to measure the effect of artificially modifying the sequence or expression level of each gene to test their specific function (Gaj et al. 2013; Mohr et al. 2014).

GENERAL EVOLUTIONARY PATTERNS OF OBPs IN *S. INVICTA*

The evolution of the OBP gene family is generally thought to follow the birth-and-death model, where gene duplication is followed by either the pseudogenization or the rapid functional divergence of the duplicate gene (Nei and Rooney 2005; Vieira et al. 2007). The *S. invicta* OBPs are organized in clusters along the genome, as in other insect species (Xu et al. 2003; Foret and Maleszka 2006; Vieira et al. 2007). However, none of these clusters appear to be monophyletic (Fig. 1). This is consistent with the birth–death model, where the fast evolution of genes can mask their true phylogenetic relationship (Vieira et al. 2007; Gotzek et al. 2011; Vieira and Rozas 2011). Alternative explanations include translocations affecting the OBPs during or after duplication, or ectopic gene conversion across different clusters after duplication (Arguello and Connallon 2011). Another argument in support of the birth-and-death model is that we find evidence of expansions in OBP number. One example is the putative ant-specific OBP expansion reported previously (the OBP cluster including SiOBP14 in Fig. 1; Gotzek et al. 2011). We found no one-to-one orthologous sequences for these genes in other ants or in other arthropods (the 11 genes in this group of OBPs have BLAST similarity to only three genes in the ant *Monomorium pharaonis*; phylogenetic group 1 in Table S7). A cluster with several novel genes identified in our study (the group including *SiOBP7* and *SiOBP8* in Fig. 1) follows a similar pattern (five OBPs have BLAST similarity to one

M. pharaonis gene, two have BLAST similarity to one *Pogonomyrmex barbatus* gene; phylogenetic group 2 in Table S7). These groups of genes may have expanded in the lineage leading to *S. invicta* and *S. geminata*, although this conclusion would require the exhaustive identification of OBPs in the present study to be replicated for other ant species. An example of a putatively recent expansion is *SiOBPZ6*, which seems to be present in multiple copies both in *S. invicta* and in *S. geminata*. Lack of heterozygosity in the region suggests that the gene copies have been recently affected by ectopic gene conversion (Arguello and Connallon 2011). Furthermore, finding that the *S. invicta* *SiOBPZ6* quadruplication is in a region that has a signature of a recent selective sweep makes it tempting to speculate that *SiOBPZ6* is involved in a recent adaptive process (Kondrashov 2012)—for example, to the invasive range of this species (Ascunce et al. 2011).

Conclusion

Previous studies have focused on how the evolution of the social chromosomes has been affected by restricted recombination (Wang et al. 2013; Pracana et al. 2017), whereas the work presented here focuses on the putative mechanisms by which these chromosomes control social organization. In summary, our analyses provide a comprehensive overview of OBPs in the fire ant genome, describing patterns of differentiation and expression that are consistent with the predicted roles of OBPs in determining social organization in this species. Our study highlights the need for tissue-specific expression profiles, as well as for broader taxonomic sampling to understand OBP evolution during the origin of the multiple-queen colony organization. Finally, our work provides a starting point for future functional studies on the roles of OBPs in the social chromosome system.

AUTHOR CONTRIBUTIONS
YW, RP, and IL conceived and designed the study; IL and RP performed the majority of genome-level analyses of OBPs; ES sequenced *S. geminata*; AP performed automated gene prediction; CMR analyzed gene expression; YW, RP, and IL drafted the manuscript and all authors contributed to later versions of the manuscript.

ACKNOWLEDGMENTS
We thank K. G. Ross, R. A. Nichols, C. Eizaguirre, L. Henry, E. Favreau, T. Colgan, two anonymous reviewers, the editor and the associate editor for advice and comments on the manuscript, and QMUL's SBCS Evolution group for support and stimulating discussion. We thank A. Devenish for supplying *Solenopsis geminata* samples. This work was supported by the Biotechnology and Biological Sciences Research Council (grant BB/K004204/1), the Natural Environment Research Council (grant NE/L00626X/1), NERC EOS Cloud, the Deutscher Akademischer Austauschdienst (DAAD) Postdoc-Programm (570704 83), Marie Curie Actions (PIEF-GA-2013-623713), and QMUL Research-IT and Mid-

Plus computational facilities (The Engineering and Physical Sciences Research Council grant EP/K000128/1).

- Illumina sequences from 15 fire ant males: NCBI SAMN00014755 and SRP017317.
- Fire ant reference genome assembly: GCA_000188075.1.

We deposited the genomic reads of the *Solenopsis geminata* sample on NCBI SRA (SRX3045159). We manually produced gene models for 24 OBPs, which we deposited to NCBI. Additionally, all data is available at wurmlab.github.io/data.

LITERATURE CITED

Arguello, J. R., and T. Connallon. 2011. Gene duplication and ectopic gene conversion in *Drosophila*. Genes 2:131–151.

Ascunce, M. S., C.-C. Yang, J. Oakey, L. Calcaterra, W.-J. Wu, C.-J. Shih, J. Goudet, K. G. Ross, and D. Shoemaker. 2011. Global invasion history of the fire ant *Solenopsis invicta*. Science 331:1066–1068.

Bastian, F., G. Parmentier, J. Roux, S. Moretti, V. Laudet, and M. Robinson-Rechavi. 2008. Bgee: integrating and comparing heterogeneous transcriptome data among species. Pp. 124–131 in A. Bairoch, S. Cohen-Boulakia, and C. Froidevaux, eds. Data integration in the life sciences, Lecture notes in computer science. Springer, Berlin, Heidelberg.

Benton, R., K. S. Vannice, C. Gomez-Diaz, and L. B. Vosshall. 2009. Variant ionotropic glutamate receptors as chemosensory receptors in *Drosophila*. Cell 136:149–162.

Bourke, A. F. G., and J. Heinze. 1994. The ecology of communal breeding: the case of multiple-queen Leptothoracine ants. Proc. R. Soc. B 345:359–372.

Bray, N. L., H. Pimentel, P. Melsted, and L. Pachter. 2016. Near-optimal probabilistic RNA-seq quantification. Nat. Biotechnol. 34:525–527.

Buechel, S. D., Y. Wurm, and L. Keller. 2014. Social chromosome variants differentially affect queen determination and the survival of workers in the fire ant *Solenopsis invicta*. Mol. Ecol. 23:5117–5127.

Camacho, C., G. Coulouris, V. Avagyan, N. Ma, J. Papadopoulos, K. Bealer, and T. L. Madden. 2009. BLAST+: architecture and applications. BMC Bioinformatics 10:421.

Charif, D., and J. R. Lobry. 2007. SeqinR 1.0-2: a contributed package to the R project for statistical computing devoted to biological sequences retrieval and analysis. Pp. 207–232 in U. Bastolla, M. Porto, H. E. Roman, and M. Vendruscolo, eds. Structural approaches to sequence evolution, Biological and medical physics, biomedical engineering. Springer, Berlin, Heidelberg.

Chevreux, B., T. Wetter, and S. Suhai. 1999. Genome sequence assembly using trace signals and additional sequence information. Pp. 45–56 in Computer science and biology: Proceedings of the German Conference on Bioinformatics GCB'99, Heidelberg,.

Corona, M., R. Libbrecht, Y. Wurm, O. Riba-Grognuz, R. A. Studer, and L. Keller. 2013. Vitellogenin underwent subfunctionalization to acquire caste and behavioral specific expression in the harvester ant *Pogonomyrmex barbatus*. PLoS Genet. 9:e1003730.

DeHeer, C. J. 2002. A comparison of the colony-founding potential of queens from single- and multiple-queen colonies of the fire ant *Solenopsis invicta*. Anim. Behav. 64:655–661.

DeHeer, C. J., M. A. D. Goodisman, and K. G. Ross. 1999. Queen dispersal strategies in the multiple-queen form of the fire ant *Solenopsis invicta*. Am. Nat. 153:660–675.

Dillies, M.-A., A. Rau, J. Aubert, C. Hennequet-Antier, M. Jeanmougin, N. Servant, et al. 2013. A comprehensive evaluation of normalization methods for Illumina high-throughput RNA sequencing data analysis. Brief. Bioinform. 14:671–683.

Eliyahu, D., K. G. Ross, K. L. Haight, L. Keller, and J. Liebig. 2011. Venom alkaloid and cuticular hydrocarbon profiles are associated with social organization, queen fertility status, and queen genotype in the fire ant *Solenopsis invicta*. J. Chem. Ecol. 37:1242–1254.

Finn, R. D., A. Bateman, J. Clements, P. Coggill, R. Y. Eberhardt, S. R. Eddy, et al. 2014. Pfam: the protein families database. Nucleic Acids Res. 42:D222–D230.

Foret, S., and R. Maleszka. 2006. Function and evolution of a gene family encoding odorant binding-like proteins in a social insect, the honey bee (*Apis mellifera*). Genome Res. 16:1404–1413.

Gaj, T., C. A. Gersbach, and C. F. III Barbas. 2013. ZFN, TALEN, and CRISPR/Cas-based methods for genome engineering. Trends Biotechnol. 31:397–405.

Garrison, E., and G. Marth. 2012. Haplotype-based variant detection from short-read sequencing. arXiv:1207.3907 [q-bio.GN].

Goodisman, M. A. D., C. J. DeHeer, and K. G. Ross. 2000. Unusual behavior of polygyne fire ant queens on nuptial flights. J. Insect Behav. 13:455–468.

Gotzek, D., H. M. Robertson, Y. Wurm, and D. Shoemaker. 2011. Odorant binding proteins of the red imported fire ant, *Solenopsis invicta*: an example of the problems facing the analysis of widely divergent proteins. PLoS One 6:e16289.

Gotzek, D., and K. G. Ross. 2007. Genetic regulation of colony social organization in fire ants: an integrative overview. Q. Rev. Biol. 82:201–226.

———. 2009. Current status of a model system: the gene *Gp-9* and its association with social organization in fire ants. PLoS One 4:e7713.

Helmkampf, M., E. Cash, and J. Gadau. 2015. Evolution of the insect desaturase gene family with an emphasis on social Hymenoptera. Mol. Biol. Evol. 32:456–471.

Huang, Y.-C., H. Yu-Ching, and J. Wang. 2014. Did the fire ant supergene evolve selfishly or socially? Bioessays 36:200–208.

Ingram, K. K., A. Kutowoi, Y. Wurm, D. Shoemaker, R. Meier, and G. Bloch. 2012. The molecular clockwork of the fire ant *Solenopsis invicta*. PLoS One 7:e45715.

Iovinella, I., F. R. Dani, A. Niccolini, S. Sagona, E. Michelucci, A. Gazzano, et al. 2011. Differential expression of odorant-binding proteins in the mandibular glands of the honey bee according to caste and age. J. Proteome Res. 10:3439–3449.

Jasper, W. C., T. A. Linksvayer, J. Atallah, D. Friedman, J. C. Chiu, and B. R. Johnson. 2015. Large-scale coding sequence change underlies the evolution of postdevelopmental novelty in honey bees. Mol. Biol. Evol. 32:334–346.

Johnson, B. R., J. Atallah, and D. C. Plachetzki. 2013. The importance of tissue specificity for RNA-seq: highlighting the errors of composite structure extractions. BMC Genomics 14:586.

Johnson, B. R., and T. A. Linksvayer. 2010. Deconstructing the superorganism: social physiology, groundplans, and sociogenomics. Q. Rev. Biol. 85:57–79.

Katoh, K., and H. Toh. 2008. Recent developments in the MAFFT multiple sequence alignment program. Brief. Bioinform. 9:286–298.

Keller, L., and K. G. Ross. 1995. Gene by environment interaction: effects of a single gene and social environment on reproductive phenotypes of fire ant queens. Funct. Ecol. 9:667–676.

———. 1998. Selfish genes: a green beard in the red fire ant. Nature 394:573–575.

———. 1999. Major gene effects on phenotype and fitness: the relative roles of *Pgm-3* and *Gp-9* in introduced populations of the fire ant *Solenopsis invicta*. J. Evol. Biol. 12:672–680.

Kondrashov, F. A. 2012. Gene duplication as a mechanism of genomic adaptation to a changing environment. Proc. R. Soc. B 279:5048–5057.

Krieger, M. J. B. 2004. To b or not to b: a pheromone-binding protein regulates colony social organization in fire ants. Bioessays 27:91–99.

Krieger, M. J. B., and K. G. Ross. 2002. Identification of a major gene regulating complex social behavior. Science 295:328–332.

———. 2005. Molecular evolutionary analyses of the odorant-binding protein gene Gp-9 in fire ants and other Solenopsis species. Mol. Biol. Evol. 22:2090–2103.

Kulmuni, J., Y. Wurm, and P. Pamilo. 2013. Comparative genomics of chemosensory protein genes reveals rapid evolution and positive selection in ant-specific duplicates. Heredity 110:538–547.

Langfelder, P., and S. Horvath. 2008. WGCNA: an R package for weighted correlation network analysis. BMC Bioinformatics 9:559.

Langmead, B., and S. L. Salzberg. 2012. Fast gapped-read alignment with Bowtie 2. Nat. Methods 9:357–359.

Lassance, J.-M., A. T. Groot, M. A. Liénard, B. Antony, C. Borgwardt, F. Andersson, E. Hedenström, D. G. Heckel, and C. Löfstedt. 2010. Allelic variation in a fatty-acyl reductase gene causes divergence in moth sex pheromones. Nature 466:486–489.

Lawson, L. P., R. K. Vander Meer, and D. Shoemaker. 2012. Male reproductive fitness and queen polyandry are linked to variation in the supergene Gp-9 in the fire ant Solenopsis invicta. Proc. R. Soc. B 279:3217–3222.

Leal, W. S. 2013. Odorant reception in insects: roles of receptors, binding proteins, and degrading enzymes. Annu. Rev. Entomol. 58:373–391.

Li, H., B. Handsaker, A. Wysoker, T. Fennell, J. Ruan, N. Homer, G. Marth, G. Abecasis, and R. Durbin. 2009. The sequence alignment/map format and SAMtools. Bioinformatics 25:2078–2079.

Li, S., J.-F. Picimbon, S. Ji, Y. Kan, Q. Chuanling, J.-J. Zhou et al. 2008. Multiple functions of an odorant-binding protein in the mosquito Aedes aegypti. Biochem. Biophys. Res. Commun. 372:464–468.

Linksvayer, T. A., J. W. Busch, and C. R. Smith. 2013. Social supergenes of superorganisms: do supergenes play important roles in social evolution? Bioessays 35:683–689.

Love, M. I., H. Wolfgang, and A. Simon. 2014. Moderated estimation of fold change and dispersion for RNA-seq data with DESeq2. Genome Biol. 15:550.

Löytynoja, A., and N. Goldman. 2005. An algorithm for progressive multiple alignment of sequences with insertions. Proc. Natl. Acad. Sci. USA 102:10557–10562.

Mohr, S. E., J. A. Smith, C. E. Shamu, R. A. Neumüller, and N. Perrimon. 2014. RNAi screening comes of age: improved techniques and complementary approaches. Nat. Rev. Mol. Cell Biol. 15:591–600.

Montgomery, S. H., and J. E. Mank. 2016. Inferring regulatory change from gene expression: the confounding effects of tissue scaling. Mol. Ecol. 25:5114–5128.

Morandin, C., M. M. Y. Tin, S. Abril, C. Gómez, L. Pontieri, M. Schiøtt, M. Schiøtt, L. Sundström, K. Tsuji, J. S. Pedersen, et al. 2016. Comparative transcriptomics reveals the conserved building blocks involved in parallel evolution of diverse phenotypic traits in ants. Genome Biol. 17:43.

Moreau, C. S., and C. D. Bell. 2013. Testing the museum versus cradle tropical biological diversity hypothesis: phylogeny, diversification, and ancestral biogeographic range evolution of the ants. Evolution 67:2240–2257.

Nachman, M. W. 2001. Single nucleotide polymorphisms and recombination rate in humans. Trends Genet. 17:481–485.

Nei, M. 1987. Molecular evolutionary genetics. Columbia Univ. Press, New York.

Nei, M., and A. P. Rooney. 2005. Concerted and birth-and-death evolution of multigene families. Annu. Rev. Genet. 39:121–152.

Niehuis, O., J. Buellesbach, J. D. Gibson, D. Pothmann, C. Hanner, N. S. Mutti, A. K. Judson, J. Gadau, J. Ruther, and T. Schmitt. 2013. Behavioural and genetic analyses of Nasonia shed light on the evolution of sex pheromones. Nature 494:345–348.

Nipitwattanaphon, M., J. Wang, M. B. Dijkstra, and L. Keller. 2013. A simple genetic basis for complex social behaviour mediates widespread gene expression differences. Mol. Ecol. 22:3797–3813.

Nipitwattanaphon, M., J. Wang, K. G. Ross, O. Riba-Grognuz, Y. Wurm, C. Khurewathanakul, et al. 2014. Effects of ploidy and sex-locus genotype on gene expression patterns in the fire ant Solenopsis invicta. Proc. R. Soc. B. 281:20141776

Pelosi, P., I. Iovinella, A. Felicioli, and F. R. Dani 2014. Soluble proteins of chemical communication: an overview across arthropods. Front. Physiol. 5:320.

Pelosi, P., J.-J. Zhou, L. P. Ban, and M. Calvello. 2006. Soluble proteins in insect chemical communication. Cell. Mol. Life Sci. 63:1658–1676.

Pracana, R., A. Priyam, I. Levantis, R. Nichols, and Y. Wurm. 2017. The fire ant social chromosome supergene variant Sb shows low diversity but high divergence from SB. Mol. Ecol. 26:2864–2879.

Privman, E., Y. Wurm, and L. Keller. 2013. Duplication and concerted evolution in a master sex determiner under balancing selection. Proc. R. Soc. B 280:20122968.

Priyam, A., B. J. Woodcroft, V. Rai, A. Munagala, I. Moghul, F. Ter, et al. 2015. Sequenceserver: a modern graphical user interface for custom BLAST databases. bioRxiv 033142. https://doi.org/10.1101/033142.

Robertson, H. M., C. G. Warr, and J. R. Carlson. 2003. Molecular evolution of the insect chemoreceptor gene superfamily in Drosophila melanogaster. Proc. Natl. Acad. Sci. USA 100(Suppl. 2):14537–14542.

Robinson, G. E., C. M. Grozinger, and C. W. Whitfield. 2005. Sociogenomics: social life in molecular terms. Nat. Rev. Genet. 6:257–270.

Robinson, S. W., P. Herzyk, J. A. T. Dow, and D. P. Leader. 2013. FlyAtlas: database of gene expression in the tissues of Drosophila melanogaster. Nucleic Acids Res. 41:D744–D750.

Ross, K. G. 1997. Multilocus evolution in fire ants: effects of selection, gene flow and recombination. Genetics 145:961–974.

Ross, K. G., and L. Keller. 1995. Ecology and evolution of social organization: insights from fire ants and other highly eusocial insects. Annu. Rev. Ecol. Syst. 26:631–656.

———. 1998. Genetic control of social organization in an ant. Proc. Natl. Acad. Sci. USA 95:14232–14237.

———. 2002. Experimental conversion of colony social organization by manipulation of worker genotype composition in fire ants (Solenopsis invicta). Behav. Ecol. Sociobiol. 51:287–295.

Stamatakis, A. 2006. RAxML-VI-HPC: maximum likelihood-based phylogenetic analyses with thousands of taxa and mixed models. Bioinformatics 22:2688–2690.

Schwander, T., R. Libbrecht, and L. Keller. 2014. Supergenes and complex phenotypes. Curr. Biol. 24:R288–R294.

Smith, J. M., and J. Haigh. 1974. The hitch-hiking effect of a favourable gene. Genet. Res. 23:23–35.

Sun, Y.-L., L.-Q. Huang, P. Pelosi, and C.-Z. Wang. 2012. Expression in antennae and reproductive organs suggests a dual role of an odorant-binding protein in two sibling Helicoverpa species. PLoS One 7:e30040.

Thompson, M. J., and C. D. Jiggins. 2014. Supergenes and their role in evolution. Heredity 113:1–8.

Trible, W., and K. G. Ross. 2016. Chemical communication of queen supergene status in an ant. J. Evol. Biol. 29:502–513.

Tschinkel, W. R. 2006. The fire ants. The Belknap Press of Harvard Univ. Press, Cambridge, Massachusetts, London, England.

UniProt Consortium. 2015. UniProt: a hub for protein information. Nucleic Acids Res. 43:D204–D212.

Vieira, F. G., and J. Rozas. 2011. Comparative genomics of the odorant-binding and chemosensory protein gene families across the Arthropoda: origin and evolutionary history of the chemosensory system. Genome Biol. Evol. 3:476–490.

Vieira, F. G., A. Sánchez-Gracia, and J. Rozas. 2007. Comparative genomic analysis of the odorant-binding protein family in 12 *Drosophila* genomes: purifying selection and birth-and-death evolution. Genome Biol. 8:R235.

Wang, J., S. Jemielity, P. Uva, Y. Wurm, J. Gräff, and L. Keller. 2007. An annotated cDNA library and microarray for large-scale gene-expression studies in the ant *Solenopsis invicta*. Genome Biol. 8:R9.

Wang, J., K. G. Ross, and L. Keller. 2008. Genome-wide expression patterns and the genetic architecture of a fundamental social trait. PLoS Genet. 4:e1000127.

Wang, J., Y. Wurm, M. Nipitwattanaphon, O. Riba-Grognuz, Y.-C. Huang, D. Shoemaker, and L Keller. 2013. A Y-like social chromosome causes alternative colony organization in fire ants. Nature 493:664–668.

Ward, P. S., S. G. Brady, B. L. Fisher, and T. R. Schultz. 2015. The evolution of myrmicine ants: phylogeny and biogeography of a hyperdiverse ant clade (Hymenoptera: Formicidae). Syst. Entomol. 40:61–81.

West, S. A., and A. Gardner. 2010. Altruism, spite, and greenbeards. Science 327:1341–1344.

Wurm, Y., J. Wang, O. Riba-Grognuz, M. Corona, S. Nygaard, B. G. Hunt, et al. 2011. The genome of the fire ant *Solenopsis invicta*. Proc. Natl. Acad. Sci. USA 108:5679–5684.

Xu, P. X., L. J. Zwiebel, and D. P. Smith. 2003. Identification of a distinct family of genes encoding atypical odorant-binding proteins in the malaria vector mosquito, *Anopheles gambiae*. Insect Mol. Biol. 12:549–560.

Xu, Y.-L., P. He, L. Zhang, S.-Q. Fang, S.-L. Dong, Y.-J. Zhang, and F. Li. 2009. Large-scale identification of odorant-binding proteins and chemosensory proteins from expressed sequence tags in insects. BMC Genomics 10:632.

Zhang, W., A. Wanchoo, A. Ortiz-Urquiza, Y. Xia, and N. O. Keyhani. 2016. Tissue, developmental, and caste-specific expression of odorant binding proteins in a eusocial insect, the red imported fire ant, *Solenopsis invicta*. Sci. Rep. 6:35452.

Zhou, X., J. D. Slone, A. Rokas, S. L. Berger, J. Liebig, A. Ray, D. Reinberg, and L. J. Zwiebel, 2012. Phylogenetic and transcriptomic analysis of chemosensory receptors in a pair of divergent ant species reveals sex-specific signatures of odor coding. PLoS Genet. 8:e1002930.

11

Environmental variation causes different (co) evolutionary routes to the same adaptive destination across parasite populations

Stuart K. J. R. Auld[1,2] and June Brand[1]

[1]*Biological and Environmental Sciences, University of Stirling, Stirling, United Kingdom*

[2]*E-mail: s.k.auld@stir.ac.uk*

Epidemics are engines for host-parasite coevolution, where parasite adaptation to hosts drives reciprocal adaptation in host populations. A key challenge is to understand whether parasite adaptation and any underlying evolution and coevolution is repeatable across ecologically realistic populations that experience different environmental conditions, or if each population follows a completely unique evolutionary path. We established twenty replicate pond populations comprising an identical suite of genotypes of crustacean host, *Daphnia magna*, and inoculum of their parasite, *Pasteuria ramosa*. Using a time-shift experiment, we compared parasite infection traits before and after epidemics and linked patterns of parasite evolution with shifts in host genotype frequencies. Parasite adaptation to the sympatric suite of host genotypes came at a cost of poorer performance on foreign genotypes across populations and environments. However, this consistent pattern of parasite adaptation was driven by different types of frequency-dependent selection that was contingent on an ecologically relevant environmental treatment (whether or not there was physical mixing of water within ponds). In unmixed ponds, large epidemics drove rapid and strong host-parasite coevolution. In mixed ponds, epidemics were smaller and host evolution was driven mainly by the mixing treatment itself; here, host evolution and parasite evolution were clear, but coevolution was absent. Population mixing breaks an otherwise robust coevolutionary cycle. These findings advance our understanding of the repeatability of (co)evolution across noisy, ecologically realistic populations.

KEY WORDS: Adaptation, coevolution, experimental evolution, host–parasite interactions.

Impact Summary

Over time, populations often become better suited to their local environment as poor performing individuals are removed by natural selection. For most parasites, environment is determined mainly by the hosts they infect. As the composition of different host types change over time, so does the selection on parasite populations. Changes in host composition will also be shaped by the parasite, as hosts and parasites are locked in a coevolutionary cycle. Here, we examine the consistency of host-parasite (co)evolution and parasite adaptation across replicate seminatural populations. Ordinarily, natural populations vary so much that it difficult to examine the repeatability of host-parasite interactions. We used a novel approach that combines the benefits of controlled experimental manipulation with ecological realism. We established 20 freshwater crustacean populations, each of which had the same genetic composition, and exposed them to isolates of the same original population of a sterilizing bacterial parasite; half of the populations experienced a stirring treatment. At the end of the season, after all the populations suffered epidemics, we sampled each of the parasite populations and exposed them to a test set of hosts. This experiment allowed us to demonstrate that parasites adapted to host genetic types that were present in the pond populations–both in terms of ability to infect the host and within-host parasite growth. However, the consistent pattern of adaptation masked very different dynamics that depended on the stirring treatment: in unstirred populations, parasites adapted to previously resistant hosts that became common (coevolution), whereas in stirred populations, hosts evolved, but not in terms of their resistance, and parasites adapted to hosts that were neither common nor rare (host evolution and parasite evolution). We therefore demonstrate that the relationship between adaptation and (co)evolution can vary according to environmental treatment, but is nevertheless not completely unique to individual populations.

Introduction

Parasites commonly exert strong selection on their host populations and vice versa, driving rapid coevolutionary change (Jaenike 1978; Brockhurst et al. 2003; Koskella and Lively 2009; Paterson et al. 2010; Schulte et al. 2011; Thrall et al. 2012; Lenski and Levin 2015). These coevolutionary interactions provide a window through which to observe host and parasite adaptation to local conditions, where parasites adapt to better infect local hosts than hosts from other populations or hosts adapt to better resist local parasites relative to foreign parasites (Lively 1989; Ebert 1994; Imhoof and Schmid Hempel 1998; Kaltz and Shykoff 1998; Oppliger et al. 1999; Koskella 2014). The relative magnitude of host and parasite local adaptation depends on both the adaptive genetic variation within each population and the strength of selection from the antagonist relative to other selection pressures (Kawecki and Ebert 2004). Populations that have different histories of selection and adaptive genetic variation can exhibit similar patterns of local adaptation, and vice versa. So, while parasite local adaptation demonstrates the potential for parasite-mediated selection on the host population (Gandon and Nuismer 2009), it cannot tell us whether that host population has sufficient genetic variation to respond to such selection. To understand the workings of the coevolutionary engine that drives local adaptation, we must examine replicate ecologically realistic populations over both space and time (Blanquart and Gandon 2013; Koskella 2014).

Patterns of host-parasite coevolution fall along a continuum. On one extreme, arms race evolution (ARE) of increased general host resistance and parasite infectivity strips genetic variation from populations, slowing coevolution (Obbard et al. 2011) until the arrival of new genotypes as a result of mutation or immigration. On the other extreme, fluctuating selection (FS), which occurs when the likelihood of infection depends on the precise combination of host and parasite genotypes, can drive Red Queen

dynamics and maintain genetic diversity in both host and parasite populations over the long term (provided the parasite is virulent; Howard and Lively 1994). Under FS, adaptation is fuelled by standing genetic variation and does not require a supply of new mutations (Hamilton 1980; Howard and Lively 1994). Theory tells us that the position of coevolutionary dynamics on the ARE-FS spectrum is governed largely by the genetics underlying host–parasite interactions (Agrawal and Lively 2002), with implications for local adaptation. When infection depends on the precise combination of host and parasite genotypes (termed genotypic specificity), FS can emerge (Hamilton 1980) and local adaptation is strong, because increased fitness on local hosts comes at an automatic cost of reduced fitness on foreign hosts. When infection is not genotype specific, that is when parasites can infect a broad range of host genotypes, ARE is more likely and local adaptation is weaker; this is because selection for increased performance on local hosts leads to correlated selection for increased performance on other foreign hosts (Morgan et al. 2005).

Infection genetics is not the only determinant of host-parasite coevolution. Both the nature of coevolution and its strength can depend on environmental conditions (Lazzaro and Little 2009; Wolinska and King 2009; Mostowy and Engelstädter 2011). For example, increased physical flux (mixing) within populations results in increased contact rate between bacteria and their phage parasites, accelerating coevolution (Brockhurst et al. 2003) and selecting on the phage to infect a broader range of host genotypes; this causes shifts coevolutionary dynamics from FS to ARE (Gómez et al. 2015). Environmental variation among populations means the mode and tempo of coevolution can potentially vary between ARE to FS, or break down into host evolution and/or parasite evolution occurring in isolation (Blanford et al. 2003; Mostowy and Engelstädter 2011; Harrison et al. 2013), leading to different patterns of local adaptation (Laine 2007, 2008). If coevolution shifts from FS to ARE, one expects increases in host range would mean parasites perform better on both local and foreign hosts, reducing the strength of local adaptation. However, to adequately test this theory, we require a better understanding of how (co)evolution and adaptation are linked across replicate populations in more ecologically complex, that is more natural, settings (Laine 2007; Thrall et al. 2012; Koskella 2014; Bankers et al. 2017; Gibson et al. 2017).

We established twenty replicate outdoor pond populations of the crustacean host, *Daphnia magna*, and its natural bacterial parasite, *Pasteuria ramosa*. In this system, infection depends on genotypic specificity (Luijckx et al. 2011), so there is the potential for FS dynamics and parasite local adaptation to emerge in these populations (Decaestecker et al. 2007). Each pond was seeded with the same suite of host genotypes and isolates from the same genetically diverse parasite population. Ponds experienced natural environmental variation over space and time, and half-experienced a physical flux (population mixing) treatment (known to reduce infection prevalence in this system; Auld and Brand 2017) to extend the test for mixing-mediated shifts in (co)evolution beyond phage-bacteria systems. We then dissected the relationship between parasite adaptation and host evolution, parasite evolution, and host-parasite coevolution using a time-shift experiment (Gaba and Ebert 2009), where a test set of host genotypes were exposed to ancestral parasite isolates and to isolates collected from the pond populations at the end of the epidemic (similar to Auld et al. 2014a). By combining laboratory experimental data on changes in infection traits with outdoor experimental data on shifts in host genotype frequencies, we dissected host evolution of resistance, parasite evolution of infectivity, and within-host growth and the overall change in parasitism due to coevolution over the course of the epidemic. We found consistent parasite local adaptation—in terms of ability to infect and grow within the host—across populations. This adaptation was underpinned by strong coevolution in unmixed ponds, but separate host evolution and parasite evolution in mixed ponds.

Methods

STUDY ORGANISMS

Pasteuria ramosa is a Gram positive bacterial endoparasite. *Pasteuria* transmission spores are ingested by filter-feeding *Daphnia magna*, and cause infection when they bind to and penetrate the host gut epithelium (Duneau et al. 2011; Auld et al. 2012a). Once inside the host, the spores grow and sporulate (Auld et al. 2014bc), causing host sterilization (Cressler et al. 2014). The parasite is an obligate killer, and millions of transmission spores are then released into the environment upon the death of the infected host (Ebert et al. 1996). *Daphnia magna* is a cyclically parthenogenetic freshwater crustacean that inhabits shallow ponds and lakes across Europe and commonly suffers infection with *Pasteuria*. Infection is easy to diagnose: *Pasteuria*-infected *Daphnia* have obvious red-brown bacterial growth in their haemolymph, lack developed ovaries or offspring in their brood chamber and often exhibit gigantism.

OUTDOOR POND EXPERIMENT

We established twenty 1000 liter artificial ponds in August 2014 and allowed them to naturally fill with rainwater over an eight-month period (Auld and Brand 2017). On the 2nd April 2015, we seeded each pond with an identical suite of 12 unique *Daphnia* genotypes (determined using microsatellite genotyping; see Auld and Brand 2017). There were ten *Daphnia* per genotype (total = 120 *Daphnia* per pond) and a genetically diverse inoculum of 1×10^8 *Pasteuria* spores. This *Pasteuria* starting popula-

tion was generated by exposing sediment samples to 21 genotypes of local *Daphnia*, harvesting the infected hosts, and re-exposing transmission spores to healthy hosts for multiple rounds of infection (Auld and Brand 2017). After a two-week establishment period, we estimated the density of *Daphnia* life stages (juveniles, healthy adults, *Pasteuria*-infected adults) in each pond on a weekly basis. We did this by passing a 0.048 m² pond net across the diameter of the mesocosm (1.51 m) and counting the resulting *Daphnia*. All *Daphnia* were returned to the ponds after counting. Half of the ponds experienced a weekly population mixing (physical flux) treatment, where mixed ponds were stirred once across the middle and once around the circumference with a 0.35 m² paddle submerged halfway into the pond (the exception to this was on day one of the experiment, when all ponds experienced the mixing treatment to ensure hosts and parasites were distributed throughout the ponds).

At the end of the season, after disease epidemics had peaked in all of the populations (November 17th, 2015) we sampled 90 infected *Daphnia* from each pond population; these samples were individually homogenized and the resulting spore solutions were pooled into three isolates per pond (i.e., where each isolate consisted of 30 homogenized infected hosts), and frozen at −20°C for use in the laboratory experiment. We also sampled 20–30 *Daphnia* from 16 of the 20 ponds (10 unmixed and six mixed) for population genetic analysis (low population densities prevented us from sampling all 20 ponds). These hosts were stored individually in 70% EtOH and later genotyped at 15 microsatellite loci (Auld and Brand 2017).

LABORATORY EXPERIMENT

We maintained replicates of a test set of fifteen *Daphnia* host genotypes with which we examined changes in *Pasteuria* infectivity and within-host growth with respect to the corresponding ancestral isolate. First, we established maternal lines for these host genotypes. Twelve of the genotypes were the same as those hosts used to establish the pond populations (named 11A, 12A, 4A, 5B, 6A, 7A, 8A, 9A, K2B, K3A, M1B, and M2A) and three genotypes were not present in the ponds but were from the same natural host population (named K1A, M3A, and M4A). There were three replicates per genotype; each replicate consisted of eight adult *Daphnia* in 100 mL of artificial media (Klüttgen et al. 1994). Hosts were maintained in a state of clonal reproduction for three generations to minimize variation due to maternal effects, and were fed 0.5 ABS chemostat-grown *Chlorella vulgaris* algae per *Daphnia* per day (ABS refers to the optical absorbance of 650 nm white light by the *C. vulgaris* culture). Jars were incubated at 20°C on a 12L:12D light cycle, and their media was changed three times per week. Second clutch neonates formed the experimental replicates.

The experimental design consisted of a factorial manipulation of these hosts and parasites. We crossed the 15 host genotypes with parasite isolates collected from each of the 20 ponds at the end of the outdoor experiment, plus isolates of the ancestral parasite used to inoculate the ponds at the beginning of the season. On the day of treatment exposure neonates from each maternal line were allocated to parasite treatments following a split-clutch design. There were three replicate parasite isolates per host genotype and thus a total of 945 experimental jars (7560 *Daphnia*). Each jar received a dose of 2×10^5 *Pasteuria* spores and kept under identical conditions as the maternal lines. After 48 hours exposure to the *Pasteuria* spores, the experimental *Daphnia* were transferred into fresh media. The infection status of each *Daphnia* was determined by eye 25 days post exposure, and infected *Daphnia* were then stored at –20°C. Counts of *Pasteuria* transmission spores were later determined with a haemocytometer.

ANALYSIS

Both the outdoor experiment data (host genotype frequency, epidemic size) and laboratory experiment data (parasite infectivity, within-host growth) were analyzed using the R statistical package (R Core Team 2013). First, we tested for rapid parasite adaptation over the course of the season, where parasites perform better on host genotypes with which they share a recent coevolutionary history. We did this by calculating the change in infectivity and spore burdens between the ancestral parasites and the parasite samples collected at the end of the season and then fitting linear-mixed effects models (LMMs) to these data; host type (sympatric/allopatric), and population flux treatment (mixed/unmixed) were fitted as fixed effects, and the pond population-by-host genotype interactions were fitted as random effects with separate intercepts for each host type. Then we fitted a LMM to test whether the change in parasite spore burden was associated with the change in infectivity; once again, pond population-by-host genotype interactions were fitted as random effects with separate intercepts for each host type.

Next, we examined how parasite adaptation covaried with host genotype frequencies at the end of the season (for the 16 ponds for which we had data on host genotype frequencies). Once again, we fitted LMMs to the data for changes in parasite traits, but this time we included final host genotype frequency (along with second-order polynomial), physical flux treatment (mixed/unmixed) and their interaction as fixed effects; here, the pond population-by-host genotype interactions were fitted as random effects with separate intercepts for each population flux treatment. For all LMMs, we applied a Satterthwaite approximation to account for different variances across treatment groups.

By combining data on change in parasite infectivity and spore burden with multilocus genotype frequency data, we dissected the effects of epidemic on: (1) how host populations evolved resistance to the ancestral parasite population (in terms of infectivity and spore burden); (2) how the parasite evolved infectivity and the capacity to proliferate within infected hosts of the ancestral host population; and (3) how coevolution shaped the proportion of infected hosts and spore burdens within evolved host and parasite populations. First, we estimated the relative fitness of host genotypes in each pond population by determining their relative frequency (eq. (1)):

$$\bar{w}_{h,\,t} = P_{h,t}.n_h, \tag{1}$$

where $P_{h,t}$ refers to the frequency of host genotype h at time t, and n_h is the total number of host genotypes used to seed the population (in this case, $n_h = 12$). So, at the beginning of the epidemic ($t = 0$), all host genotypes have a relative fitness of 1. At the end, the relative fitness of each host genotype varies within and across populations.

For each pond population, we calculated the mean change in host susceptibility to infection (Δi_h), and change in the parasite burden in infected hosts (Δs_h) that were exposed to the ancestral parasite (eq. (2a), (2b)):

$$\Delta i_{\,h} = \frac{1}{n}.\sum_h ((i_{h,t=0}.\bar{w}_{h,t=1}) - i_{h,t=0}), \tag{2a}$$

$$\Delta s_h = \frac{1}{n}.\sum_h ((s_{h,t=0}.\bar{w}_{h,t=1}) - s_{h,t=0}), \tag{2b}$$

where $i_{h,t}$ is the proportion of hosts h in each pond that suffer infection at time t (Note that the parasite isolates are identical across populations when $t = 0$.) n is the number of pond populations and $s_{h,t}$ refers to the spore burden on infected hosts h in each pond. Then, we calculated the change in parasite infectivity (Δi_p), and within host spore burdens (Δs_p) for parasite populations that were exposed to the ancestral host population (eq. (3a), (3b)):

$$\Delta i_{\,p} = \frac{1}{n}.\sum_h (i_{h,t=1} - i_{h,t=0}), \tag{3a}$$

$$\Delta s_h = \frac{1}{n}.\sum_h (s_{h,t=1} - s_{h,t=0}), \tag{3b}$$

Next, we calculated the population-level change in infectivity and spore burden that would result from host-parasite coevolution, by weighting the change in parasite traits by the change in host genotype frequencies (eq. (4a), (4b)).

$$\Delta i_{\,hp} = \frac{1}{n}.\sum_h ((i_{h,t=1}.\bar{w}_{h,t=1}) - i_{h,t=0}), \tag{4a}$$

$$\Delta s_{hp} = \frac{1}{n}.\sum_h ((s_{h,t=1}.\bar{w}_{h,t=1}) - s_{h,t=0}). \tag{4b}$$

Finally, we dissected the population-level effects of epidemic size on host evolution of susceptibility to infection and parasite within-host growth (Δi_h, Δs_h), parasite evolution of infectivity and within-host growth (Δi_p, Δs_p), and the coevolutionary outcomes in terms of likelihood of infection and parasite burdens (Δi_{hp}, Δs_{hp}). We did this by fitting linear models (LMs) to each of the six response variables with epidemic size included as a fixed effect.

Results

PARASITE ADAPTATION TO SYMPATRIC HOST POPULATIONS

After just a single epidemic, a striking pattern of rapid parasite adaption emerged across populations. Findings from our laboratory experiment revealed parasite isolates from the end of the season evolved to be more infectious than the ancestral parasite population when exposed to the host genotypes present in the pond populations (sympatric hosts), but significantly less infectious when exposed to novel host genotypes with which they had not interacted with (allopatric hosts) (linear-mixed effects model, LMM: $F_{1, 228.94} = 110.92$, $P < 0.0001$; Fig. 1A; Fig. S1A; Table S1). These changes in parasite infectivity did not differ according to physical flux treatment (LMM: $F_{1, 158.8} = 0.003$, $P = 0.96$; Fig. 1A; Fig. S1A; Table S1).

There was also an increase in the number of parasite spores per infected host in sympatric hosts over the season, but no change in spore burdens in allopatric hosts (LMM: $F_{1, 103.32} = 6.24$, $P < 0.014$; Fig. 1B; Fig. S1B; Table S2); again, this did not differ across physical flux treatments (LMM: $F_{1, 295.68} = 0.003$, $P = 0.99$; Fig. 1B; Fig. S1B; Table S2). The variance explained by the parasite population by host genotype interaction was higher for parasites exposed to sympatric hosts (0.89) than for parasites exposed to allopatric hosts (0.55). Finally, we found a strong positive relationship between the change in spore burden and change in infectivity (Fig. S2, LMM: $F_{1,550.58} = 221.39$, $P < 0.0001$).

PARASITE EVOLUTION AND HOST GENOTYPE FREQUENCIES

Despite near-identical local adaption across physical flux treatments, we uncovered different underlying patterns of host-parasite (co)evolution. After pairing data from both the outdoor and laboratory experiment, we uncovered a strong positive relationship between the change in parasite infectivity over the season and final host genotype frequency in unmixed ponds; in mixed populations, increases in parasite infectivity were highest on host genotypes at intermediate final frequencies (quadratic effect of host genotype frequency × physical flux interaction; LMM: $F_{1, 163.91} = 4.19$, $P = 0.017$; Fig. 2A; Table S3). The proportion of the variance in change in infectivity explained by the parasite population by host

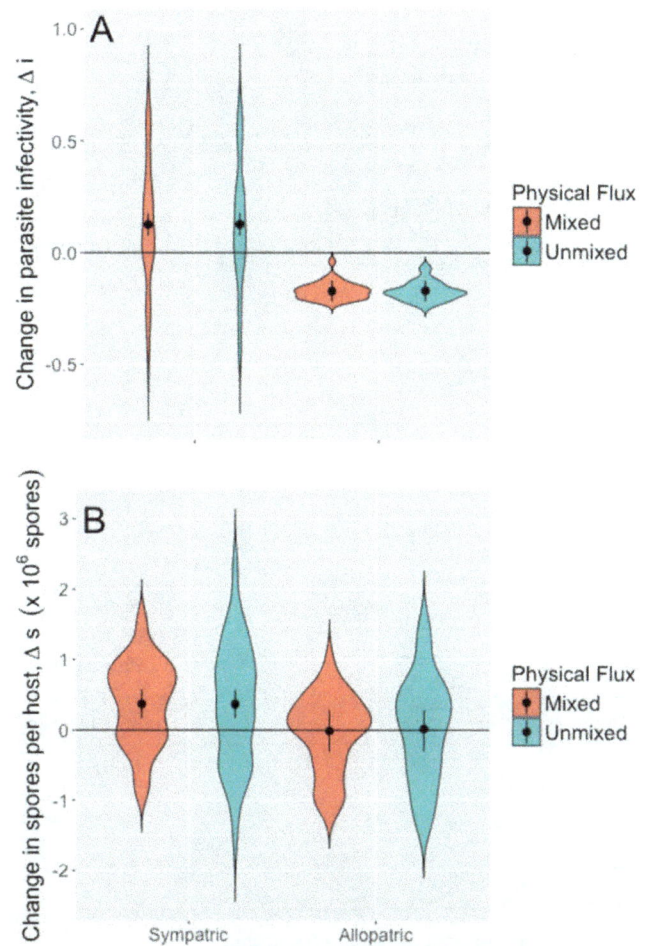

Figure 1. Change in (A) parasite infectivity; and (B) within-host growth in parasites taken from mixed ($n = 10$) and unmixed ($n = 10$) populations when exposed to sympatric hosts (genotypes that were present in the pond populations) and allopatric hosts (related genotypes that were not present in the pond populations). Violin plots show the distribution of the raw data; points and bars denote the means and 95% confidence intervals predicted by the LMM.

genotype interaction was similar in unmixed (39%) and mixed populations (44%). We found a strong positive relationship between the change in the number of parasite spores per infected host over the season and final host genotype frequency in unmixed ponds, and a very weak positive relationship in mixed ponds (host genotype frequency × physical flux interaction; LMM: $F_{1, 189.25} = 5.25$, $P = 0.023$; Fig. 2B; Table S4). The proportion of the variance in change in within-host parasite burden explained by the parasite population by host genotype interaction was higher in unmixed (32%) than in mixed populations (23%).

DISSECTING HOST EVOLUTION, PARASITE EVOLUTION, AND COEVOLUTION

By combining data on infectivity and spore burden in both ancestral and evolved parasite populations with multilocus genotype

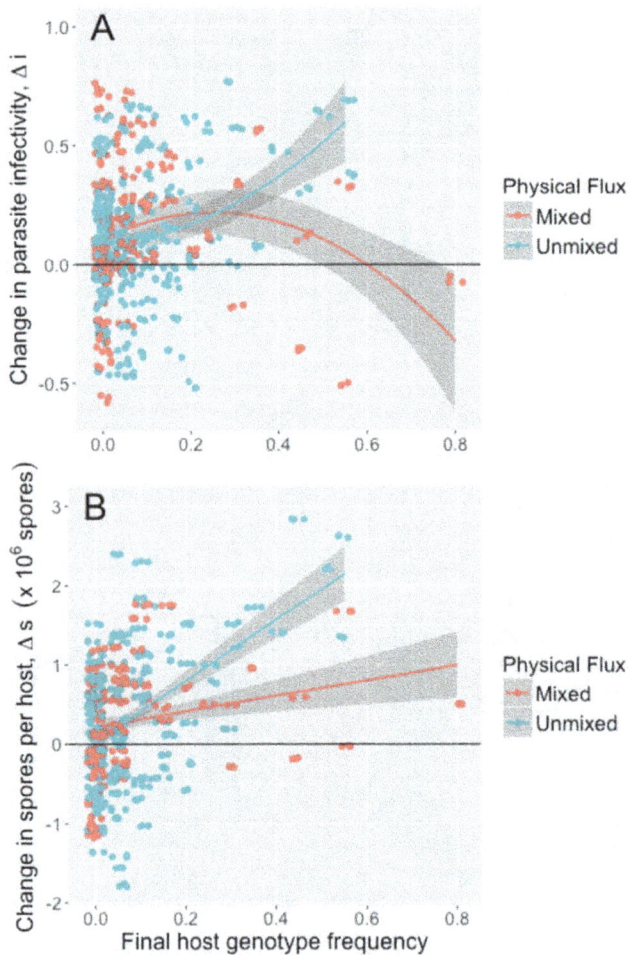

Figure 2. Associations between change in infection traits and host genotype frequencies in sympatric host populations at the end of the epidemic. (A) change in parasite infectivity, and (B) change in within-host growth (between parasite isolates collected after epidemics and corresponding ancestral isolates) taken from mixed ($n = 6$) and unmixed ($n = 10$) populations. Lines and shaded bands denote the slopes and 95% confidence intervals predicted by the LMM; points are jittered for clarity.

Figure 3. Effect of epidemic size on population-level coevolution of infectivity ($n = 16$). (A) Host susceptibility (susceptibility to the ancestral parasite weighted by relative shifts in host genotype frequencies), (B) parasite infectivity (mean infectivity of postepidemic parasite isolates assuming fixed host genotype frequencies), and (C) coevolution of overall infection risk (mean infectivity of postepidemic parasite isolates weighted by relative shifts in host genotype frequencies). Lines and shaded bands denote the predicted slopes and 95% confidence intervals predicted by each LM.

frequency data, we dissected the population-level effects of epidemic size on: (1) how host populations evolved resistance to the ancestral parasite population; (2) how the parasite evolved infectivity and the capacity to proliferate within infected hosts of the ancestral host population (here host genotype frequencies are fixed); and (3) how coevolution shaped the proportion of infected hosts and spore burdens within evolved host and parasite populations. We found larger epidemics were associated with the evolution of reduced host susceptibility to infection (i.e. increased resistance) with the ancestral parasite (Δi_h, linear model, LM: $F_{1,14} = 5.97$, $P = 0.028$; Fig. 3A). The evolution of increased parasite infectivity over the season was not associated with epidemic size (Δi_p, LM: $F_{1,14} = 0.02$, $P = 0.88$; Fig. 3B). The overall change in infection risk (i.e., the change in parasite infectivity when weighted

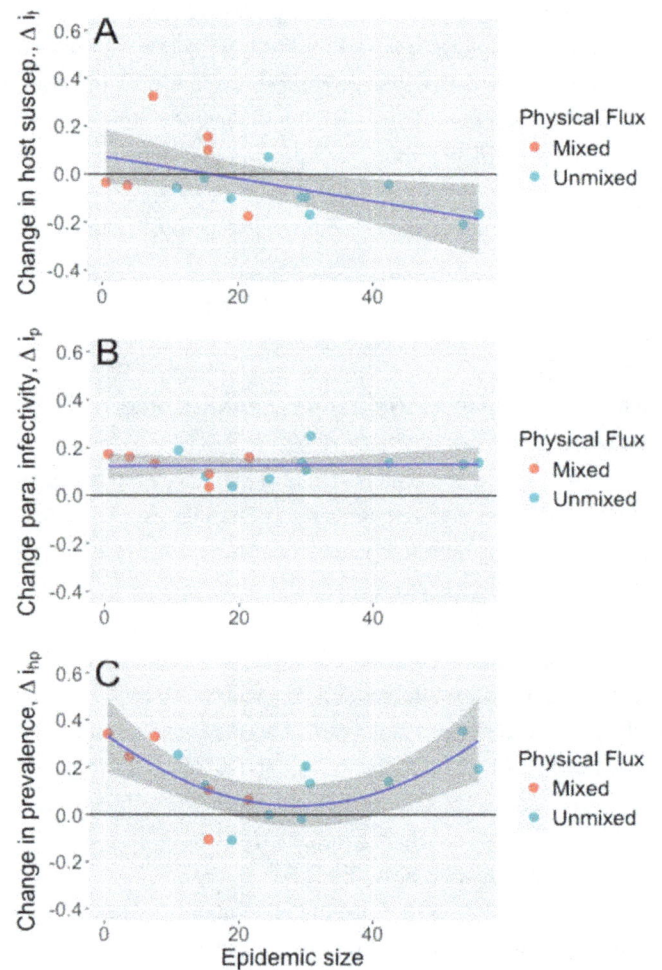

by the final host genotype frequency) exhibited a quadratic relationship with epidemic size: ponds that experienced either small or large epidemics showed an increase in infection risk, whereas ponds that experienced epidemics of an intermediate size showed no overall change in infection risk (Δi_{hp}, LM: $F_{2,13} = 6.40$, $P = 0.012$; Fig. 3C).

Epidemic size had a subtly different effect on the coevolutionary patterns of parasite within-host growth and host resistance to it. Similar to the infectivity data, larger epidemics were associated with the evolution of reduced spore burdens in hosts infected with the ancestral parasite (Δs_h, LM: $F_{1,14} = 7.31$, $P = 0.017$; Fig. 4A). The evolution of increased parasite within-host growth

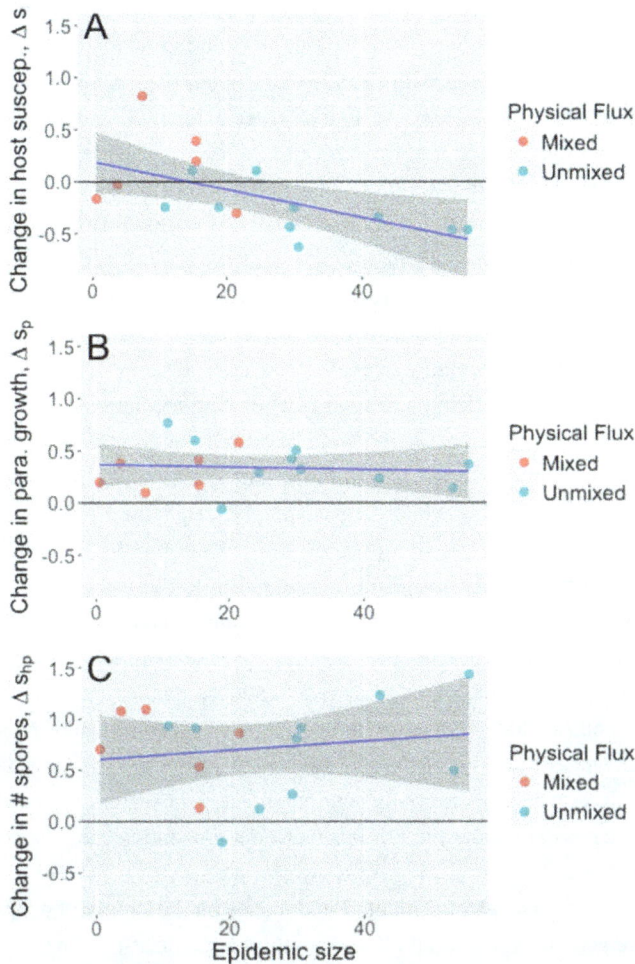

Figure 4. Effect of epidemic size on population-level coevolution of parasite within-host growth ($n = 16$). (A) Host susceptibility (mean spore burdens from the ancestral parasite weighted by relative shifts in host genotype frequencies), (B) parasite within-host growth (mean spore burdens from postepidemic parasite isolates assuming fixed host genotype frequencies), and (C) coevolution of overall parasite burden (mean spore burdens from postepidemic parasite isolates weighted by relative shifts in host genotype frequencies). Lines and shaded bands denote the predicted slopes and 95% confidence intervals predicted by each LM.

though there is the added complication that their principal environment, the host, is also evolving in response to selection (Kawecki and Ebert 2004; Schulte et al. 2011). Moreover, in many host-parasite systems, evolution and adaptation occurs in rapid bursts during disease epidemics (Duffy et al. 2009; Penczykowski et al. 2016). We examined the changes in parasite infection traits across replicate semi-natural populations of *Daphnia magna* and their castrating parasite *Pasteuria ramosa*, and tested if any signatures of adaptation and (co)evolution were either consistent over space, dependent on physical flux, or specific to each individual population. We identified consistent signatures of parasite local adaptation that emerged after just a single epidemic. However, this adaptation was driven by coevolution in unmixed ponds and parasite evolution only in mixed ponds.

Pasteuria populations evolved to better infect sympatric host genotypes at a cost of being poorer at infecting allopatric host genotypes (i.e., hosts which they have not shared a recent coevolutionary history) over the epidemic (Fig. 1A). The emergence of rapid local adaptation and foreign maladaptation is expected, given the known infection genetics in *Daphnia-Pasteuria* systems: alleles that allow a parasite to infect one set of host genotypes lead to an inability to infect other host genotypes (Carius et al. 2001; Auld et al. 2012b, but see Luijckx et al. 2011). The infection genetics in this system is therefore consistent with the matching allele (MA) model of infection (Grosberg and Hart 2000; Bento et al. 2017) and is known to exhibit FS dynamics in the long term (Decaestecker et al. 2007).

The parasite burden data revealed a different pattern: there was evidence of parasite local adaptation to sympatric hosts, but this did not come at a cost of maladaptation to allopatric hosts (Fig. 1B). Previous work has shown that within-host parasite growth varies across parasite genotypes, but does not exhibit genotypic specificity (Vale and Little 2009). So, while infectivity is governed by MA genetics, within-host parasite growth is likely to be a quantitative trait that depends mainly on parasite genotype and its interactions with other environmental conditions (though change in infectivity and change in within-host growth are correlated; Fig. S2). Whether or not infection occurs is by far the most important step in the parasite transmission process, because infected hosts are sterilized and failure of the parasite to bind to the host gut leads to a failure of parasite replication. Therefore, the most intense fluctuating selection occurs for resistant/infectivity alleles at this initial step; alleles for parasite replication will experience less host-mediated selection, because any variation in fitness among infected hosts is minimal when compared to variation between infected and healthy hosts (Ebert et al. 2004).

We expected the physical flux treatment to increase contact rate between hosts and parasites (May and Anderson 1979) leading to larger epidemics, stronger parasite-mediated selection and

over the season was not associated with epidemic size (Δs_p, LM: $F_{1,14} = 0.12$, $P = 0.74$; Fig. 4B). Finally, there was an increase in the overall change in spore burden (i.e., the change in spores per infected host when weighted by the final host genotype frequency) over the season, but this was not associated with epidemic size (Δs_{hp}, LM: $F_{1,14} = 0.387$, $P = 0.54$; Fig. 4C).

Discussion

Natural selection commonly drives populations to become adapted to their local environment. Parasites are no different, al-

greater parasite adaptation (Morgan et al. 2005), as found in a phage-bacteria system (Gómez et al. 2015). However, adaptation was equally strong across physical flux treatments even though epidemics were smaller in mixed ponds than in unmixed ponds (Fig. 1A). This is because the host populations in mixed ponds experience direct selection from the mixing treatment (Auld and Brand 2017), with downstream consequences for parasite evolution and adaptation. An examination of coevolutionary patterns across 16 of the 20 populations (i.e., populations for which there was host genotype frequency data) revealed that physical flux changed the nature of host-mediated selection on the parasite population. Previous work has found that coevolutionary paths are idiosyncratic to individual populations (Schulte et al. 2011), whereas we find that coevolution is consistent across the unmixed populations, and that parasite evolution and host evolution is consistent across the mixed populations. Host genotypes either increase or decrease in frequency in a generally consistent manner across populations of each mixing treatment, and measures of drift are comparatively low (Auld and Brand 2017). Also, in unmixed populations, the parasite evolved increased infectivity and within-host growth when exposed to host genotypes that became more common, but not when exposed to hosts that became rare (Fig. 2A, B). By contrast, in mixed populations, the parasite best adapted to infect host genotypes that achieved intermediate frequency and grew best within hosts that became rare (Fig. 2A, B).

Although the (co)evolutionary paths differ according to physical flux treatment, we find evidence that FS on parasite infectivity is operating across the board: the proportion of random effects variance explained by a host genotype by population interaction is consistent across physical flux treatments, which tells us there is no overall shift in host range (as would be predicted by ARE models: Betts et al. 2014). The same is not true for the change in parasite burden: the variance explained by the host genotype by population interaction is much lower in mixed ponds than in unmixed ponds, indicating the parasite is more of a generalist in mixed ponds in terms of within-host growth (Gómez et al. 2015). Both ARE and FS models of coevolution predict that parasite populations should adapt to common host genotypes after a lag, and that parasites should generally perform best on hosts from the recent past (Jaenike 1978; Hutson and Law 1981; Nee 1989; Sasaki 2000; Lively 2016). Empirical data from invertebrate-trematode (Dybdahl and Lively 1998; Koskella and Lively 2009), bacteria-phage (Koskella 2014) and *Daphnia-Pasteuria* (Decaestecker et al. 2007) systems support these theoretical predictions; our findings further demonstrate that this adaptation can occur extremely rapidly, within a single epidemic.

By weighting the changes in parasite infectivity and within-host growth by shifts in host genotype frequency, we were able to dissect host evolution, parasite evolution and host-parasite coevolution. Our previous study demonstrated that unmixed populations suffered larger epidemics than mixed populations (Auld and Brand 2017). Here, we found larger epidemics (in unmixed ponds) selected for host resistance to the ancestral parasite, both in terms of infectivity and within-host growth (Fig. 3A, Fig. 4A), consistent with previous studies (Duncan et al. 2006; Duffy et al. 2012). By contrast, epidemic size had no effect on the mean change in parasite infectivity of within-host growth (Fig. 3B, Fig. 4B). The change in overall infection risk as a result of host-parasite coevolution revealed how coevolution varied across epidemics of differing size, and that parasite-mediated selection and host-mediated selection were often not equal in magnitude. In mixed populations that suffered small epidemics, hosts remained susceptible while parasites evolved increased infectivity and growth, leading to higher overall infection risk (Fig. 3A); host evolution did not result in adaptation to the ancestral parasite, whereas parasite evolution did result in adaptation to the local suite of host genotypes. In unmixed populations, infection risk increased with epidemic size: large epidemics and strong parasite-mediated selection for host resistance was outweighed by parasite evolution of increased infectivity. Moreover, the parasite evolved to be proportionally better at infecting host genotypes that were resistant to the ancestral parasite, demonstrating coevolution (Fig. 2A).

Host and parasite adaptation depends on the supply of adaptive genetic variation and the strength of antagonist-mediated selection relative to other selective forces. Host-mediated selection is equally strong across populations irrespective of epidemic size and environmental conditions, whereas parasite-mediated selection was much stronger in unmixed populations because they experienced larger epidemics. The host generally provides the principal environment and thus the main selective force acting on parasite infectivity and growth, whereas host populations experience a wide range of different selective forces in addition to selection for resistance to parasitism. Unmixed ponds, with their large epidemics and symmetric parasite- and host-mediated selection, were "coevolutionary hotspots." By contrast, mixed ponds with their smaller epidemics, weak parasite-mediated selection and strong host-mediated selection were "coevolutionary coldspots" (Thompson 2005). Nevertheless, these environment-dependent signatures of evolution and coevolution result in similar patterns of adaptation across parasite populations, demonstrating a certain level of repeatability across noisy environments.

AUTHOR CONTRIBUTIONS

S.K.J.R.A. planned the project. S.K.J.R.A. and J.B. collected the outdoor experiment data. S.K.J.R.A. conducted the laboratory infection experiment, analyzed the data, and wrote the manuscript. Both S.K.J.R.A. and J.B. approved the final version of the manuscript.

ACKNOWLEDGMENTS
A. Hayward, M. Tinsley and three anonymous reviewers provided useful manuscript comments. Work was funded by a NERC Fellowship to S.K.J.R.A (NE/L011549/1).

LITERATURE CITED

Agrawal, A., and C. M. Lively. 2002. Infection genetics: gene-for-gene versus matching-alleles models and all points in between. Evol. Ecol. Res. 4:91–107.

Auld, S. K. J. R., and J. Brand. 2017. Simulated climate change, epidemic size and host evolution across host-parasite populations. Global Change Biol. Early View.

Auld, S. K. J. R., K. H. Edel, and T. J. Little. 2012a. The cellular immune response of *Daphnia magna* under host–parasite genetic variation and variation in initial dose. Evolution 66:3287–3293.

Auld, S. K. J. R., P. J. Wilson, and T. J. Little. 2014a. Rapid change in parasite infection traits over the course of an epidemic in a wild host–parasite population. Oikos 123:232–238.

Auld, S. K. J. R., S. R. Hall, and M. A. Duffy. 2012b. Epidemiology of a *Daphnia*—multiparasite system and its implications for the red queen. PLoS One 7:e39564.

Auld, S. K. J. R., S. R. Hall, J. H. Ochs, M. Sebastian, and M. A. Duffy. 2014b. Predators and patterns of within-host growth can mediate both among-host competition and evolution of transmission potential of parasites. Am. Nat. 184:S77–S90.

Bankers, L., P. Fields, K. E. McElroy, J. L. Boore, J. M. J. Logsdon, and M. Neiman. 2017. Genomic evidence for population-specific responses to co-evolving parasites in a New Zealand freshwater snail. Mol. Ecol. 26:3663–3675.

Bento, G., J. Routtu, P. D. Fields, Y. Bourgeois, L. Du Pasquier, and D. Ebert. 2017. The genetic basis of resistance and matching-allele interactions of a host-parasite system: the *Daphnia magna-Pasteuria ramosa* model. PLoS Genet. 13:e1006596.

Betts, A., O. Kaltz, and M. E. Hochberg. 2014. Contrasted coevolutionary dynamics between a bacterial pathogen and its bacteriophages. Proc. Natl Acad. Sci. 111:11109–11114.

Blanford, S., M. B. Thomas, C. Pugh, and J. K. Pell. 2003. Temperature checks the Red Queen? Resistance and virulence in a fluctuating environment. Ecol. Lett. 6:2–5.

Blanquart, F., and S. Gandon. 2013. Time-shift experiments and patterns of adaptation across time and space. Ecol. Lett. 16:31–38.

Brockhurst, M. A., A. D. Morgan, P. B. Rainey, and A. Buckling. 2003. Population mixing accelerates coevolution. Ecol. Lett. 6:975–979.

Carius, H. J., T. J. Little, and D. Ebert. 2001. Genetic variation in a host-parasite association: potential for coevolution and frequency-dependent selection. Evolution 55:1136–1145.

Cressler, C. E., W. A. Nelson, T. Day, and E. McCauley. 2014. Starvation reveals the cause of infection-induced castration and gigantism. Proc. R Soc. B 281:20141087.

Decaestecker, E., S. Gaba, J. A. M. Raeymaekers, R. Stoks, L. Van Kerckhoven, D. Ebert, and L. De Meester. 2007. Host-parasite "Red Queen" dynamics archived in pond sediment. Nature 450:870–873.

Duffy, M. A., J. H. Ochs, R. M. Penczykowski, D. J. Civitello, C. A. Klausmeier, and S. R. Hall. 2012. Ecological context influences epidemic size and parasite-driven evolution. Science 335:1636–1638.

Duffy, M. A., S. R. Hall, C. E. Cáceres, and A. R. Ives. 2009. Rapid evolution, seasonality, and the termination of parasite epidemics. Ecology 90:1441–1448. Ecological Society of America.

Duncan, A. B., S. E. Mitchell, and T. J. Little. 2006. Parasite-mediated selection and the role of sex and diapause in *Daphnia*. J. Evol. Biol. 19:1183–1189.

Duneau, D., P. Luijckx, F. Ben Ami, C. Laforsch, and D. Ebert. 2011. Resolving the infection process reveals striking differences in the contribution of environment, genetics and phylogeny to host-parasite interactions. BMC Biol. 9:11.

Dybdahl, M. F., and C. M. Lively. 1998. Host-parasite coevolution: evidence for rare advantage and time-lagged selection in a natural population. Evolution 52:1057.

Ebert, D. 1994. Virulence and local adaptation of a horizontally transmitted parasite. Science 265:1084–1086.

Ebert, D., H. Joachim Carius, T. Little, and E. Decaestecker. 2004. The evolution of virulence when parasites cause host castration and gigantism. Am. Nat. 164:S19–S32.

Ebert, D., P. Rainey, T. M. Embley, and D. Scholz. 1996. Development, life cycle, ultrastructure and phylogenetic position of *Pasteuria ramosa* Metchnikoff 1888: rediscovery of an obligate endoparasite of *Daphnia magna* straus. Phil. Trans. R Soc. B 351:1689–1701.

Gaba, S., and D. Ebert. 2009. Time-shift experiments as a tool to study antagonistic coevolution. Trends Ecol. Evol. 24:226–232.

Gandon, S., and S. L. Nuismer. 2009. Interactions between genetic drift, gene flow, and selection mosaics drive parasite local adaptation. Am. Nat. 173:212–224.

Gibson, A. K., L. F. Delph, and C. M. Lively. 2017. The two-fold cost of sex: experimental evidence from a natural system. Evol. Lett. 1:6–15.

Gómez, P., Ben Ashby, and A. Buckling. 2015. Population mixing promotes arms race host–parasite coevolution. Proc. R Soc. B 282:20142297.

Grosberg, R. K., and M. W. Hart. 2000. Mate selection and the evolution of highly polymorphic self/nonself recognition genes. Science 289:2111–2114.

Hamilton, W. D. 1980. Sex versus non-sex versus parasite. Oikos 35:282.

Harrison, E., A.-L. Laine, M. Hietala, and M. A. Brockhurst. 2013. Rapidly fluctuating environments constrain coevolutionary arms races by impeding selective sweeps. Proc. R Soc. B 280:20130937.

Howard, R. S., and C. M. Lively. 1994. Parasitism, mutation accumulation and the maintenance of sex. Nature 367:554–557.

Hutson, V., and R. Law. 1981. Evolution of recombination in populations experiencing frequency-dependent selection with time delay. Proc. R Soc. B 213:345–359.

Imhoof, B., and P. Schmid Hempel. 1998. Patterns of local adaptation of a protozoan parasite to its bumblebee host. Oikos 82:59.

Jaenike, J. 1978. A hypothesis to account for the maintenance of sex within populations. Evol. Theory 3:191–194.

Kaltz, O., and J. A. Shykoff. 1998. Local adaptation in host-parasite systems. Heredity 81:361–370.

Kawecki, T. J., and D. Ebert. 2004. Conceptual issues in local adaptation. Ecology Letters 7:1225–1241.

Klüttgen, B., U. Dülmer, M. Engels, and H. T. Ratte. 1994. ADaM, an artificial freshwater for the culture of zooplankton. Water Res. 28:743–746.

Koskella, B. 2014. Bacteria-phage interactions across time and space: merging local adaptation and time-shift experiments to understand phage evolution*. Am. Nat. 184:S9–S21.

Koskella, B., and C. M. Lively. 2009. Evidence for negative frequency-dependent selection during experimental coevolution of a freshwater snail and a sterilizing trematode. Evolution 63:2213–2221.

Laine, A.-L. 2007. Detecting local adaptation in a natural plant–pathogen metapopulation: a laboratory vs. field transplant approach. J. Evol. Biol. 20:1665–1673.

———. 2008. Temperature-mediated patterns of local adaptation in a natural plant–pathogen metapopulation. Ecol. Lett. 11:327–337.

Lazzaro, B. P., and T. J. Little. 2009. Immunity in a variable world. Phil. Trans. R. Soc. B 364:15–26.

Lenski, R. E., and B. R. Levin. 2015. Constraints on the coevolution of bacteria and virulent phage: a model, some experiments, and predictions for natural communities. Am. Nat. 125:585–602.

Lively, C. M. 1989. Adaptation by a parasitic trematode to local populations of its snail host. Evolution 43:1663.

———. 2016. Coevolutionary epidemiology: disease spread, local adaptation, and sex. Am. Nat. 187:E77–E82. https://doi.org/10.1086/684626.

Luijckx, P., F. Ben Ami, L. Mouton, L. Du Pasquier, and D. Ebert. 2011. Cloning of the unculturable parasite *Pasteuria ramosa* and its Daphnia host reveals extreme genotype–genotype interactions. Ecol. Lett. 14:125–131.

May, R. M., and R. M. Anderson. 1979. Population biology of infectious diseases: part II. Nature 280:455–461.

Morgan, A. D., S. Gandon, and A. Buckling. 2005. The effect of migration on local adaptation in a coevolving host-parasite system. Nature 437:253–256.

Mostowy, R., and J. Engelstädter. 2011. The impact of environmental change on host–parasite coevolutionary dynamics. Proc. R Soc. B 278:2283–2292.

Nee, S. 1989. Antagonistic co-evolution and the evolution of genotypic randomization. J. Theor. Biol. 140:499–518.

Obbard, D. J., F. M. Jiggins, N. J. Bradshaw, and T. J. Little. 2011. Recent and recurrent selective sweeps of the antiviral RNAi gene argonaute-2 in three species of Drosophila. Mol. Biol. Evol. 28:1043–1056.

Oppliger, A., R. Vernet, and M. Baez. 1999. Parasite local maladaptation in the Canarian lizard *Gallotia galloti* (Reptilia: Lacertidae) parasitized by haemogregarian blood parasite. J. Evol. Biol. 12:951–955.

Paterson, S., T. Vogwill, A. Buckling, R. Benmayor, A. J. Spiers, N. R. Thomson, M. Quail, F. Smith, D. Walker, B Libberton, A. Fenton, N. Hall, and M. A. Brockhurst. 2010. Antagonistic coevolution accelerates molecular evolution. Nature 464:275–278.

Penczykowski, R. M., A.-L. Laine, and B. Koskella. 2016. Understanding the ecology and evolution of host–parasite interactions across scales. Evol. Appl. 9:37–52.

R Core Team. 2013. R: A language and environment for statistical computing. Vienna, Austria.

Sasaki, A. 2000. Host-parasite coevolution in a multilocus gene-for-gene system. Proc. R Soc. B 267:2183–2188.

Schulte, R. D., C. Makus, B. Hasert, N. K. Michiels, and H. Schulenburg. 2011. Host–parasite local adaptation after experimental coevolution of *Caenorhabditis elegans* and its microparasite *Bacillus thuringiensis*. Proc. R Soc. B 278:2832–2839.

Thompson, J. N. 2005. The geographic mosaic of coevolution. Chicago Univ. Press, Chicago.

Thrall, P. H., A.-L. Laine, M. Ravensdale, A. Nemri, P. N. Dodds, L. G. Barrett, and J. J. Burdon. 2012. Rapid genetic change underpins antagonistic coevolution in a natural host-pathogen metapopulation. Ecol. Lett. 15:425–435.

Vale, P. F., and T. J. Little. 2009. Measuring parasite fitness under genetic and thermal variation. Heredity 103:102–109.

Wolinska, J., and K. C. King. 2009. Environment can alter selection in host–parasite interactions. Trends Parasitol. 25:236–244.

12

Feminizing *Wolbachia* endosymbiont disrupts maternal sex chromosome inheritance in a butterfly species

Daisuke Kageyama,[1,2] (iD) Mizuki Ohno,[3] Tatsushi Sasaki,[3] Atsuo Yoshido,[3] Tatsuro Konagaya,[4] Akiya Jouraku,[1] Seigo Kuwazaki,[1] Hiroyuki Kanamori,[5] Yuichi Katayose,[5] Satoko Narita,[1,6] Mai Miyata,[7] Markus Riegler,[8] (iD) and Ken Sahara[3,9] (iD)

[1] *Institute of Agrobiological Sciences, National Agriculture and Food Research Organization, Tsukuba, Ibaraki 305–0854, Japan*

[2] *E-mail: kagymad@affrc.go.jp*

[3] *Laboratory of Applied Entomology, Faculty of Agriculture, Iwate University, Morioka 020–8550, Japan*

[4] *Graduate School of Science, Kyoto University, Kyoto 606–8502, Japan*

[5] *Institute of Crop Science, National Agriculture and Food Research Organization, Tsukuba, Ibaraki 305–0854, Japan*

[6] *Tsukuba Primate Research Center, National Institute of Biomedical Innovation, Hachimandai, Tsukuba, Ibaraki 305–0843, Japan*

[7] *Graduate School of Horticulture, Chiba University, Matsudo, Chiba 271–8510, Japan*

[8] *Hawkesbury Institute for the Environment, Western Sydney University, Penrith, New South Wales 2751, Australia*

[9] *E-mail: sahara@iwate-u.ac.jp*

Wolbachia is a maternally inherited ubiquitous endosymbiotic bacterium of arthropods that displays a diverse repertoire of host reproductive manipulations. For the first time, we demonstrate that *Wolbachia* manipulates sex chromosome inheritance in a sexually reproducing insect. *Eurema mandarina* butterfly females on Tanegashima Island, Japan, are infected with the *w*Fem *Wolbachia* strain and produce all-female offspring, while antibiotic treatment results in male offspring. Fluorescence in situ hybridization (FISH) revealed that *w*Fem-positive and *w*Fem-negative females have Z0 and WZ sex chromosome sets, respectively, demonstrating the predicted absence of the W chromosome in *w*Fem-infected lineages. Genomic quantitative polymerase chain reaction (qPCR) analysis showed that *w*Fem-positive females lay only Z0 eggs that carry a paternal Z, whereas females from lineages that are naturally *w*Fem-negative lay both WZ and ZZ eggs. In contrast, antibiotic treatment of adult *w*Fem females resulted in the production of Z0 and ZZ eggs, suggesting that this *Wolbachia* strain can disrupt the maternal inheritance of Z chromosomes. Moreover, most male offspring produced by antibiotic-treated *w*Fem females had a ZZ karyotype, implying reduced survival of Z0 individuals in the absence of feminizing effects of *Wolbachia*. Antibiotic treatment of *w*Fem-infected larvae induced male-specific splicing of the *doublesex* (*dsx*) gene transcript, causing an intersex phenotype. Thus, the absence of the female-determining W chromosome in Z0 individuals is functionally compensated by *Wolbachia*-mediated conversion of sex determination. We discuss how *Wolbachia* may manipulate the host chromosome inheritance and that *Wolbachia* may have acquired this coordinated dual mode of reproductive manipulation first by the evolution of female-determining function and then cytoplasmically induced disruption of sex chromosome inheritance.

KEY WORDS: Butterfly, chromosome inheritance, sex determination, *Wolbachia*.

Impact Summary

Genomes are vulnerable to selfish genetic elements that enhance their own transmission often at the expense of host fitness. Examples are cytoplasmic elements such as maternally inherited bacteria that cause feminization, male-killing, parthenogenesis, and cytoplasmic incompatibility. We demonstrate, for the first time, that the inheritance of a chromosome can be hampered by the ubiquitous endosymbiotic bacterium *Wolbachia*. For *Eurema mandarina* butterfly lineages with a Z0 sex chromosome constitution, we provide direct and conclusive evidence that *Wolbachia* induces production of all-female progeny by a dual role: the compensation for the female-determining function that is absent in Z0 lineages and the prevention of maternal sex chromosome inheritance to offspring. Therefore, our findings highlight that both sex determination and chromosome inheritance—crucially important developmental processes of higher eukaryotes—can be manipulated by cytoplasmic parasites.

Genomes of sexually reproducing organisms are exposed to genetic conflicts. For example, some genes bias reproduction toward male offspring while other genes within the same genome may favor reproduction of more daughters. Selfish genetic elements (SGEs), such as meiotic drivers, cytoplasmic sex ratio distorters and transposons, are extreme examples, which enhance their own transmission often at the expense of their hosts' fitness (Burt and Trivers 2006; Werren 2011). There is growing evidence that SGEs, and their genetic conflict with host genomes, trigger important evolutionary change and innovation in eukaryotes (Werren 2011).

Meiotic drive, also referred to as segregation distortion (SD), is a violation of Mendelian law as it leads to the more frequent inheritance of one copy of a gene than the expected 50% (Jaenike 2001; Lindholm et al. 2016). A segregation distorter that sits on a sex chromosome biases the sex ratio. For example, X-linked segregation distorter (X drive) and Y-linked segregation distorter (Y drive) in flies (Diptera), result in female-biased and male-biased sex ratios, respectively (Lindholm et al. 2016). In male-heterogametic species, X and Y drivers are expected to be encoded in the nuclear genome. In female-heterogametic species, however, W chromosome and cytoplasm behave as a single linkage group and thus distortion of sex chromosome inheritance in female-heterogametic species can theoretically also be caused by cytoplasmic elements. Although this possibility has previously been proposed (Hurst 1993; Beukeboom and Perrin 2014), lack of empirical evidence questions whether it is mechanistically possible for cytoplasmic elements to cause meiotic drive.

Wolbachia pipientis (Alphaproteobacteria), simply referred to as *Wolbachia*, attracts significant interest in evolutionary and developmental biology but also in applied fields such as pest management because it can manipulate reproduction of arthropods in various ways such as cytoplasmic incompatibility, parthenogenesis induction, feminization, and male-killing (Werren et al. 2008). Here we demonstrate for the first time that *Wolbachia* is responsible for the disruption of sex chromosome inheritance. We do this by providing multifaceted and conclusive evidence that in the butterfly *Eurema mandarina Wolbachia*-induced disruption of chromosome inheritance, which may be the result of SD, constitutes the underlying mechanism for the production of all-female progeny. In most populations, *E. mandarina* is infected with the cytoplasmic-incompatibility (CI)-inducing *Wolbachia* strain *w*CI at a high prevalence of close to 100% (Hiroki et al. 2005; Narita et al. 2006). Hiroki et al. (2002 and 2004) first reported all-female offspring production in *E. mandarina* (then known as *Eurema hecabe* yellow type), which was considered to be due to the feminization of genetic males (ZZ) by coinfections with the *Wolbachia* strain *w*Fem (hereafter referred to as double infection CF while single infection with *w*CI is referred to as C). Three observations about CF lineages supported this view, that is (a) antibiotic treatment of adult females led to the production of all-male offspring (Hiroki et al. 2002), (b) antibiotic treatment of larvae resulted in intersex adults (Narita et al. 2007a) and (c) females did not have the W chromatin body (Hiroki et al. 2002; Narita et al. 2007a). This has recently been challenged, because it was demonstrated that CF females have only one Z chromosome and that this Z chromosome always derived from their fathers implying that a disruption of chromosome inheritance may be in place although it was not clear whether *Wolbachia* induced the disruption (Kern et al. 2015). As a consequence, two novel (yet untested) hypotheses were formed, namely, that CF females have either a Z0 or a W'Z sex chromosome set (whereby W' cannot be visualized in W chromatin assays and does not have a female-determining function), and that the disruption of Z chromosome inheritance occurs in CF lineages due to *Wolbachia* or another factor, such as those encoded by the host nucleus. Moreover, the intensity of chromosome disruption was not known, and therefore, killing of ZZ males may complement the incomplete chromosome disruption to achieve all-female production.

In a multifaceted approach, by combining fluorescence in situ hybridization (FISH), genome sequencing, quantitative PCR, reverse transcription PCR and antibiotic treatment, we have tested these hypotheses and revealed that CF females have Z0, and that *Wolbachia* is the cause for both the 100% disruption of Z chromosome inheritance and the female sex determination of Z0 individuals. Our results demonstrate, for the first time, *Wolbachia* as the agent that is responsible for distorted sex chromosome inheritance, and thereby highlight that cytoplasmic elements can have profound effects on oogenesis, sex chromosome inheritance, and sex determination–fundamental biological processes of eukaryotes.

Methods

COLLECTION AND REARING OF *E. MANDARINA*

Female adults of *E. mandarina* (Lepidoptera: Pieridae) were collected on Tanegashima Island, Kagoshima, Japan (Fig. S1). In the laboratory, each female was allowed to lay embryos on fresh leaves of *Lespedeza cuneata* (Fabales: Fabaceae) in a plastic cup with absorbent cotton immersed with 5% honey solution. The artificial diet for larvae was prepared by mixing leaf powder of *Albizia julibrissin* (Fabales: Fabaceae) in the custom-made Silkmate devoid of mulberry leaves (Nihon-Nosan, Yokohama, Japan). Insects were reared under the 16 h/8 h light/dark photoperiod at 25°C.

ANTIBIOTIC TREATMENT

We performed antibiotic treatment of two different stages (larval stage and adult stage) of *E. mandarina*. For larval antibiotic treatment, larvae were fed with the artificial diet (shown above) containing 0.05% tetracycline hydrochloride (tet). For adult antibiotic treatment, female adults were fed with 5% honey solution containing 0.1% tet. Specifically, CF females were mated to antibiotic-treated male offspring of C females. Antibiotic treatment of these males was to avoid a possible occurrence of CI. After mating, each CF female was allowed to lay eggs on fresh leaves of *L. cuneata* in a plastic cup with absorbent cotton immersed with 5% honey solution containing 0.1% tet. Fresh leaves of *L. cuneata* and cotton with tet-containing honey solution were exchanged daily.

DIAGNOSIS OF *WOLBACHIA* STRAINS

To diagnose *Wolbachia* strains in *E. mandarina*, several legs of each adult were homogenized in STE buffer (10 mM Tris-HCl (pH 8.0), 1 mM EDTA (pH 8.0), 150 mM NaCl) and incubated at 56°C for 30 min followed by 92°C for 5 min. After centrifugation at 15,000 rpm for 2 min, the supernatant was used for polymerase chain reaction (PCR) using different primer pairs. The primer pair wsp81F (5′–TGGTCCAATAAGTGATGAAGAAAC–3′) and wsp691R (5′–AAAAATTAAACGCTACTCCA–3′) amplifies a ca. 610-bp fragment of the *Wolbachia* wsp gene (Braig et al. 1998). The primer pair wsp81F and HecCIR (5′–ACTAAC GTCGTTTTTGTTTAG–3′) amplifies a 232-bp fragment of the *wsp* gene of *w*CI, while the primer pair HecFemF (5′– TTACTCACAATTGGCTAAAGAT–3′) and the wsp691R amplifies a 398-bp fragment of *wsp* gene of *w*Fem (Hiroki et al. 2004; Narita et al. 2007b).

WHOLE GENOME SEQUENCING AND *DE NOVO* ASSEMBLY

We performed whole genome sequencing for three types of *E. mandarina* individuals (CF females, C females, and C males) that were collected on Tanegashima Island, Japan (Fig. S1). Six genomic DNA libraries (two libraries for each sample type derived from two individuals) were constructed following manufacturer's instructions (http://www.illumina.com). The average insert size of the libraries was approximately 350 bp and each library was multiplexed using a single indexing protocol. The genomic DNA libraries were sequenced by Illumina MiSeq using MiSeq Reagent Kit v3 (600-cycle) (Illumina, San Diego, CA). Generated raw reads (8.31 Gb, 5.34 Gb, and 6.94 Gb for CF females, C females and C males, respectively) were filtered by Trimmomatic (Bolger et al. 2014) and then mapped to the complete genome of *Wolbachia* strain wPip (GenBank: NC_010981.1) by Bowtie2 (Langmead and Salzberg 2012). Mapped reads were discarded (to eliminate *Wolbachia* sequences) and then remaining reads of the three samples were merged and *de novo* assembled by SGA assembler (Simpson and Durbin 2012). Generated genome contig sequences were used for further analysis.

ANALYSIS OF MAPPED READ COUNTS ON CHROMOSOMES

To verify that CF and C females have one Z chromosome, we compared normalized mapped read counts of the three samples on Z chromosomes and remaining chromosomes. The filtered reads of each sample were mapped to the genome contigs by Bowtie2 (only concordantly and uniquely mapped reads were counted) and then normalized mapped read count of each sample on each contig was calculated based on the ratio of the number of total mapped reads between the three samples. Nucleotide sequences of relatively long genome contigs (length is 2 kb or more) with enough coverage (20 or more mapped reads) were extracted and compared with the gene set A of *B. mori* (Suetsugu et al. 2013) by blastx search (cutoff e-value is 1e-50). Genome contigs with blastx hits were extracted and classified into 28 chromosomes based on the location of the homologous *B. mori* genes. For each chromosome, the average number of relative normalized mapped read counts was calculated for each sample (the number of C males was normalized to 1) using the normalized mapped read counts in the classified genome contigs, respectively.

SANGER SEQUENCING

To genotype Z chromosomes, a highly variable intron of Z-linked triosephosphate isomerase (*Tpi*) gene was PCR amplified using the primers, 5′–GGTCACTCTGAAAGGAGAACCACTTT– 3′ and 5′–CACAACATTTGCCCAGTTGTTGCAA–3′, located in coding regions (Jiggins et al. 2001). The PCR products were treated with ExoSAP-IT® (Affymetrix Inc., Santa Clara, CA) and subjected to direct sequencing at Eurofins Genomics K.K. (Tokyo, Japan). No indels or SNPs were observed in sequence chromatograms of females; some males were heterozygous due to detected double peaks and shifts of sequence reads. By sequencing from both sides, it was possible to obtain the genotypes of males and females (Fig. S4).

FISH ANALYSIS

In most lepidopteran species, a conspicuous heterochromatic body is exclusively found in female polyploid nuclei. Since W derived-BAC as well as genomic probes have highlighted the W chromosomes and heterochromatin bodies in *B. mori* (Sahara et al. 2003a,b), there is no doubt that the bodies consist of the W chromosomes. The diagnosis however remains unreliable if a species of interest carries a W–autosomal translocation and/or partial deletion of the W (Traut and Marec 1996; Abe et al. 2008). Hiroki et al. (2002) as well as Narita et al. (2007a) relied on the W-body diagnosis for C and CF females and concluded that they have WZ and ZZ sex chromosome constitutions, respectively. However, Kern et al. (2015) has recently found that, on the basis of genomic qPCR designed to amplify Z-linked gene sequences (*Tpi* and *Ket*) relative to an autosomal gene (*EF-1α*), both CF and C females have only one Z chromosome while males have two Z chromosomes. This finding rejected the previous conclusion that the sex chromosome constitution of CF females is ZZ (Hiroki et al. 2002; Narita et al. 2007a) but was inconclusive about whether CF females have a Z0 or W'Z system (with W' as a modified W that has lost the female-determining function and cannot be detected by the W-body assay). Hence, we carried out more extensive chromosome analysis (other than just the W-body) to directly prove whether CF females carry the W or not.

In Lepidoptera, the W chromosome can be highlighted by FISH using probes prepared from whole genomic DNA of males or females. The capability of FISH probes in detecting the W chromosome is due to the numerous repetitive short sequences occupying the W chromosome, which is then prone to be hybridized by random sequences. Genomic probes also paint repetitive regions scattered across other chromosomes, albeit at a lower density (autosomes and Z chromosome). Here we made mitotic and pachytene chromosome preparations from wing discs and gonads, respectively, in the last instar larvae of C and CF individuals of *E. mandarina* (see Yoshido et al. (2014) for details). Genomic DNA was extracted from tet-treated C female larvae. Insect telomeric repeats were amplified by nontemplate PCR (Sahara et al. 1999). *Kettin* (*Ket*) gene fragments were amplified from adult cDNA synthesized by PrimeScript™ RT reagent Kit (TaKaRa, Otsu, Japan) and cloned by TOPO® TA Cloning® Kit (Thermo Fisher Scientific, Waltham, MA). We used four pairs of primers, Em_kettin_F1: 5′–AGGTAATCCAACGCCAGTCG–3′ and Em_kettin_R1: 5′–TGCTTGCCCTAAGGCATTGT–3′,

Em_kettin_F2: 5′–ACAATGCCTTAGGGCAAGCA–3′ and Em_kettin_R2: 5′–TGGGCAAAGCCTCTTCATGT–3′, Em_kettin_F3: 5′–AGATTCCGCACTACGCATGA–3′ and Em_kettin_R3: 5′–TAAATTGTGGTGGGACGGCA–3′, Em_kettin_F5: 5′–ACATGAAGAGGCTTTGCCCA–3′ and Em_kettin_R5: 5′–TCATGCGTAGTGCGGAATCT–3′, for PCR amplification with 94°C for 5 min followed by 35 cycles of 94°C for 30 s, 60°C for 30 s and 72°C for 3 min finalized by 72°C for 10 min. Probe labeling was done by using the Nick Translation Kit (Abbott Molecular, Des Plaines, IL). We selected Green-dUTP, Orange-dUTP (Abbott Molecular Inc.) and Cy5-dUTP (GE Healthcare Japan, Tokyo) fluorochromes for genomic DNA, *Ket* and insect telomeric repeat (TTAGG)n probes respectively. Hybridizations were carried out according to protocols described elsewhere (Yoshido et al. 2014). Signal and chromosome images were captured with a DFC350FX CCD camera mounted on a DM 6000B microscope (Leica Microsystems Japan, Tokyo) and processed with Adobe Photoshop CS2. We applied green, red and yellow pseudocolors to signals from Green, Orange, and Cy5, respectively.

QUANTITATIVE POLYMERASE CHAIN REACTION (QPCR)

Embryos of mated females were sampled 48 h after the oviposition and stored at −80°C until DNA extraction. Embryos were individually subjected to DNA extraction using DNeasy® Blood & Tissue Kit (Qiagen, Tokyo, Japan). Real-time fluorescence detection quantitative PCR (qPCR) was performed using SYBR Green and a LightCycler® 480 System (Roche Diagnostics K.K., Tokyo, Japan). Z-linked *Tpi* was amplified using TPI-F (5′–GGCCTCAAGGTCATTGCCTGT–3′) and TPI-R (5′–ACACGACCTCCTCGGTTTTACC–3′), Z-linked *Ket* was amplified using Ket-F (5′–TCAGTTAAGGCTATTAACGCTCTG–3′) and Ket-R (5′–ATACTACCTTTTGCGGTTACTGTC–3′), and autosomal *EF-1α* was amplified using EF-1F (5′–AAATCGGTGGTATCGGTACAGTGC–3′) and EF-1R (5′–ACAACAATGGTACCAGGCTTGAGG–3′) (Kern et al. 2015). For each qPCR, a standard dilution series of PCR products (10^8, 10^7, 10^6, 10^5, 10^4, and 10^3 copies per microliter) was included in order to estimate the absolute copy numbers of the target sequence in the samples. To prepare standard samples, PCR products were gel-excised and purified by Wizard® SV (Promega). Copy numbers of the standard samples were estimated by the concentration measured by a spectrophotometer, considering that the molecular weight of a nucleotide is 309 g/mol. For each qPCR, two replicates were performed that delivered similar results. All qPCRs were performed using a temperature profile of 40 cycles of 95°C for 5 s, 60°C for 10 s, and 72°C for 10 s. The qPCR data were analyzed by the Absolute Quantification analysis using the Second

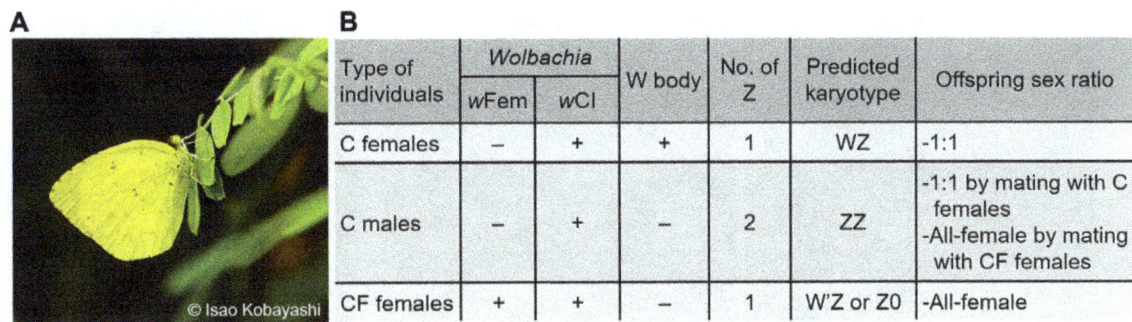

Figure 1. *E. mandarina* butterflies used in this study. (A) A photo of *E. mandarina* taken in Tanegashima Island. (B) Characteristics of three types of *E. mandarina* individuals inhabiting Tanegashima Island.

Derivative Maximum method implemented in the LightCycler® 480 Instrument Operator Software Version 1.5 (Roche).

RT-PCR

RNA was extracted from adult abdomens that were stored at –80°C using RNeasy® Mini Kit (Qiagen, Tokyo, Japan). The cDNA synthesized by using Superscript™ III (Invitrogen) and Oligo(dT) was used as a template for RT-PCR. A partial sequence of *dsx* that contains alternative splicing sites was amplified using a primer pair, E520F (5′–GCAACGAC CTCGACGAGGCTTCGCGGA–3′) and EhdsxR4 (5′–AGG GGCAGCCAGTGCGACGCGTACTCC–3′) and a temperature profile of 94°C for 2 min, 30 cycles of 94°C for 1 min, 57°C for 1 min and 72°C for 1 min 30 s, followed by 72°C for 7 min. The sequences of seven dsx^F isoforms and a dsx^M isoform were deposited in DDBJ/EMBL/Genbank (LC215389-LC215396).

Results

ALL-FEMALE-PRODUCING CF FEMALES HAVE A Z0 SEX CHROMOSOME CONSTITUTION

We performed FISH on *E. mandarina* chromosomes prepared from CF females, C females, and C males collected on Tanegashima Island (Fig. 1; Fig. S1). In the mitotic complement of C females, which harbor a $2n = 62$ karyotype, genomic probes highlighted the W chromosome, with scattered signals on the other chromosomes (Fig. 2A; see Materials and Methods for technical details). A probe for the Z-linked gene *Kettin* (*Ket*) identified the single Z chromosome in C females (Fig. 2A), and also hybridized to the Z chromosome paired with the W chromosome in pachytene bivalents (Fig. 2J). The *Ket* probe identified two Z chromosomes in the mitotic complement of C males (Fig. 2B; $2n = 62$). No painted W chromosome was observed in interphase nuclei (Fig. 2H,I), the mitotic complement (Fig. 2C), and pachytene complement (Fig. 2L) of CF females, but the *Ket* signal appeared on the single Z chromosome in the mitotic complement (Fig. 2C) and Z univalent in the pachytene complement (Fig. 2L). Based on the

relative read counts homologous to *Bombyx mori* Z-linked and autosomal genes in females and males, our genome sequencing data support the notion that CF and C females have one Z chromosome (Fig. 2M–O; Fig. S2), which is consistent with genomic qPCR data based on two loci, *Triosephosphate isomerase* (*Tpi*) and *Ket*, relative to the autosomal gene *EF-1α* (Kern et al. 2015). Thus, our results directly reveal the sex chromosome constitution of C females, C males, and CF females as WZ, ZZ, and Z0, respectively. This confirms one of two previously suggested sex chromosome constitution of CF females (Kern et al. 2015) while it disproves another previous interpretation based on W-body diagnosis that CF females are ZZ (Hiroki et al. 2002; Narita et al. 2007a).

ALL EMBRYOS OVIPOSITED BY CF FEMALES ARE Z0

We performed real-time genomic qPCR (to detect Z-linked *Tpi* or *Ket* relative to autosomal *EF-1α*) on individual fertilized eggs (48-h embryos), and found that embryos oviposited by C females had either one or two Z chromosomes at nearly equal frequencies (Fig. 3A left; Fig. S3). In contrast, all embryos oviposited by CF females were single Z carriers (Fig. 3A middle; Fig. S3). These findings indicate that the progeny of CF females are exclusively Z0 individuals, supporting the previous data that the maternal Z chromosomes are not inherited in CF lineages (Kern et al. 2015).

WOLBACHIA CAUSES THE EXCLUSIVE PRODUCTION OF Z0 EMBRYOS BY CF FEMALES

To abolish the effects of *Wolbachia*, tetracycline (tet) was administered to adult CF females previously inseminated by antibiotic-treated male offspring of C females. The Z-linked gene dose of embryos laid by these tet-treated females ranged from approximately 0.5–1.0, indicating that some embryos are Z0 and others are ZZ (Fig. 3A right; Fig. S3). This suggests that the *Wolbachia* strain *w*Fem in CF females causes the exclusive production of gametes without sex chromosomes that then develop as Z0 embryos after fertilization. Therefore, our finding is the first empirical evidence that in a female-heterogametic species the sex-specific linkage disequilibrium can be caused by cyto-

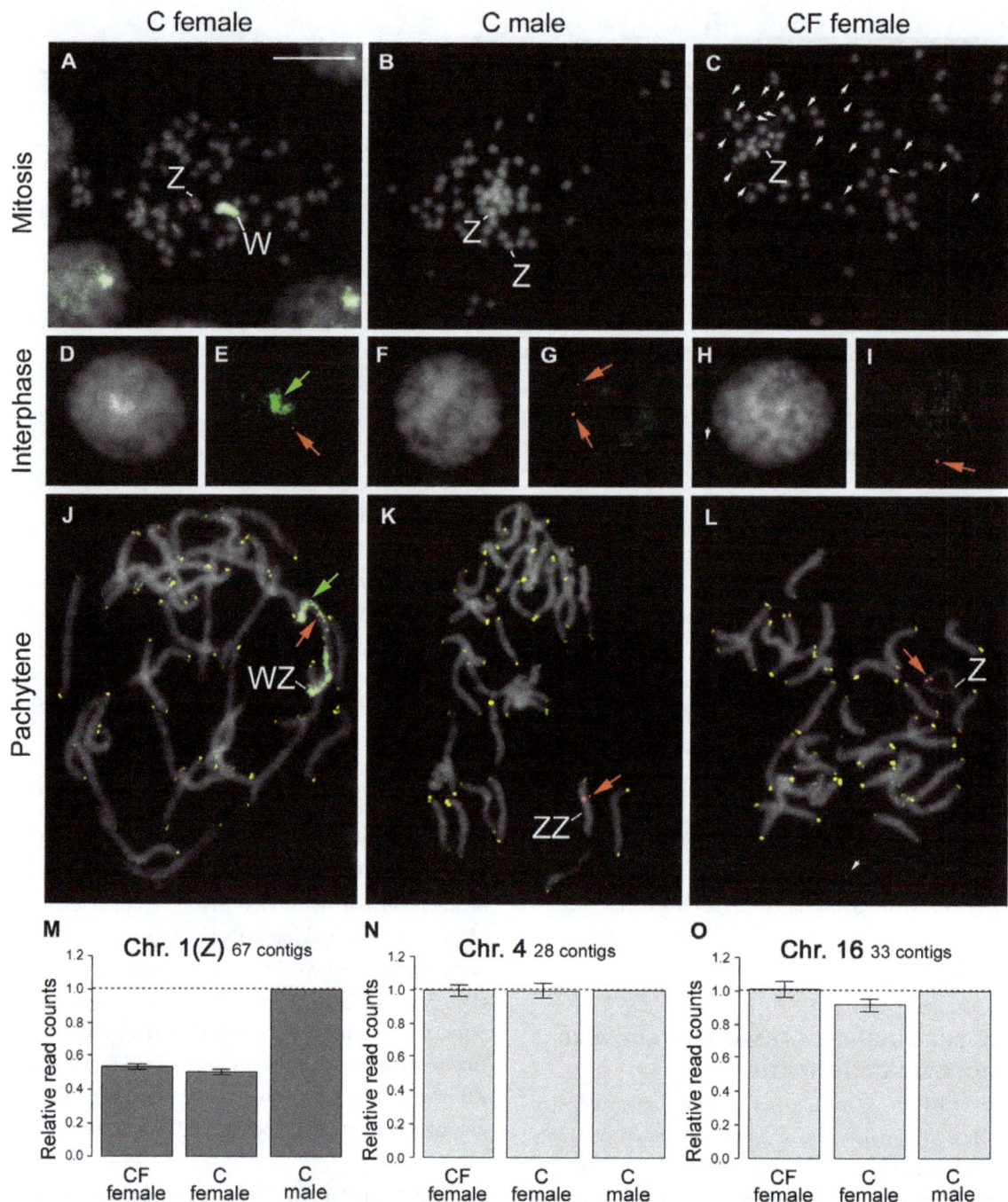

Figure 2. Fluorescence in situ hybridization and sequence read counts for a C female, C male, and CF female *E. mandarina*. (A–C) Mitotic complements hybridized with a genomic probe (green; green arrows) and a Z-linked *Ket* probe (red; red arrows) in a C female (2n = 62) (A), C male (2n = 62) (B), and CF female (2n = 61) (C). (D–I) Genomic in situ hybridization (GISH) and FISH with a Z-linked *Ket* probe performed on interphase nuclei of *E. mandarina* C females (D, E), C males (F, G), and CF females (H, I). (J–L) GISH, telomere-FISH, and FISH with *Ket* probe performed on pachytene complements of *E. mandarina* C females (G, n = 31), C males (H, n = 31), and CF females (I, n = 31). Green paint signals in A, E, and J revealed that C females have the W chromosome. The *Ket* probe signals (red) appeared on the Z pairing to the W in C females (J), the ZZ bivalent in C males (K), and the Z univalent of CF females (L). The single signals were observed both in C and CF female nuclei. The signals in C females (J) and males (K) clearly showed their respective WZ and ZZ chromosome sets, and a Z0 chromosome set in CF females (L). W: W chromosome; Z: Z chromosome; white arrows: *Wolbachia*-like structures. A bar represents 10 μm. M–O: Relative normalized sequence read counts in CF females, C females, and C males for 67 contigs homologous to *Bombyx mori* loci on chromosome 1 (Z chromosome; M), 28 contigs homologous to *B. mori* loci on chromosome 4 (N), and 33 contigs homologous to *B. mori* loci on chromosome 16 (O), with relative read counts set to 1 (males). Details about genome sequencing are provided in Materials and Methods.

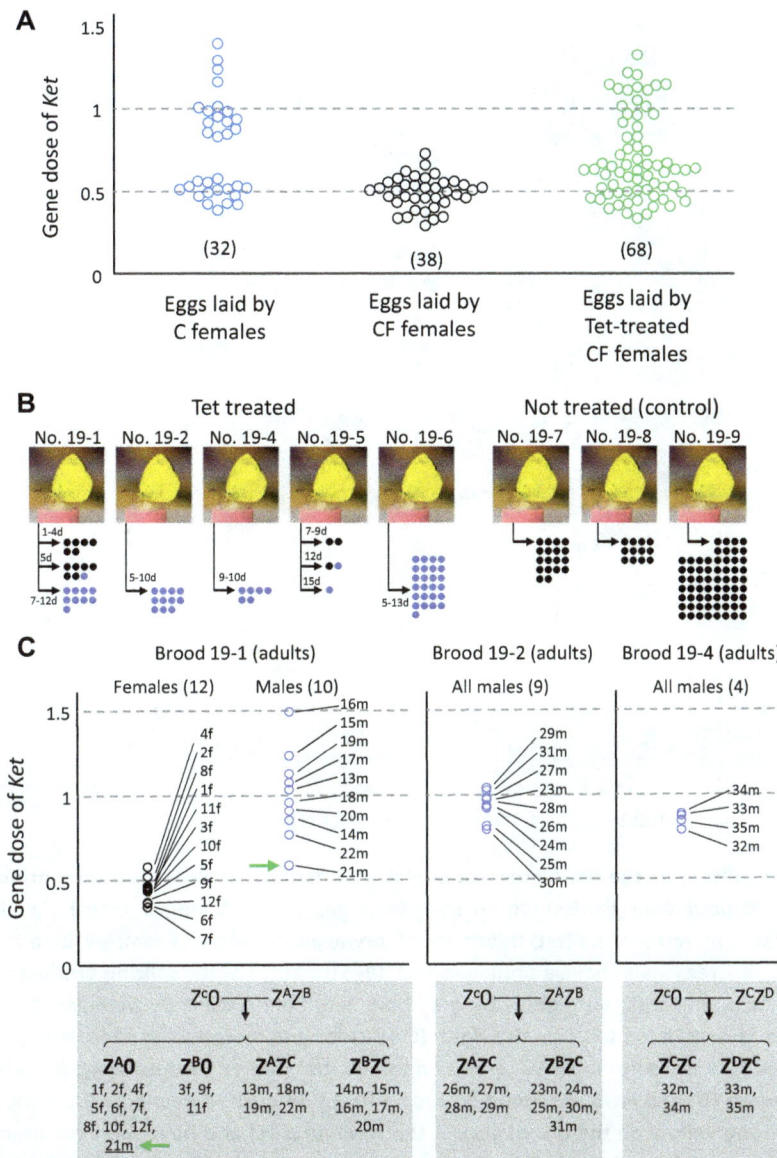

Figure 3. Effects of wFem on Z-linked gene dose in *E. mandarina* offspring. (A) Estimate of the gene dose of *Ket* (relative gene copies per copy of *EF-1α*) by genomic quantitative polymerase chain reaction (qPCR) analysis in each of the fertilized eggs laid by C females, CF females, and tetracycline (tet)-treated CF females. Each colored circle represents a single fertilized egg. Sample sizes are given in parentheses. (B) Offspring sex ratio of five females tet-treated prior to oviposition and three nontreated CF females. Numbers to the left of the arrows represent duration (days) of tet treatment. Blue dots and red dots represent males and females, respectively. (C) Estimate of the gene dose of *Ket* (relative gene copies per copy of *EF-1α*) by genomic qPCR in each of the adult offspring produced by CF females that were tet-treated during the adult stage (prior to oviposition). Each circle represents an adult offspring. Z chromosomes of these offspring individuals were genotyped as Z^A, Z^B, Z^C, or Z^D on the basis of intron nucleotide sequence of Z-linked *Tpi*. The green arrow points to a male individual (adult) whose karyotype was considered to be Z0 but possibly ZZ' (see text for details). f: female, m: male.

plasmic elements (Hurst 1993; Beukeboom and Perrin 2014). Furthermore, *Wolbachia*-like structures were observed near the chromosomes in CF females (Fig. 2C) while less apparent in C females (Fig. 2A) and C males (Fig. 2B), and this may represent different tropism and function of wFem when contrasted with wCI. Sixty-nine adults (15 females and 54 males) were obtained from offspring produced by five tet-treated adult CF females

(Fig. 3B). Three of these tet-treated females produced only male offspring. Exclusive production of males was previously observed in tet-treated *E. mandarina* females derived from a different population on Okinawa-jima Island, Okinawa Prefecture, Japan (Hiroki et al. 2002). In this study, we obtained 15 female offspring from two broods in the first days after tet treatment; however, the mothers produced more males as the duration of tet treatment

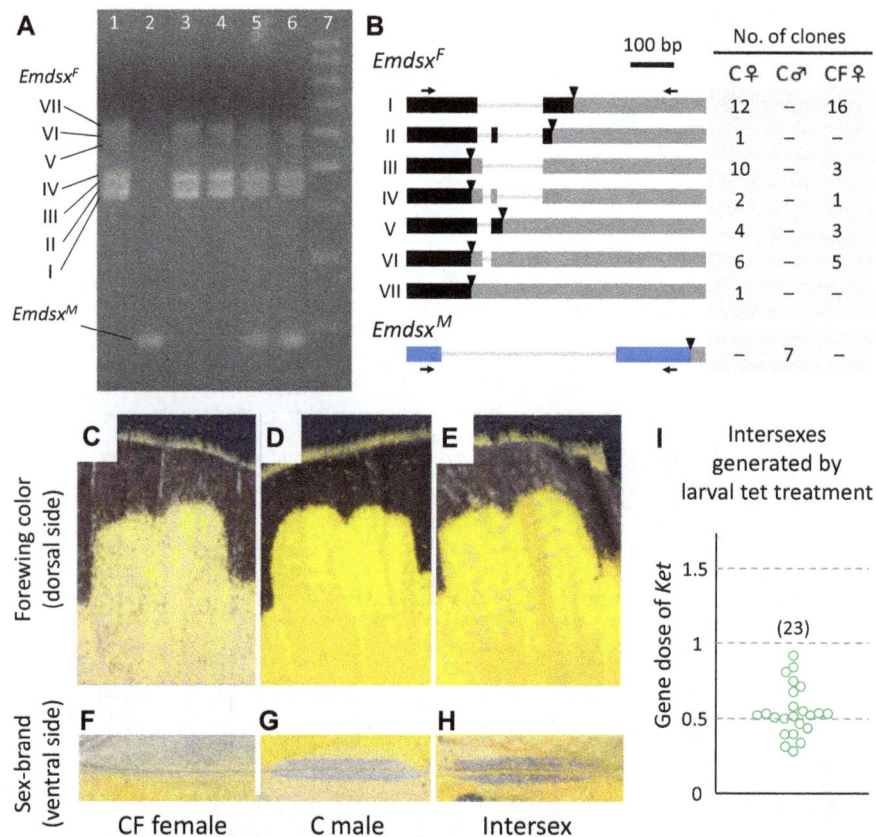

Figure 4. Effects of *w*Fem on splicing of the *doublesex* gene in *E. mandarina*. **(A)** Reverse-transcription polymerase chain reaction (RT-PCR) products of *E. mandarina doublesex* (*Emdsx*) run on an agarose gel. Lane 1: C female; lane 2: C male; lanes 3 and 4: CF females; lanes 5 and 6: intersexes generated by tetracycline (tet) treatment of larvae produced by CF females; lane 7: 100-bp ladder. Females have at least seven splicing products, whereas males have a single product. **(B)** Structures of the splicing products of *Emdsx*. Translated regions are indicated by red and blue bars, untranslated regions by gray bars, and stop codons by triangles. Numbers of clones obtained by cloning the RT-PCR products are shown in the table on the right. **(C–H)** color and morphology of forewings. Females are pale yellow on the dorsal side of the forewings (C) and do not have sex brand on the ventral side of the forewings (F), while males are intense yellow on the dorsal side of the forewings (D) and have sex brand on the ventral side of the forewings (G). Many of the intersexes generated by tet-treating CF larvae are strong yellow on the dorsal side of the forewings (E) and have faint sex brand on the ventral side of the forewings (H). **(I)** Estimate of the gene dose of *Ket* (relative gene copies per copy of *EF-1α*) by genomic qPCR in each of the intersex CF individuals that were produced by tet treatment during larval stages. Each circle represents an adult offspring.

increased, and eventually produced only males. Examination of the Z-linked gene dose of these offspring by genomic qPCR showed that the females had one Z chromosome, whereas almost all of the males had two Z chromosomes (Fig. 3C). The nucleotide sequences of the introns of the *Tpi* gene demonstrate that, in brood 19-1, all females ($n = 12$) were hemizygous and nine out of 10 males were heterozygous (Fig. 3C; Fig. S4). Curiously, one male (21m) that exhibited the lowest gene dose of *Ket* (0.588) appeared to be hemizygous (Fig. 3C). These results suggest that the emerged females had a Z0 sex chromosome constitution, whereas most males had a ZZ sex chromosome constitution, with one exception (21m) of either Z0 or ZZ' (Z' represents partial deletion/mutation in Z). These results also demonstrate that, in principle, tet-treated adult CF females can oviposit embryos with either a Z0 or ZZ sex chromosome constitutions (Fig. 3A right;

Fig. S3). However, Z0 males appear to have zero or very low survival rates.

INVOLVEMENT OF *WOLBACHIA* IN THE SEX DETERMINATION OF *E. MANDARINA*

Next, we fed CF larvae on a tet-containing diet. As previously observed (Narita et al. 2007a), all individuals treated in this way developed an intersex phenotype at the adult stage, typically represented with male-like wing color and an incomplete male-specific structure on the wing surface (Fig. 4E,H; Fig. S6). The qPCR assay to assess the Z-linked gene dose revealed that these intersexes ($n = 23$) had just one Z chromosome (Fig. 4I), and therefore a Z0 genotype. Because these Z0 individuals were destined to develop as females without tet treatment, *w*Fem is likely to be responsible for female sex determination. Further evidence in support of this idea was obtained by examining the sex-specific splicing products of *dsx* (Fig. S7), a widely conserved gene responsible for sexual differentiation (Bopp et al. 2014). Similar to *B. mori* (Ohbayashi et al. 2001), C females exhibited female-specific splicing products of *E. mandarina dsx* (*Emdsx^F*), whereas C males had a male-specific splicing product of *E. mandarina dsx* (*Emdsx^M*; Lanes 1 and 2 in Fig. 4A, respectively; Fig. 4B). Similarly to C females, CF females exhibited exclusive expression of *Emdsx^F* (Lanes 3 and 4 in Fig. 4A; Fig. 4B). Intersexual butterflies, generated by feeding the larval offspring of CF females a tet-containing larval diet, expressed both *Emdsx^F* and *Emdsx^M* (Lanes 5 and 6 in Fig. 4A; Fig. S5).

Discussion

We provide comprehensive and conclusive indirect (qPCR of Z gene dosage) and direct (W chromosome painting; genomic analyses) evidence for the absence of the W chromosome in CF individuals. Furthermore, we demonstrate that the *Wolbachia* strain *w*Fem is directly responsible for the disruption of sex chromosome inheritance in CF females of *E. mandarina*. This is the first empirical proof for previous theoretical predictions that cytoplasmic SGEs, such as *Wolbachia*, can disrupt chromosome inheritance (e.g., by meiotic drive). In *E. mandarina*, *w*Fem has a dual role in both causing disruption of maternal inheritance of Z chromosomes and converting male sex determination into female sex determination (feminization) in Z0 lineages that have lost W chromosome and its female-determining function.

WOLBACHIA DISRUPTS Z CHROMOSOME INHERITANCE IN Z0 FEMALES

Our data provides evidence that the exclusive production of Z0 embryos by CF females is due to a yet unidentified developmental process that leads to the disruption of sex chromosome inheritance in CF females, thereby the absence of maternal Z chromosome in CF offspring (Kern et al. 2015). We believe that two mutually exclusive hypotheses can account for the disruption of Z chromosome inheritance observed in CF individuals (Fig. 5A). The first assumes that a gamete without the maternal Z chromosome (or without any sex chromosome overall), is always selected to become an egg pronucleus (meiotic drive against Z-bearing gametes) (Fig. 5A left) (Pardo-Manuel de Villena and Sapienza 2001). The second assumes that meiosis itself is normal, and that maternal Z chromosomes (or sex chromosomes in general), are selectively eliminated from Z-bearing gametes during, or possibly after, meiosis (Fig. 5A right). At present, it is unclear which of the two scenarios (meiotic drive *sensu stricto* or elimination of the maternal Z) is more plausible. However, it is noteworthy that, in the moth *Abraxas grossulariata*, a matriline consisting of putative Z0 females was observed to produce only females or a great

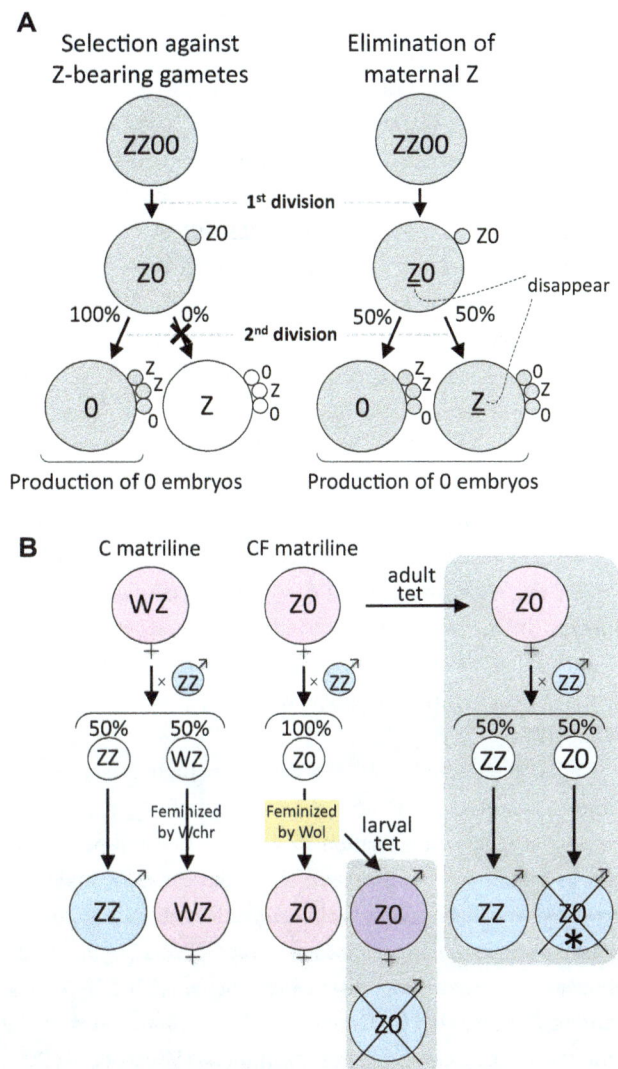

Figure 5. (A) Schematic illustration of two alternative mechanistic models of disruption of Z chromosome inheritance that explain the observed data. The "Selection against Z gametes" model assumes that Z-bearing gametes are selected against during meiosis (left). The "Elimination of maternal Z" model assumes that Z chromosomes are eliminated during or after normal meiosis, while all the autosomes being intact (right). (B) All-female production explained by *Wolbachia*–host interaction. Effects of *w*Fem on the development and sex determination of *E. mandarina*, and outcomes of larval versus adult tet treatment are illustrated. Asterisk: The majority of Z0 males die, but a few survived.

excess of females, and the underlying mechanism was considered to be the selective elimination of Z chromosomes (Doncaster 1913, 1914, 1915, 1922). However, the presence of cytoplasmic bacteria such as *Wolbachia* has not yet been examined for this moth species. If we assume that the elimination of the maternal Z chromosome occurs in *E. mandarina*, the exceptional individual 21m (Fig. 3C) could be viewed as ZZ' rather than Z0, wherein Z' is a maternal Z chromosome that was only partially deleted in the position including *Tpi* and *Ket* by the incomplete action of *w*Fem. It is possible to further speculate that the presence of *w*Fem results in the elimination of sex chromosomes in general (Z or W chromosomes) and, therefore, the absence of W chromosomes in CF females may also be a direct effect of *w*Fem.

THE FEMINIZING EFFECT OF *WOLBACHIA* COMPENSATES FOR THE LOSS OF THE W CHROMOSOME IN Z0 INDIVIDUALS

In general, lepidopteran species with Z0/ZZ sex chromosome constitution are considered to determine their sexes by Z-counting mechanisms, wherein ZZ is male and Z0 is female (Traut et al. 2007; Sahara et al. 2012). However, the appearance of the male phenotype in Z0 individuals of *E. mandarina* after antibiotic treatment suggests that *w*Fem in Z0 individuals compensates for the loss of W and its female-determining function (Fig. 5B). We speculate that the W chromosome of *E. mandarina* acts as an epistatic feminizer. In *B. mori*, the W chromosome–more specifically, a piRNA located on the W chromosome–acts as an epistatic feminizer by silencing *Masculinizer* on the Z chromosome (Kiuchi et al. 2014).

Reduced survival of Z0 individuals or their offspring after antibiotic treatment of larvae or adults, respectively, may suggest improper dosage compensation in Z0 males. Improper dosage compensation was also proposed to be the cause of male- and female-specific lethality in *Wolbachia*-infected and cured lines of *Ostrinia* moths (Kageyama and Traut 2004; Sugimoto and Ishikawa 2012; Fukui et al. 2015; Sugimoto et al. 2015). Another symbiont *Spiroplasma poulsonii* kills *Drosophila* males by targeting the dosage compensation machinery (Veneti et al. 2005; Cheng et al. 2016). It was recently found that *S. poulsonii* causes DNA damage on the male X chromosome interacting with the male-specific lethal (MSL) complex. The DNA damage leads to male killing through apoptosis via p53-dependent pathways (Harumoto et al. 2016).

HOW DID THE COORDINATED DUAL EFFECTS OF *WOLBACHIA* EVOLVE?

We demonstrated that *w*Fem causes the disruption of Z chromosome inheritance and the conversion of sex determination in *E. mandarina* in two steps (Fig. 5B). This is similar to the dual role of *Wolbachia* and *Cardinium* in haplodiploid parasitoid wasps where they induce thelytokous parthenogenesis in a two-step mechanism, comprising diploidization of the unfertilized egg followed by feminization (Giorgini et al. 2009; Ma et al. 2015). Here, we develop the potential evolutionary scenario that led to the appearance of both effects in *E. mandarina* (Fig. 6).

A WZ female *Eurema* butterfly may have acquired *w*Fem that exerted a feminizing effect on ZZ males. The feminizing effect was lethal to ZZ individuals because of improper dosage

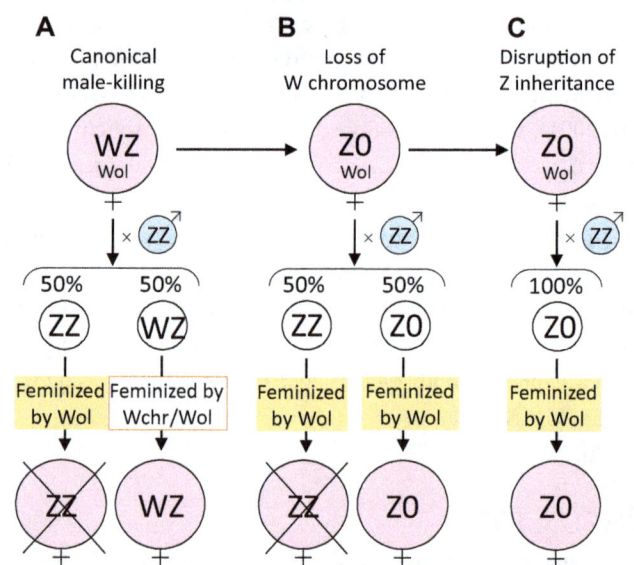

Figure 6. Hypothetical evolutionary trajectory of the *Wolbachia*–host interaction in *E. mandarina*. See Discussion for details.

compensation, as evident in *Wolbachia*-infected *Ostrinia* moths (Fig. 6A) (Fukui et al. 2015; Sugimoto et al. 2015). This could be viewed as a manipulation similar to a male-killing phenotype (Dyson et al. 2002; Harumoto et al. 2016). However, the feminizing effect of *w*Fem was redundant in WZ females where the W chromosome acted as a female determiner (Kiuchi et al. 2014). Conversely, the function of W had also become redundant in CF individuals and this could have led to the loss of the W chromosome and the rise of a Z0 lineage (Fig. 6B). Similarly, in *Ostrinia* moths, a female-determining function is thought to have been lost from the W chromosome in *Wolbachia*-infected matrilines (Sugimoto and Ishikawa 2012). Spontaneous loss of a nonfunctional W chromosome may be easier than expected: in a wild silkmoth *Samia cynthia*, the W chromosome does not have a sex-determining function, and Z0 females are frequently obtained in experimental crosses between subspecies (Yoshido et al. 2016). *Wolbachia* has previously been found to be associated with the loss and birth of W chromosomes in the woodlouse *Armadillidium vulgare* (Rigaud et al. 1997; Leclercq et al. 2016). However, in *A. vulgare* it has not yet been tested whether *Wolbachia* interferes with chromosome segregation and inheritance as we have mechanistically demonstrated it for *E. mandarina*; that is after the loss of the W chromosome in CF lineages, *Wolbachia* then acquired a novel function that affected female oogenesis and resulted in the disruption of Z chromosome inheritance (Fig. 6C). It is unlikely that the disruption arose prior to the appearance of female-determining function of *Wolbachia*: if the appearance of the disruption of Z chromosome inheritance were to precede the loss of the W chromosome, the feminizing or female-determining function would become unnecessary for *Wolbachia* because there would be no males.

In the short term, disruption of Z chromosome inheritance in females in a female-heterogametic species represents a great advantage to cytoplasmic symbionts because all vertically transmitted symbionts gain the opportunity to survive. However, males are still required for fertilization, and fixation of the symbionts in the host population will inevitably lead to the extinction of both the symbionts and the hosts (Hatcher et al. 1999). In the long term, suppressors against sex ratio distortion, as has been observed for the male-killing phenotypes in the butterfly *Hypolimnas bolina* or a ladybird beetle (Charlat et al. 2007; Majerus and Majerus 2010), can be expected to evolve in the host. However, the evolutionary outcomes of the suppression of a combined action of *Wolbachia* in *E. mandarina* would be different from that of male-killing suppression, because it would lead to all-male progeny, resulting in the loss of the matriline that inherits the feminizing and sex-distorting *Wolbachia*. This process thereby selects for an increased frequency of WZ females.

CONCLUDING REMARKS

In summary, we demonstrate for the first time that the manipulation of sex chromosome inheritance and cytoplasmically induced disruption of chromosome inheritance, which would either be the result of meiotic drive against Z-bearing gametes or elimination of Z chromosome, can be added to the repertoire of host manipulations induced by *Wolbachia*. Therefore, the host effects of this bacterium are far more diverse and profound than previously appreciated. Disentangling these complex interactions between insects and *Wolbachia* may provide further exciting discoveries in the areas of host–parasite interactions, endosymbiosis as well as cell and chromosome biology in years to come, and perhaps also provide new avenues for pest population control.

AUTHOR CONTRIBUTIONS

D.K., K.S., designed the research; D.K., M.O., T.S., A.Y., T.K., S.K., H.K., Y.K., S.N., M.M., M.R., K.S., performed the research; D.K., A.J., K.S., analyzed the data; D.K., M.R., K.S., wrote the article with input from A.Y.

ACKNOWLEDGMENTS

We thank Isao Kobayashi for help in collecting butterflies and Ranjit Kumar Sahoo for stimulating discussion. This work was funded by Japan Society for the Promotion of Science (JSPS) to D.K. and K.S. (No. 16K08106), and to K.S. (16H05050). A part of this study is supported by NIAS technical support system at NARO.

LITERATURE CITED

Abe, H., T. Fujii, N. Tanaka, T. Yokoyama, H. Kakehashi, M. Ajimura, K. Mita, Y. Banno, Y. Yasukochi, T. Oshiki, et al. 2008. Identification of the female-determining region of the W chromosome in *Bombyx mori*. Genetica 133:269–282. https://doi.org/10.1007/s10709-007-9210-1

Beukeboom, L. W., and N. Perrin. 2014. The Evolution of Sex Determination. Oxford Univ. Press, Oxford.

Bolger, A. M., M. Lohse, and B. Usadel. 2014. Trimmomatic: a flexible trimmer for Illumina sequence data. Bioinformatics 30:2114–2120. https://doi.org/10.1093/bioinformatics/btu170

Bopp, D., G. Saccone, and M. Beye. 2014. Sex determination in insects: variations on a common theme. Sex. Dev. 8:20–28. https://doi.org/10.1159/000356458

Braig, H. R., W. Zhou, S. L. Dobson, and S. L. O'Neill. 1998. Cloning and characterization of a gene encoding the major surface protein of the bacterial endosymbiont *Wolbachia pipientis*. J. Bacteriol. 180:2373–2378. https://doi.org/10.1099/0022-1317-69-1-35

Burt, A., and R. Trivers. 2006. Genes in Conflict: The Biology of Selfish Genetic Elements. Harvard Univ. Press, Cambridge, MA.

Charlat, S., E. A. Hornett, J. H. Fullard, N. Davies, G. K. Roderick, N. Wedell, and G. D. D. Hurst. 2007. Extraordinary flux in sex ratio. Science 317:214. https://doi.org/10.1126/science.1143369

Cheng, B., N. Kuppanda, J. C. Aldrich, O. S. Akbari, and P. M. Ferree. 2016. Male-killing *Spiroplasma* alters behavior of the dosage compensation complex during *Drosophila melanogaster* embryogenesis. Curr. Biol. 26:1339–1345. https://doi.org/10.1016/j.cub.2016.03.050.

Doncaster, L. 1913. On an inherited tendency to produce purely female families in *Abraxas grossulariata*, and its relation to an abnormal chromosome number. J. Genet. 3:1–10. https://doi.org/10.1007/BF02981560

———. 1914. On the relations between chromosomes, sex-limited transmission and sex determination in *Abraxas grossulariata*. J. Genet. 4:1–21. https://doi.org/10.1007/BF02981560

———. 1915. The relation between chromosomes and sex-determination in "*Abraxas grossulariata*." Nature 95:395. https://doi.org/10.1038/095395a0

———. 1922. Further observations on chromosomes and sex-determination in *Abraxas grossulariata*. Q. J. Microsc. Sci. 66:397–406. https://doi.org/10.1038/095395a0

Dyson, E. A., M. K. Kamath, and G. D. D. Hurst. 2002. *Wolbachia* infection associated with all-female broods in *Hypolimnas bolina* (Lepidoptera: Nymphalidae): evidence for horizontal transmission of a butterfly male killer. Heredity 88:166–171. https://doi.org/10.1038/sj.hdy.6800021

Fukui, T., M. Kawamoto, K. Shoji, T. Kiuchi, S. Sugano, T. Shimada, Y. Suzuki, and S. Katsuma. 2015. The endosymbiotic bacterium *Wolbachia* selectively kills male hosts by targeting the masculinizing gene. PLoS Pathog. 11:1–14. https://doi.org/10.1371/journal.ppat.1005048

Giorgini, M., M. M. Monti, E. Caprio, R. Stouthamer, and M. S. Hunter. 2009. Feminization and the collapse of haplodiploidy in an asexual parasitoid wasp harboring the bacterial symbiont *Cardinium*. Heredity 102:365–371. https://doi.org/10.1038/hdy.2008.135

Harumoto, T., H. Anbutsu, B. Lemaitre, and T. Fukatsu. 2016. Male-killing symbiont damages host's dosage-compensated sex chromosome to induce embryonic apoptosis. Nat. Commun. 7:12781. https://doi.org/10.1038/ncomms12781

Hatcher, M. J., D. E. Taneyhill, A. M. Dunn, and C. Tofts. 1999. Population dynamics under parasitic sex ratio distortion. Theor. Pop. Biol. 56:11–28. https://doi.org/10.1006/tpbi.1998.1410

Hiroki, M., Y. Kato, T. Kamito, and K. Miura. 2002. Feminization of genetic males by a symbiotic bacterium in a butterfly, *Eurema hecabe* (Lepidoptera: Pieridae). Naturwissenschaften 89:167–170. https://doi.org/10.1007/s00114-002-0303-5

Hiroki, M., Tagami, Y., Miura, K., and Kato, Y. 2004. Multiple infection with *Wolbachia* inducing different reproductive manipulations in the butterfly *Eurema hecabe*. Proc. Biol. Sci. 271:1751–1755. https://doi.org/10.1098/rspb.2004.2769

Hiroki, M., Y. Ishii, and Y. Kato. 2005. Variation in the prevalence of cytoplasmic incompatibility-inducing *Wolbachia* in the butterfly *Eurema hecabe* across the Japanese archipelago. Evol. Ecol. Res. 7:931–942.

Hurst, L. D. 1993. The incidences mechanisms and evolution of cytoplasmic sex ratio distorters in animals. Biol. Rev. 68:121–194. https://doi.org/10.1111/j.1469-185X.1993.tb00733.x

Jaenike, J. 2001. Sex chromosome meiotic drive. Annu. Rev. Ecol. Syst. 32:25–49. https://doi.org/10.1146/annurev.ecolsys.32.081501.113958

Jiggins, C. D., M. Linares, R. E. Naisbit, C. Salazar, Z. H. Yang, and J. Mallet. 2001. Sex-linked hybrid sterility in a butterfly. Evolution 55:1631–1638. https://doi.org/10.1111/j.0014-3820.2001.tb00682.x

Kageyama, D., and W. Traut. 2004. Opposite sex-specific effects of *Wolbachia* and interference with the sex determination of its host *Ostrinia scapulalis*. Proc. Biol. Sci. 271:251–258. https://doi.org/10.1098/rspb.2003.2604

Kern, P., J. M. Cook, D. Kageyama, and M. Riegler. 2015. Double trouble: combined action of meiotic drive and *Wolbachia* feminization in *Eurema* butterflies. Biol. Lett. 11:20150095. https://doi.org/10.1098/rsbl.2015.0095

Kiuchi, T., H. Koga, M. Kawamoto, K. Shoji, H. Sakai, Y. Arai, G. Ishihara, S. Kawaoka, S. Sugano, T. Shimada, et al. 2014. A single female-specific piRNA is the primary determiner of sex in the silkworm. Nature 509:4–6. https://doi.org/10.1038/nature13315

Langmead, B., and S. L. Salzberg. 2012. Fast gapped-read alignment with Bowtie 2. Nat. Meth. 9:357–359. https//doi.org/10.0.4.14/nmeth.1923

Leclercq, S., J. Thézé, M. A. Chebbi, I. Giraud, B. Moumen, L. Ernenwein, P. Grève, C. Gilbert, and R. Cordaux. 2016. Birth of a W sex chromosome by horizontal transfer of *Wolbachia* bacterial symbiont genome. PNAS 113:201608979. https://doi.org/10.1073/pnas.1608979113

Lindholm, A. K., K. A. Dyer, R. C. Firman, L. Fishman, W. Forstmeier, L. Holman, H. Johannesson, U. Knief, H. Kokko, A. M. Larracuente, et al. 2016. The ecology and evolutionary dynamics of meiotic drive. Trends Ecol. Evol. 31:315–326. https://doi.org/10.1016/j.tree.2016.02.001

Ma, W.-J., B. A. Pannebakker, L. van de Zande, T. Schwander, B. Wertheim, and L. W. Beukeboom. 2015. Diploid males support a two-step mechanism of endosymbiont-induced thelytoky in a parasitoid wasp. BMC Evol. Biol. 15:84. https://doi.org/10.1186/s12862-015-0370-9

Majerus, T. M. O., and M. E. N. Majerus. 2010. Intergenomic arms races: detection of a nuclear rescue gene of male-killing in a ladybird. PLoS Pathog. 6:1–7. https://doi.org/10.1371/journal.ppat.1000987

Narita, S., M. Nomura, Y. Kato, and T. Fukatsu. 2006. Genetic structure of sibling butterfly species affected by *Wolbachia* infection sweep: evolutionary and biogeographical implications. Mol. Ecol. 15:1095–1108. https://doi.org/10.1111/j.1365-294X.2006.02857.x

Narita, S., D. Kageyama, M. Nomura, and T. Fukatsu. 2007a. Unexpected mechanism of symbiont-induced reversal of insect sex: feminizing *Wolbachia* continuously acts on the butterfly *Eurema hecabe* during larval development. Appl. Environ. Microbiol. 73:4332–4341. https://doi.org/10.1128/AEM.00145-07

Narita, S., M. Nomura, and D. Kageyama. 2007b. Naturally occurring single and double infection with *Wolbachia* strains in the butterfly *Eurema hecabe*: transmission efficiencies and population density dynamics of each *Wolbachia* strain. FEMS Microbiol. Ecol. 61:235–245. https://doi.org/10.1111/j.1574-6941.2007.00333.x

Ohbayashi, F., M. G. Suzuki, K. Mita, K. Okano, and T. Shimada. 2001. A homologue of the *Drosophila doublesex* gene is transcribed into sex-specific mRNA isoforms in the silkworm, *Bombyx mori*. Comp. Biochem. Physiol. B Biochem. Mol. Biol. 128:145–158. https://doi.org/10.1016/S1096-4959(00)00304-3

Pardo-Manuel de Villena, F., and C. Sapienza. 2001. Nonrandom segregation during meiosis: the unfairness of females. Mamm. Genome 12:331–339. https://doi.org/10.1007/s003350040003

Rigaud, T., P. Juchault, and J.-P. Mocquard. 1997. The evolution of sex determination in isopod crustaceans. BioEssays 19:409–416. https://doi.org/10.1002/bies.950190508

Sahara, K., F. Marec, and W. Traut. 1999. TTAGG telomeric repeats in chromosomes of some insects and other arthropods. Chromosom. Res. 7:449–460. https://doi.org/10.1023/A:1009297729547

Sahara, K., A. Yoshido, N. Kawamura, A. Ohnuma, H. Abe, K. Mita, T. Oshiki, T. Shimada, S. Asano, H. Bando, et al. 2003a. W-derived BAC probes as a new tool for identification of the W chromosome and its aberrations in *Bombyx mori*. Chromosoma 112:48–55. https://doi.org/10.1007/s00412-003-0245-5

Sahara, K., F. Marec, U. Eickhoff, and W. Traut. 2003b. Moth sex chromatin probed by comparative genomic hybridization (CGH). Genome 46:339–342. https://doi.org/10.1139/g03-003

Sahara, K., A. Yoshido, and W. Traut. 2012. Sex chromosome evolution in moths and butterflies. Chromosom. Res. 20:83–94. https://doi.org/10.1007/s10577-011-9262-z

Simpson, J. T., and R. Durbin. 2012. Efficient de novo assembly of large genomes using compressed data structures. Genome Res. 22:549–556. https://doi.org/10.1101/gr.126953.111

Suetsugu, Y., R. Futahashi, H. Kanamori, K. Kadono-Okuda, S. Sasanuma, J. Narukawa, M. Ajimura, A. Jouraku, N. Namiki, M. Shimomura, et al. 2013. Large scale full-length cDNA sequencing reveals a unique genomic landscape in a lepidopteran model insect, *Bombyx mori*. G3 Genes|Genomes|Genetics 3:1481–1492. https://doi.org/10.1534/g3.113.006239

Sugimoto, T. N., and Y. Ishikawa. 2012. A male-killing *Wolbachia* carries a feminizing factor and is associated with degradation of the sex-determining system of its host. Biol. Lett. 8:412–415. https://doi.org/10.1098/rsbl.2011.1114

Sugimoto, T. N., T. Kayukawa, T. Shinoda, Y. Ishikawa, and T. Tsuchida. 2015. Misdirection of dosage compensation underlies bidirectional sex-specific death in *Wolbachia*-infected *Ostrinia scapulalis*. Insect Biochem. Mol. Biol. 66:72–76. https://doi.org/10.1016/j.ibmb.2015.10.001

Traut, W., and F. Marec. 1996. Sex chromatin in Lepidoptera. Q. Rev. Biol. 71:239–256. https://doi.org/10.1086/419371

Traut, W., K. Sahara, and F. Marec. 2007. Sex chromosomes and sex determination in Lepidoptera. Sex Dev. 1:332–346. https://doi.org/10.1159/000111765

Veneti, Z., J. K. Bentley, T. Koana, H. R. Braig, and G. D. D. Hurst. 2005. A functional dosage compensation complex required for male killing in *Drosophila*. Science 307:1461–1463. https://doi.org/10.1126/science.1107182

Werren, J. H., L. Baldo, and M. E. Clark. 2008. *Wolbachia*: master manipulators of invertebrate biology. Nat. Rev. Microbiol. 6:741–751. https://doi.org/10.1038/nrmicro1969

Werren, J. H. 2011. Selfish genetic elements, genetic conflict, and evolutionary innovation. PNAS 108:10863–10870. https://doi.org/10.1073/pnas.1102343108

Yoshido, A., K. Sahara, and Y. Yasukochi. 2014. Chapter 6; Silkmoths (Lepidoptera). Pp. 219–256 *in* I Sharakhov, ed. Protocols for cytogenetic mapping of arthropod genomes. CRC Press, Boca Raton, USA.

Yoshido, A., F. Marec, and K. Sahara. 2016. The fate of W chromosomes in hybrids between wild silkmoths, *Samia cynthia* ssp.: no role in sex determination and reproduction. Heredity 116:424–433. https://doi.org/10.1038/hdy.2015.110

13

Inconsistent reproductive isolation revealed by interactions between *Catostomus* fish species

Elizabeth G. Mandeville,[1,2] Thomas L. Parchman,[3] Kevin G. Thompson,[4] Robert I. Compton,[5]

Kevin R. Gelwicks,[5] Se Jin Song,[6] and C. Alex Buerkle[1] (ID)

[1] Department of Botany and Program in Ecology, University of Wyoming, Laramie Wyoming 82071
 [2] E-mail: emandevi@uwyo.edu
[3] Department of Biology, University of Nevada, Reno Nevada 89557
[4] Colorado Parks and Wildlife, Montrose Colorado 81401
[5] Wyoming Game and Fish Department, Laramie Wyoming 82070
[6] Department of Ecology and Evolutionary Biology, University of Colorado, Boulder Colorado 80309

Interactions between species are central to evolution and ecology, but we do not know enough about how outcomes of interactions between species vary across geographic locations, in heterogeneous environments, or over time. Ecological dimensions of interactions between species are known to vary, but evolutionary interactions such as the establishment and maintenance of reproductive isolation are often assumed to be consistent across instances of an interaction between species. Hybridization among *Catostomus* fish species occurs over a large and heterogeneous geographic area and across taxa with distinct evolutionary histories, which allows us to assess consistency in species interactions. We analyzed hybridization among six *Catostomus* species across the Upper Colorado River basin (US mountain west) and found extreme variation in hybridization across locations. Different hybrid crosses were present in different locations, despite similar species assemblages. Within hybrid crosses, hybridization varied from only first generation hybrids to extensive hybridization with backcrossing. Variation in hybridization outcomes might result from uneven fitness of hybrids across locations, polymorphism in genetic incompatibilities, chance, unidentified historical contingencies, or some combination thereof. Our results suggest caution in assuming that one or a few instances of hybridization represent all interactions between the focal species, as species interactions vary substantially across locations.

KEY WORDS: Admixture, ancestry, *Catostomus*, hybridization, introduced species, reproductive isolation.

Impact Summary

Species occupy variable environments over large geographic areas, where they interact with a range of other species, including closely related species. Outcomes of interactions between species can vary across locations, depending on factors like the environmental context and which other species are in close proximity. While ecological components of species interactions are known to vary substantially, many studies of evolution employ a simplifying assumption that evolutionary processes are relatively consistent in all places where a pair of species interacts. In this article, we considered how reproductive isolation between species (the mechanisms that keep species from interbreeding with closely related species) varies when the same pairs of species interact repeatedly in a range of environments. We used large genetic datasets to study six species of *Catostomus* fish ("suckers") in the US mountain west. From previous work, we know that these species sometimes interbreed, producing hybrid individuals. In this study, we found that occurrence and extent of hybridization vary dramatically across many replicate locations where species come into contact. This pattern holds across multiple pairs of species. From these results, we conclude that these fish species are maintained as separate species (prevented from interbreeding) by different factors in different locations, with variable effectiveness of barriers to reproduction between different species. If these fish species are typical, this means that evolutionary biologists might need to incorporate a greater expectation of geographic variation into studies of evolutionary processes like hybridization. Our results are also extremely applicable to conservation of native *Catostomus* fish species that are threatened by hybridization, since this research suggests that successful conservation strategies might also need to be tailored to individual rivers.

Species interactions can produce variable ecological outcomes (e.g., Brooks and Dodson 1965; Paine 1966; Carpenter and Kitchell 1988; Valone and Brown 1995; Brown et al. 2001), such that contingency and variability are commonly expected, even if deterministic processes also contribute (Hubbell 2001; Jackson et al. 2009). While genetic and phenotypic variance is a central subject of study in evolutionary biology, simpler, more deterministic models are appealing and are often assumed to apply to the history of organisms and the evolution and genetics of their traits (Weiss 2008; Hewitt 2011; Rockman 2012). Theory predicts and empirical studies show that species' histories and trait architectures are partly, and sometimes largely, idiosyncratic, so some combination of contingency and determinism is a plausible expectation in evolutionary genetics (Taylor and McPhail 2000; Losos 2010; Kaeuffer et al. 2012; Rockman 2012; Soria-Carrasco et al. 2014). Our study characterized the extent to which species interactions, in this case hybridization, result in consistent outcomes as predicted by simple models of reproductive isolation between species, and to what extent hybridization is variable as suggested by empirical examples.

Simple models for isolating barriers have played a central role in the conceptualization of speciation (e.g., two locus Dobzhansky-Muller incompatibilites) and in some, possibly exceptional, cases have empirical support from trait mapping (Rieseberg and Blackman 2010; Wolf et al. 2010; Nosil and Schluter 2011; Yuan et al. 2013). Despite the appeal of simple models, isolation between species is expected to arise from multiple, polygenic traits that are expressed at multiple stages of the life history (Ramsey et al. 2003; Lindtke et al. 2014). As is true for most quantitative traits, one would expect the relevant phenotypes to be shaped by functional genetic polymorphisms that vary across a species' range and to be influenced by environmental variation. Additionally, similar phenotypic outcomes of independent evolution might be evident (Losos 1998; Mahler

et al. 2013), but develop as a result of different underlying mutations (Natarajan et al. 2016) or processes (Stayton 2015). Consequently, the genetics of speciation and dynamics of incompletely isolated species should vary among locations where species potentially hybridize. But this variance has rarely been quantified across several natural populations. If appreciable variation exists in the evolutionary outcomes of contact between species, these species are unlikely to be isolated as a result of traits with simple genetic architectures that are shared and independent of environment across the species' ranges. Instead, in these cases, speciation and reproductive isolation are more contingent, as a result of variable genetics or environments in which traits are expressed.

Our previous work on hybridization among *Catostomus* fishes demonstrated how reproductive isolation can vary geographically (Mandeville et al. 2015). We found substantial differences in the outcomes of hybridization in 785 fish from three sites, involving five different parental species (Mandeville et al. 2015). There were hybrid fish in each of the three rivers, but the species that produced hybrid offspring varied by river, as did the extent of hybridization and backcrossing. If we had sampled any one of these rivers in isolation, we would have been misled about the dynamics of reproductive isolation for these species (as was initially the case for McDonald et al. 2008). With samples from only three sites, we established that there was variability, but we were not able to characterize the extent and nature of variation in reproductive isolation and outcomes of hybridization.

To accurately characterize species interactions and hybridization, it was necessary to work at a broader spatial scale, with informative genomic data. In this study, we employed extensive geographic and taxonomic sampling to determine the consistency of species interactions. Our primary goal was to measure variation in hybridization among *Catostomus* fishes in Wyoming and Colorado on a broad geographic scale, using genomic data to estimate ancestry and distinguish among distinct hybridization outcomes. Our study addressed two specific questions: **(1)** To what extent are outcomes of interactions between *Catostomus* species consistent across sites and rivers?, and **(2)** What relationships exist between intraspecific genetic variation and outcomes of contact and hybridization between species? Broad geographic and genomic sampling allowed us to characterize hybridization in an unusually comprehensive manner, leading to greater understanding of how hybridization outcomes vary in natural populations across the shared range of the hybridizing species. We sought to detect the consequences of variation in traits affecting reproductive isolation (Sweigart et al. 2007; Good et al. 2008; Cutter 2012; Kozlowska et al. 2012) and to learn how ecological opportunity, historical contingency, and chance contribute to the evolution and maintenance of reproductive isolation (Taylor and McPhail 2000; Seehausen 2007; Wagner et al. 2012).

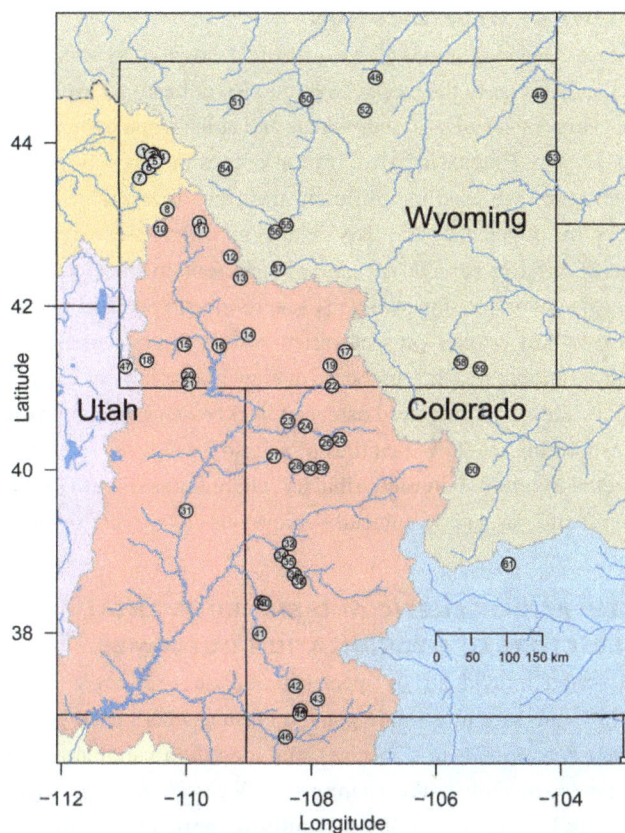

Figure 1. Map showing approximate sampling locations for populations of fish in this study. Boundaries between major river basins are shown in gray, and each major river basin is shown with a different background color. Our study of hybridization focused on the Upper Colorado River basin (UCRB; pink); other populations were sampled for reference individuals of species that were introduced in the UCRB but are native to adjacent basins. For clarity, only major rivers are shown and small streams are omitted.

Methods

Sampling for this project was accomplished through partnerships with state and federal agencies. Fin tissue was sampled nonlethally from 2932 individual fish. Samples span six species and hybrids, and represent 61 locations in the US mountain west (Fig. 1; 765 individuals originally collected for McDonald et al. (2008) and Mandeville et al. (2015) were resequenced for this study). Where possible, 20–30 individuals of each species or cross were sampled in each location. Fish were identified phenotypically by experienced field personnel, and fin clips were stored in ethanol. Six species were sampled, including *Catostomus latipinnis* (flannelmouth sucker), *C. discobolus* (bluehead sucker), *C. commersoni* (white sucker), *C. platyrhynchus* (mountain sucker), *C. catostomus* (longnose sucker), and *C. ardens* (Utah sucker), and hybrids of these species. *C. latipinnis*, *C. discobolus*, and *C. platyrhynchus* are native to the Upper Colorado River basin, our primary study area. The other three species are native to adjacent basins, and

have been introduced to the Upper Colorado River basin, probably within the past 100 years (Baxter et al. 1995; Gelwicks et al. 2009). One introduced species, *C. commersoni*, has become extremely widespread and abundant in its introduced range (Gill et al. 2007; Gelwicks et al. 2009); the other two were only sampled in a few locations in the Upper Colorado River basin.

We extracted DNA from fin clips using DNeasy 96 Blood & Tissue Kits (Qiagen, Inc.). We then prepared reduced-complexity genomic libraries for high throughput DNA sequencing, following the genotyping-by-sequencing method in Parchman et al. (2012). For this project, each lane of sequencing included 300–400 individual fish. High throughput DNA sequencing (Illumina Hiseq 2500, SR 1×100) was completed at the University of Texas, Austin, by the Genome Sequencing and Analysis Facility (UT-GSAF). Prior to sequencing, UT-GSAF used a BluePippin (Sage Science) device to size-select DNA fragments 250–300 base pairs in length from our libraries.

FILTERING AND ASSEMBLY

DNA sequencing on eight Illumina Hiseq 2500 lanes produced 427 gigabytes of raw data, representing 1.77×10^9 short DNA sequences. We filtered the data to remove common contaminant sequences (PhiX, *E. coli*) and excess Illumina primers and adaptors, and retained 1.43×10^9 reads. We then used a custom `perl` script to match barcode sequences to individual fish, and retained 1.37×10^9 sequence reads.

Since there is no draft genome sequence for *Catostomus* species, we constructed an artificial reference genome using a *de novo* assembly (as in Parchman et al. 2012; Gompert et al. 2014; Mandeville et al. 2015). We used `smng` (SeqMan NGen, DNAstar, Inc.) to assemble a subset of the data (40 million reads). We required a minimum match percentage of 90% and a depth of at least 25 reads to retain a contig. This *de novo* assembly produced an artificial reference genome with 327,388 contigs. We then completed a reference-based assembly of all sequence data using `bwa` version 0.7.5 (Li and Durbin 2009). 9.38×10^8 reads (68% of barcoded reads) assembled to the reference. To ensure that all individuals had sufficient data for downstream analyses, we removed 147 individuals with fewer than 9845 reads assembled, corresponding to the 5% quantile of assembled reads. We retained 2785 individuals.

VARIANT CALLING

From the individual assembly files (bam files), we identified single nucleotide variants using `samtools` and `bcftools` (version 0.1.19; Li et al. 2009; Li 2011). We required that >70% of all individuals (>1950 individuals) had data at a genetic site to identify a variable nucleotide and calculate genotype likelihoods at that locus. We initially identified 100,242 single nucleotide variants. To compare multiple species, we wanted sites that were polymorphic in multiple populations, so we excluded sites with minor allele frequencies <5% (Gompert et al. 2014). We also excluded sites with more than two alleles. We ensured greater independence of loci by randomly selecting one variable site per contig. We retained 11,221 SNPs that we used for all analyses of hybridization.

ANCESTRY ESTIMATION WITH ENTROPY

We used a hierarchical Bayesian model, entropy (Gompert et al. 2014), to estimate ancestry of each individual. Like structure (Pritchard et al. 2000; Falush et al. 2003), entropy does not require *a priori* information about membership of individuals in species or demes. We distinguished F_1, backcrossed, and advanced generation hybrids by combining estimates of q and Q from entropy, where q is the proportion of an individual's ancestry from each parental species, and Q is the proportion of loci in an individual that have ancestry from both parental species.

To estimate q (proportion of ancestry) we ran a model with $k = 6$ genetic clusters (one for each species; McDonald et al. 2008; Mandeville et al. 2015), using data at 11,221 loci. entropy uses Markov Chain Monte Carlo (MCMC) to estimate posterior distributions for all parameters. We ran three MCMC chains for 45,000 steps, retaining every 5th step, and discarded the first 40,000 steps as burn-in. This resulted in 1000 samples from the posterior distribution of each of three chains. For a subset of individuals, we plotted MCMC chains to check for adequate mixing and convergence of parameter estimates. We estimated posterior distributions for proportion of ancestry (q) in each fish and genotype at each locus for each individual.

We also estimated interspecific ancestry (Q), the proportion of loci in an individual with ancestry from both parental species, using entropy (Gompert et al. 2014). Estimates of Q are particularly informative about whether hybrids are the progeny of a cross involving the parental species, or are advanced generation hybrids. Interspecific ancestry is expected to be $Q = 1$ for F_1 hybrids (each locus has one allele copy from each parental species), and has an expected value of $Q = 0.5$ for F2 hybrids and backcrosses between F_1 hybrids and parental species. To achieve model convergence and to ensure that relevant local parental allele frequencies were used for local hybrids, we ran entropy separately for each hybridizing species pair in each river where >3 individuals of the same hybrid cross were sampled, using all 11,221 SNPs. For each hybrid cross in each river, we ran entropy for 200,000 steps with a $k = 2$ model and the Q model for admixture, discarded the first 150,000 steps as burn-in, and retained every 10th step, resulting in 5000 samples from the posterior distribution for each of three chains. We confirmed mixing and convergence using plots of MCMC chains for a subset of individuals and parameters. We then used the bivariate relationship of Q and q to characterize the composition and ancestry of hybrids.

EXTENT OF HYBRIDIZATION

Using estimates of q and Q, we quantified extent of hybridization for the 22 rivers in the Upper Colorado River basin for which we had sample sizes of >20 individuals and at least one hybrid cross was present. Extensive hybridization occurs in two ways in this system, and we used two different measures to compare across rivers. In some locations, many distinct hybrid crosses (combinations of different parental species) were present; in other locations, extensive hybridization involved backcrossing to parental species. Since hybrid crosses present varied among rivers, our first response variable was the number of distinct hybrid crosses sampled in a location. We measured extent of backcrossing quantitatively by estimating the 95% quantiles of ancestry estimates (q) for individual hybrid fish in each tributary where flannelmouth×white hybrids (the most geographically ubiquitous cross) were sampled.

POPULATION GENETIC AND ENVIRONMENTAL PREDICTORS OF HYBRIDIZATION OUTCOMES

We sorted individuals to species based on entropy results (>95% ancestry in a single genetic cluster, where each genetic cluster corresponds to a species), and identified variable genetic sites independently within each species. We used samtools and bcftools (version 0.1.19) to identify variants, and required that >80% of individuals within a species had data at a site for a single nucleotide variant to be called. We filtered out extremely low frequency variants (present in fewer than three individuals) and selected one SNP per contig. We then ran entropy for each species independently, for $k = 1$–5 genetic clusters. We used posterior estimates from entropy to reestimate genotype at each locus and built a genotype covariance matrix among individuals of each species, which we used in a principal components analysis (prcomp in R, R Development Core Team 2016). We then used PC scores for each species to examine the relationship between intraspecific variation in parental species and outcomes of hybridization.

We also used publicly available data to examine possible correlations (cor.test in R) between extent of hybridization in a location and environmental variables. We used data on land use, elevation, gradient, and other potentially relevant site characteristics, and tested for correlations between these variables and extent of hybridization at a location, as measured by both degree of backcrossing and number of distinct hybrid crosses. A more detailed description of this analysis is in the Supplemental material.

Results

DNA sequencing produced 1.77×10^9 reads. From these data, we identified 11,221 well-supported independent SNPs for 2785 individual fish. Mean sequence coverage for retained SNPs was

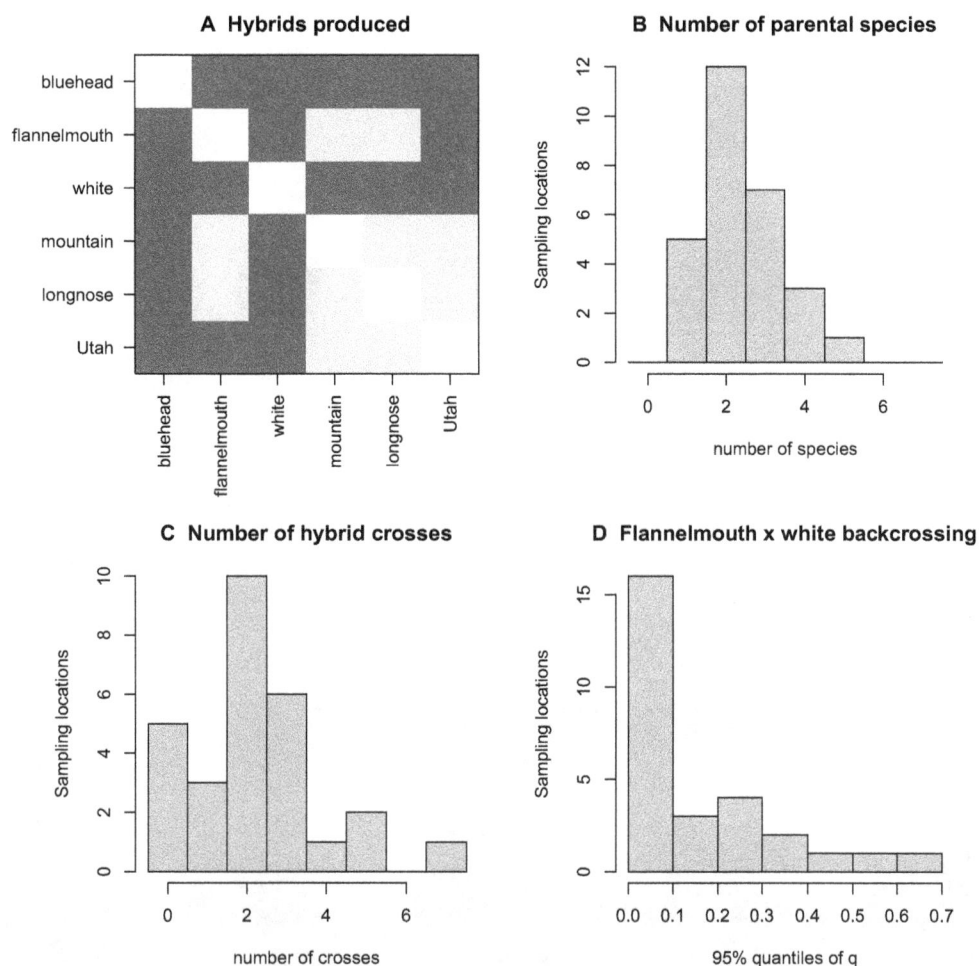

Figure 2. The identity of hybrid crosses (A; dark gray–present, light gray–absent), number of parental species (B) and number of hybrid crosses (C) varies across the 28 rivers in the Upper Colorado River where more than 20 individuals were sampled. Extent of backcrossing in flannelmouth×white hybrids, the most geographically widespread cross, also varies by river (D).

6.1 reads per locus per individual. We then sorted individuals by species, and identified polymorphic loci within each of six parental species (7672–19,797 SNPs per species) to describe the relationship between within-species genetic structure and hybridization outcomes.

HYBRIDIZATION PRODUCES DIVERSE OUTCOMES

Our analyses confirm that hybridization among *Catostomus* species is geographically widespread and variable in the Upper Colorado River basin. Of the 61 locations sampled, 38 were within the Upper Colorado River basin, while the other 23 were in adjacent river basins (Fig. 1). Populations outside the basin were sampled to provide reference populations for focal taxa independent of hybridization in the Upper Colorado River basin. For each individual fish, we estimated q, the proportion of ancestry in each genetic cluster in a $k = 6$ model using `entropy` (Gompert et al. 2014). Under this model, each cluster corresponds to a named species (Mandeville et al. 2015).

In 21 out of 28 locations within the Upper Colorado River basin with sample sizes of >20 individuals, we identified at least one type of interspecific hybrid (Fig. 2), but the identity of crosses, the number of crosses, and the extent of backcrossing and later-generation hybridization varied by river (Fig. 3 and 4). Hybrids were produced between native and nonnative species, but also between pairs of native species. Among the three most common species, bluehead, flannelmouth, and white suckers, we observed more geographic instances of hybridization and more hybrid individuals in the crosses involving nonnative white suckers (*C. commersoni*). Flannelmouth×white sucker hybrids were observed at 15 out of the 28 sites in Figure 2 (246 total individuals), and bluehead×white hybrids were observed at 12 out of those 28 sites (146 total individuals). In contrast, bluehead×flannelmouth hybrids, which involve two native parental species, were observed at 8 of the 28 sites (56 total individuals).

We observed 12 different hybrid crosses among six parental species, including hybrids that occurred outside of the Upper

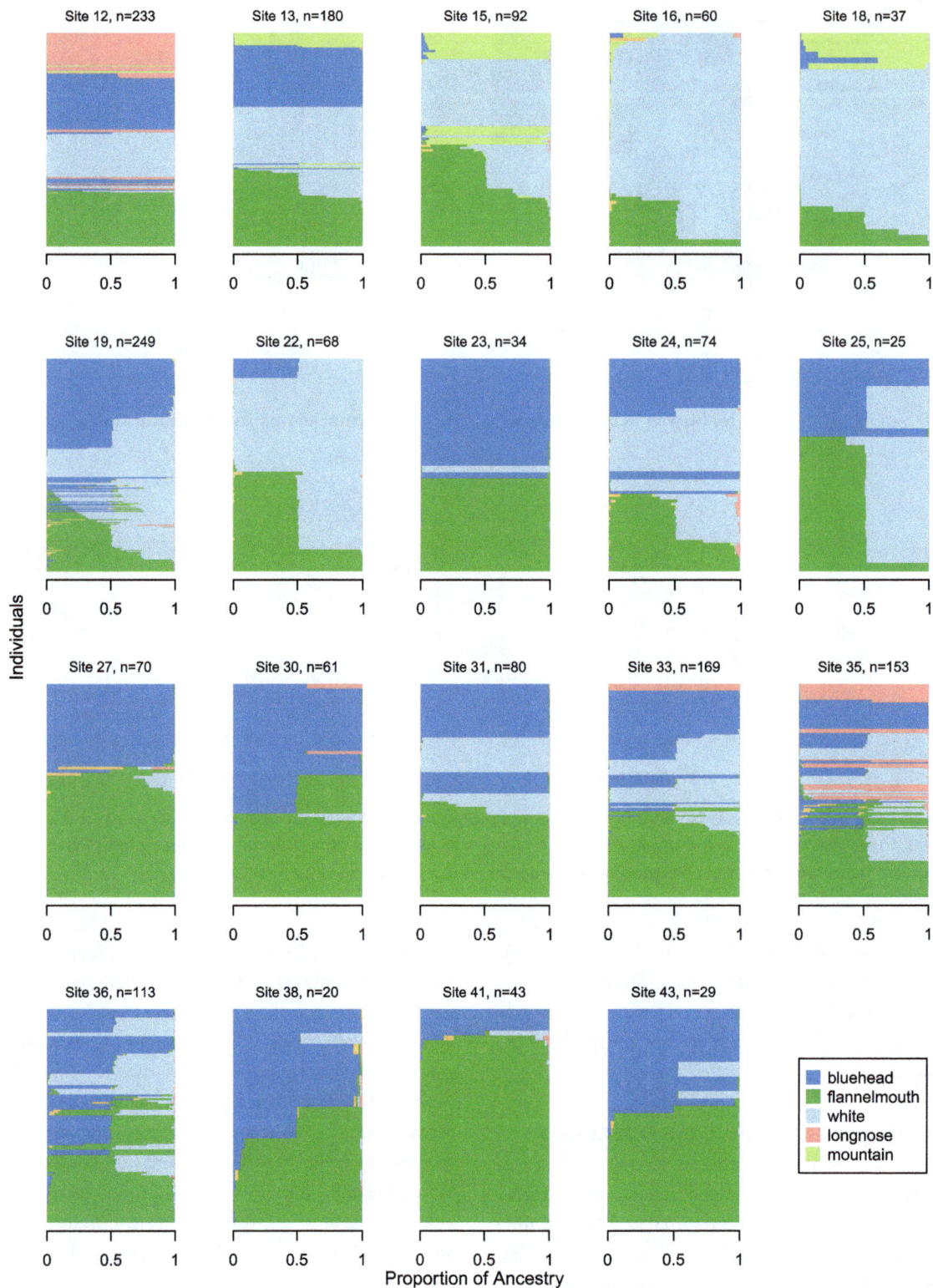

Figure 3. Outcomes of hybridization vary across locations where native and nonnative *Catostomus* species coexist in the Upper Colorado River basin. These 19 plots represent rivers in the Upper Colorado River basin with large sample sizes and multiple species of interest. Individual fish are arranged along the vertical axis of each plot, and bars are colored according to the proportion of an individual's ancestry contributed from each of six possible parental species (`entropy`, *k* = 6 model; Utah suckers were not sampled in these rivers and are excluded from the legend). Rivers are presented in north to south order.

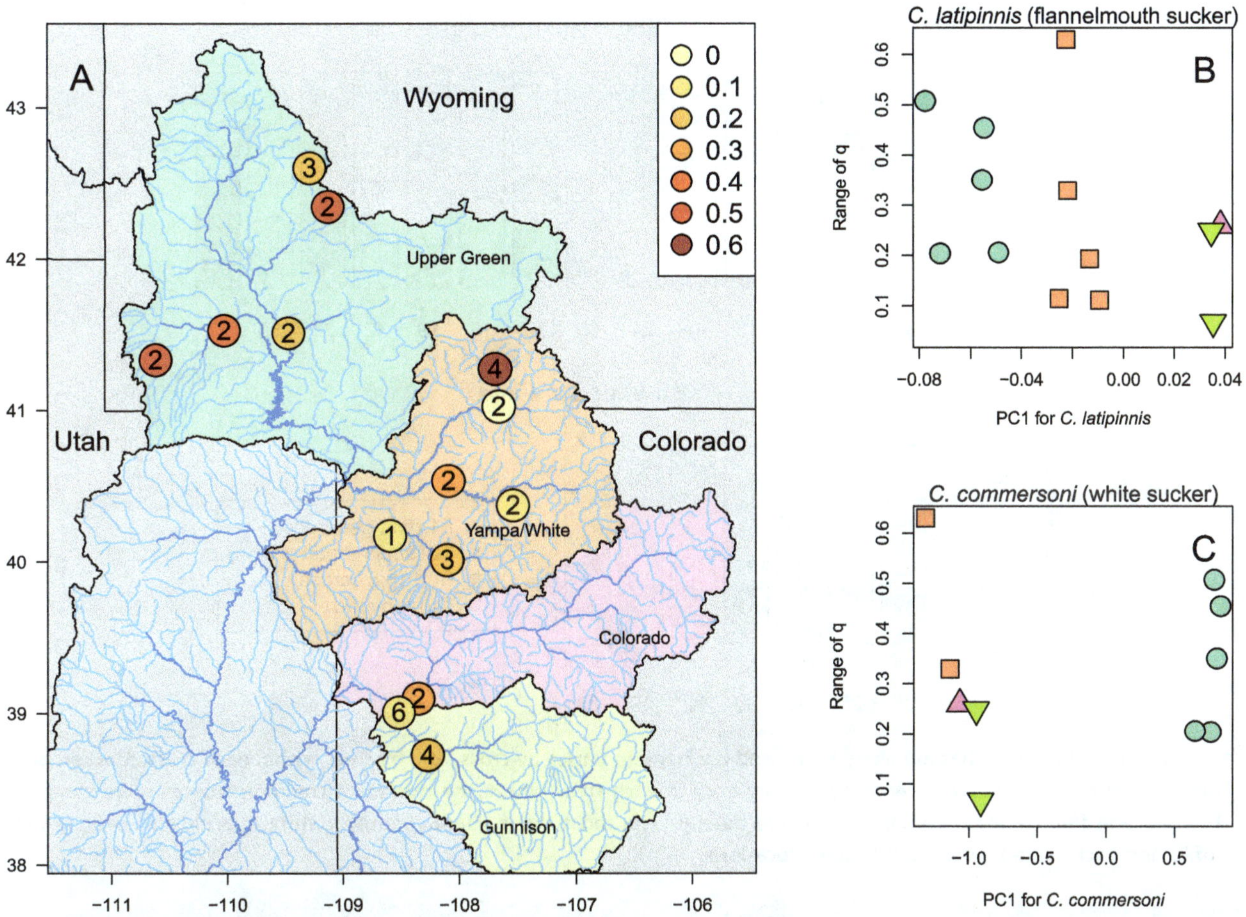

Figure 4. Hybridization outcomes are variable across geographic space (A). Within each river basin (color-coded) where flannelmouth and white suckers come into contact, there are instances of extensive hybridization and more constrained outcomes. Color of points corresponds to the range of q, or proportion of ancestry, in flannelmouth x white hybrids. The text inside the points gives the number of different hybrid combinations in each river. B, C There is no correlation between the dominant axis of genetic variation within parental species and hybridization outcomes. For B and C, the horizontal axis shows population means on PC1, the first principal component of genetic variation, for each parental species. The vertical axis shows the range of q values in a river (range between 2.5 and 97.5% quantiles of estimates of q for individuals in a population), which is a measure of how much backcrossing occurs in flannelmouth x white hybrids in a river. Points are colored according to river basin in which a sampling location lies.

Colorado River basin. To guard against inferring erroneous patterns of hybridization due to model uncertainty, this count only includes crosses for which more than one putative hybrid individual was sampled. Two types of hybrids had ancestry from three parental species (bluehead x flannelmouth x white, Muddy Creek, Colorado River, Gunnison River; flannelmouth x Utah x white, Halfmoon Lake). Of the ten crosses with two parental species, six feature a nonnative species hybridizing with a native species, while four are between two native species. Nine crosses were in the Upper Colorado River basin.

EXTENT OF HYBRIDIZATION IS VARIABLE ACROSS SPECIES AND LOCATIONS

We observed different combinations of q (proportion of ancestry) and Q (interspecific ancestry) in hybrid individuals in

different hybrid crosses, indicating different numbers of generations of hybridization and different parentage of hybrids (Fig. 6). Using q and Q, we classified hybrid individuals as likely F_1 (q 0.4–0.6; $Q > 0.75$), potential F_2 (q 0.4–0.6; Q 0.25–0.75), potential first-generation backcrosses (q 0.15–0.35 or q 0.65–0.85; Q 0.25–0.75), and other types of hybrids (Fig. 5). Number and identity of hybrid crosses varied by location. Within each hybrid cross, there was variation in what additional hybridization (if any) occurred beyond the F_1 generation (Figs. 3, 4, 5, 6). In some cases, hybridization led to later-generation recombinant hybrids and backcrosses; in other cases, outcomes of hybridization were more constrained (asymmetric backcrossing or absence of later generation hybrids).

Flannelmouth x white hybrids were most geographically widespread (Fig. 5, 6). Hybridization ranged from only F_1 hybrids

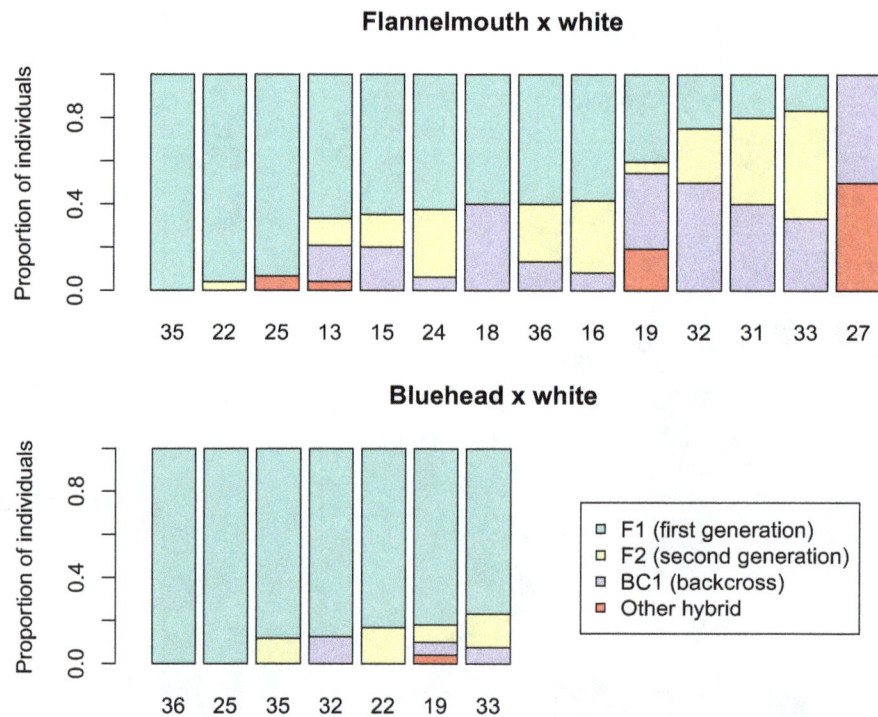

Figure 5. Proportion of hybrids that are likely F_1, F_2, and backcrosses varies by cross, and by river within each cross. Stacked bars show the proportion of hybrid individuals in each hybrid category in each river where the two most common crosses were sampled. Both bluehead×white and flannelmouth×white hybrids are mostly F_1 in most rivers. Flannelmouth×white hybrids also include moderate numbers of F_2 and backcrossed individuals in some locations.

in some locations (e.g., site 35, the mainstem of the Gunnison in Colorado) to extensive later-generation hybridization and repeated backcrossing to both parental species (e.g., site 19, Muddy Creek, Carbon County, Wyoming). In some locations, backcrossing was observed only toward one parental species (toward flannelmouth: sites 24, 32, 33, and 36; toward white: sites 16 and 31). In other locations, hybrids were formed through symmetrical backcrossing to both parental species (sites 12, 13, 15, 18, and 19). In contrast, bluehead×white hybrids were mostly first generation (F_1; Figs. 5, 6), with a small number of backcrosses in several locations (sites 19, 32, and 33) and F_2 hybrids observed in only one location (site 19).

ANALYSES OF POSSIBLE CAUSES FOR VARIABLE HYBRIDIZATION

If variation in reproductive isolation is associated with genetic differentiation of populations at a large geographic scale within a species, we expect that hybridization outcomes might be associated with intraspecific genetic structure in one or both parental species. We therefore used principal components analysis to quantify intraspecific population genetic structure and compare to hybridization outcomes for flannelmouth×white sucker hybrids. The range of q values (proportion of ancestry; larger range corresponds to more backcrossing) for flannelmouth×white hybrids

differed substantially between adjacent localities (Fig. 4, **A**), despite genetic similarity of populations within river basins (proximity in principal component space; see PC1 in Fig. 4 **B** and **C**). There was no correlation between q range and PC1 for either parental species (Fig. 4, **B** and **C**), suggesting that extent of hybridization was not correlated with major axes of genetic differentiation within parental species, although it is possible that a small number of causative loci could vary among populations independently of the major axes of variation quantified by a PCA.

If variation in reproductive isolation is connected to ecological context, we expect that hybridization outcomes might be correlated with ecological or environmental characteristics of sites. We used publicly available data quantify the association between attributes of sites (e.g., stream gradient, land usage, elevation) and hybridization outcomes. No environmental attributes of sites were strongly associated with extent of hybridization (more details are included in Supplemental Material). However, the number of parental species at a location was positively and significantly correlated with number of hybrid crosses and extent of backcrossing in flannelmouth×white hybrids (Pearson correlations = 0.56 and 0.51; $P<0.05$). Multiple parental species are needed to produce multiple hybrid crosses, but interestingly, the number of parental species was also positively correlated with extent of backcrossing in a single cross, flannelmouth×white sucker hybridization.

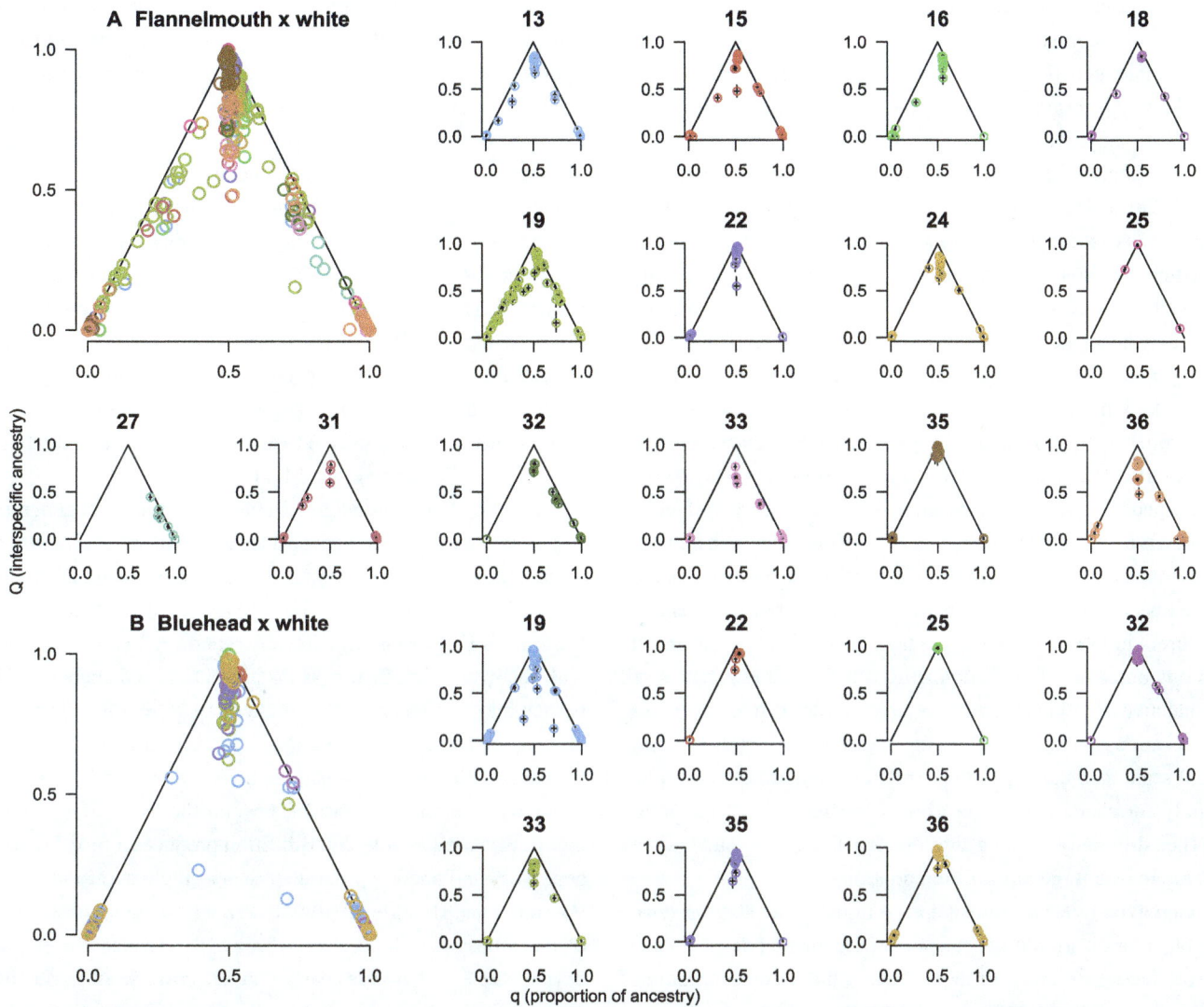

Figure 6. Estimates of proportion of ancestry (q) and interspecific ancestry (Q) for flannelmouth × white (A) and bluehead × white hybrids (B) show that the extent of hybridization and backcrossing varies among rivers. The large plots show all hybrids of each cross, color-coded by river, and the smaller plots each represent one river where hybrids are present. F_1 hybrids occupy the apex of the gray triangle in each plot ($q=0.5$, $Q=1$). F_2 hybrids have the same expected proportion of ancestry ($\bar{q} = 0.5$), but depressed expectations for interspecific ancestry ($\bar{Q}=0.5$). BC_1 hybrids have expectations of $\bar{q}=0.25$ or 0.75, and $\bar{Q}=0.5$.

Discussion

We present evidence for highly variable outcomes of interactions between *Catostomus* species, indicating that reproductive isolation is remarkably inconsistent across the geographic area where these species interact. Furthermore, variation in several pairs of *Catostomus* species suggests that variation in evolutionary interactions between species might be general rather than exceptional (Kozlowska et al. 2012; Fukami 2015). Our findings contribute to the mounting evidence that reproductive isolation and hybridization can vary substantially through space and time, and across different species pairs in secondary contact (Buerkle and Rieseberg 2001; Vines et al. 2003; Lepais et al. 2009; Nolte et al. 2009; Teeter et al. 2010; Haselhorst and Buerkle 2013; Mandeville et al.

2015). In the sections below we elaborate on possible causes and consequences of this variability.

POTENTIAL CAUSES OF VARIABLE REPRODUCTIVE ISOLATION

Variation in hybridization among populations could stem from variation in loci and traits associated with isolation, variation due to epistatic interactions among genes within individuals, or phenotype-by-phenotype interactions between potentially hybridizing individuals. Additionally, variation in hybridization might arise from variation in traits that results from genotype-by-environment interactions across heterogenous environments. We lack the evidence to fully evaluate the contributions of these potential causes in the case of *Catostomus* hybridization, but discuss their plausibility below.

Variable genetics underlying reproductive isolation have been described in many taxa (e.g., Wade et al. 1997; Rieseberg 2000; Reed and Markow 2004; Kopp and Frank 2005; Shuker et al. 2005; Vyskočilová et al. 2005; Sweigart et al. 2007; Good et al. 2008; Nolte et al. 2009; Teeter et al. 2010), indicating that genetic components of reproductive isolation can be polymorphic and inconsistent within a species. Variation could arise prezygotically from variable traits in parental taxa, both by simple trait variation or through the interactions of traits of individuals (gene-by-gene or phenotype-by-phenotype interactions; e.g., matching of phenology). Additionally, variation in isolation could arise postzygotically, from genetically variable progeny that are produced through hybridization (Bomblies and Weigel 2007; Bomblies et al. 2007). If intraspecific variation for genetic components of reproductive isolation is responsible for variation in hybridization (Cutter 2012; Kozlowska et al. 2012), hybridization outcomes could be shared by members of subspecific demes. However, population genetic structure within *Catostomus* parental species was not correlated with outcomes of hybridization (Fig. 4). If variable genetics of reproductive isolation are responsible for variation in *Catostomus* hybridization, the causal alleles are likely to vary at a different spatial scale than the population structure we detected, or not be strongly correlated with major axes of subspecific genetic variation (i.e., differences among sites due to a few loci of large effect rather than overall genomic differentiation).

Genotype-by-environment interactions could also produce variable hybridization if genetic and environmental determinants of reproductive isolation, or their efficacy, differ across locations (Seehausen et al. 1997; Taylor and McPhail 2000). For example, in African cichlids, elevated water turbidity can lead to loss of reproductive isolation between sympatric species (Seehausen et al. 1997, 2008). In trout, warming water temperatures have led to increased hybridization between native and nonnative species (Muhlfeld et al. 2014). It is also likely that time since introduction of a nonnative species or proximity to introduction site could affect hybridization outcomes (as in tiger salamanders; Fitzpatrick et al. 2010), but since the introduction history of white sucker populations is poorly characterized, it is difficult to know how much introduction history affects hybridization outcomes. However, adjacent populations sometimes have different hybridization outcomes despite probable similarity in introduction times, suggesting that time since introduction is not the primary determinant of extent of hybridization.

EVOLUTIONARY SIGNIFICANCE OF VARIABLE REPRODUCTIVE ISOLATION

Research on hybridization and speciation has typically focused on processes that maintain reproductive isolation for species as a whole (Endler 1977; Barton and Hewitt 1985; Hewitt 1988; Barton and Hewitt 1989). In contrast, our study of *Catostomus* hybridization indicates that realized reproductive isolation is extremely variable across many locations, among several pairs of *Catostomus* fishes, which implies that no single, consistent set of mechanisms is responsible for maintaining reproductive isolation between these species. Along with other studies, our findings support the idea that variability in isolating barriers might be common for incompletely isolated and potentially hybridizing taxa (Sweigart et al. 2007; Good et al. 2008; Kozlowska et al. 2012). This variation exists in the context of likely secondary contact among *Catostomus* species, whereas variation might exist in other systems that have undergone primary divergence, among populations that have evolved isolation to different extents (Riesch et al. 2017; Stuart et al. 2017).

For effective reproductive isolation to exist at all geographic locations where *Catostomus* species co-occur would require different natural selection among sites. At locations with primarily F_1 hybrids, effective isolation would arise from selection against the traits of F_1 hybrids (e.g., low F_1 fecundity). In contrast, at sites with relatively high F_1 fitness (as in trout and salamanders; Fitzpatrick and Shaffer 2007; Muhlfeld et al. 2009; Fitzpatrick et al. 2010) and viable later-generation F_n and backcrossed hybrids, effective isolation would instead require very low fertility of F_n and backcrossed hybrids. However, beyond the F_1, or if F_1 hybrids themselves are variable, hybridization produces a broad range of genotypes and phenotypes, rather than a single hybrid phenotype (Gompert and Buerkle 2016). We do not have direct observations of the fitness of *Catostomus* hybrids. If hybridization proceeds beyond the F_1, it is less likely that selection would effectively maintain isolation, and local gene flow between species would be likely (Barton and Bengtsson 1986; Gavrilets and Gravner 1997; Bank et al. 2012; Gompert et al. 2012; Lindtke and Buerkle 2015). Adaptive and neutral introgression of traits and genomic regions across species boundaries is also likely to vary geographically, potentially affecting local evolutionary and ecological dynamics. It is unclear what long-term outcomes of variable hybridization and introgression will be in *Catostomus* fishes, and more generally, what consequences heterogeneity in species boundaries has for the evolutionary cohesion of populations within parental species.

ECOLOGICAL CONSEQUENCES OF VARIABLE REPRODUCTIVE ISOLATION

If variable postzygotic selection on hybrids drives variation in hybridization among sites, ecological traits of hybrids will be important to outcomes of interactions between species. Fitness of hybrids might therefore vary geographically as a result of variation in ecologically important traits. Hybrids often have different phenotypes from either parental species, and might be able to exploit different resources (e.g., Williams and Ehleringer 2000; Lexer et al. 2004; Gompert et al. 2006; Rieseberg et al. 2007; Stelkens and Seehausen 2009). Specific to this system, we know that different *Catostomus* species have different diets and swimming abilities (Cross et al. 2013; Walsworth et al. 2013; Underwood et al. 2014), and overlap in spawning habitat and timing to varying extents (Sweet and Hubert 2010). Given their variation in admixture, and the different crosses involved, *Catostomus* hybrids are likely to express a wide range of phenotypes, perhaps including transgressive phenotypes (Stelkens and Seehausen 2009; Stelkens et al. 2009). In locations with more extensive backcrossing or later-generation hybridization, recombinant hybrids with high fitness might be formed, potentially with novel ecological traits (as in sunflowers; Rieseberg et al. 2003) or simply highly competitive relative to the parental species, leading to variable ecological outcomes of hybridization.

The patterns of hybridization described in this study are important for conservation of native *Catostomus* species in the Upper Colorado River basin. Our conclusions from this study also apply more generally to understanding when extinction via hybridization is likely to occur. Two species in this study, *C. discobolus* and *C. latipinnis*, are the focus of conservation and management in Wyoming, Colorado, and Utah (Gill et al. 2007; Gelwicks et al. 2009; Senecal et al. 2010), and population sizes are believed to be declining (Bezzerides and Bestgen 2002). Hybridization with nonnative *C. commersoni* has been viewed as a major threat to persistence of native species. Initially, based on results of McDonald et al. (2008), the primary concern for management was the potential loss of genetic identity of native taxa. However, based on results reported here and previously (Mandeville et al. 2015), it is likely that genetic homogenization of native and nonnative species will occur locally, if at all. Hybridization would lead to introgression in some locations, blurring genetic identity of parental species (Rhymer and Simberloff 1996; Wolf et al. 2001), but local genetic homogenization of these species would be unlikely in locations with little hybridization or no hybridization beyond the F_1 (Fig. 3,5). Identifying variation in effectiveness of reproductive isolation will help managers prioritize where and how to intervene. Demographic threats to the persistence of native species also exist and might be exacerbated by variation in ecological success of hybrids. Hybridization very likely results in an opportunity cost and lower population mean fitness, since heterospecific reproduction almost certainly occurs at the cost of conspecific reproduction. The question for management is how large this cost is. If the reduction in population mean fitness is large, then hybridization of any of the forms we observed might represent a substantial threat to the local persistence of species. If mean fitness is reduced only slightly as a result of hybridization, introgression and locally homogenizing gene flow would be a greater concern.

CONCLUSIONS

In this study, we provide evidence for variable genomic outcomes of hybridization among multiple *Catostomus* species pairs across a large geographic area. The variation we observed in hybridization suggests that reproductive isolation is also variable, with no single mechanism of reproductive isolation maintaining separation between species across all locations where they come into contact. Few studies have examined outcomes of hybridization across a similar geographic area. Thus, it is unclear to what extent our results represent a general pattern of variable reproductive isolation and variable evolutionary consequences of species interactions. It is possible that the variation we observed is characteristic of reproductive isolation in many taxa. Variation in outcomes of reproductive interactions between species would be consistent with what we know about the ecological outcomes of interactions between species, which are more commonly recognized to be influenced by contingency and context than are primarily evolutionary outcomes of species interactions like hybridization. If *Catostomus* fishes are indeed representative of typical dynamics of reproductive isolation across the range of an interaction between species, this suggests that our conceptual models of reproductive isolation as a consistent process operating at the species level might need to be revised to accommodate potential for substantial variation across time and space.

AUTHOR CONTRIBUTIONS
E.G.M., T.L.P., and C.A.B. designed and implemented genomic analyses of hybridization. E.G.M. prepared reduced-complexity genomic libraries. E.G.M., T.L.P., and S.J.S. completed DNA extractions. K.G.T., K.R.G., R.I.C., and S.J.S. contributed expertise in the study system, and directed sampling for this work, including selection of populations to sample. All authors contributed to writing and revising the manuscript.

ACKNOWLEDGMENTS
This work was funded by the Wyoming Game and Fish Department through State Wildlife Grant #001793 and Bureau of Land Management Cooperative Agreement 12AC20048, and by Colorado Parks and Wildlife through Species Conservation Trust Fund project SCTF001C. E.G.M. was supported by NIH INBRE funding to the University of Wyoming (NCRR P20RR016474/NIGMS P20GM103432) and by the UW Biodiversity Institute. Computing was accomplished in part with an allocation from the University of Wyoming's Advanced Research Computing Center, on its Mount Moran IBM System X cluster (http://n2t.net/ark:/85786/m4159c).

We thank the Wyoming Game and Fish Department, Colorado Parks and Wildlife, and the US Fish and Wildlife Service, who have made this work possible. Specifically, thanks to Mark Smith and Dave Zafft for arranging and administering funding. We would also like to thank the many people who collected samples or facilitated sampling for this work, including Diana Miller, Paul Attwood, Rob Keith, Bill Cleary, Hillary Walrath, Nick Walrath, Paul Gerrity, Rob Gipson, Bill Bradshaw, Craig Amadio, Pete Cavalli, Darren Rhea, Zack Klein, Jenn Logan, Kyle Battige, Jim White, Lori Martin, Dan Kowalski, Paul Jones, Mike Japhet, Sherm Hebein, Bob Burdick, Dale Ryden, and many seasonal fisheries technicians for WGFD, CPW, and USFWS. Sampling was accomplished under WGFD and CPW permits, and also under Colorado scientific permit #08AQ1026 and University of Colorado animal protocol #1103.05. This manuscript was improved by comments and suggestions from Bob Hall, Annika Walters, Cynthia Weinig, Dave McDonald, Monia Haselhorst, Katie Wagner, Jessica Rick, Jimena Golcher-Benavides, Zach Gompert, and two anonymous reviewers.

LITERATURE CITED

Bank, C., R. Buerger, and J. Hermisson. 2012. The limits to parapatric speciation: Dobzhansky-Muller incompatibilities in a continent-island model. Genetics 191:845–U345.

Barton, N., and B. Bengtsson. 1986. The barrier to genetic exchange between hybridizing populations. Heredity 57:357–376.

Barton, N. H., and G. M. Hewitt. 1985. Analysis of hybrid zones. Ann. Rev. Ecol. Syst. 16:113–148.

———. 1989. Adaptation, speciation and hybrid zones. Nature 341:497–503.

Baxter, G. T., M. D. Stone, and L. Parker. 1995. Fishes of Wyoming. Wyoming Game and Fish Department Cheyenne.

Bezzerides, N., and K. Bestgen. 2002. Status review of roundtail chub, flannelmouth sucker, and bluehead sucker in the Colorado River Basin. Tech. rep., Larval Fish Laboratory, Colorado State University, Fort Collins.

Bomblies, K., J. Lempe, P. Epple, N. Warthmann, C. Lanz, J. L. Dangl, and D. Weigel. 2007. Autoimmune response as a mechanism for a Dobzhansky-Muller-type incompatibility syndrome in plants. PLOS Biol. 5:1962–1972.

Bomblies, K., and D. Weigel. 2007. Hybrid necrosis: autoimmunity as a potential gene-flow barrier in plant species. Nat. Rev. Genet. 8:382–393.

Brooks, J. L., and S. I. Dodson. 1965. Predation, body size, and composition of plankton. Science 150:28–35.

Brown, J. H., T. G. Whitham, S. M. Ernest, and C. A. Gehring. 2001. Complex species interactions and the dynamics of ecological systems: long-term experiments. Science 293:643–650.

Buerkle, C. A., and L. H. Rieseberg. 2001. Low intraspecific variation for genomic isolation between hybridizing sunflower species. Evolution 55:684–691.

Carpenter, S. R., and J. F. Kitchell. 1988. Consumer control of lake productivity. BioScience 38:764–769.

Cross, W. F., C. V. Baxter, E. J. Rosi-Marshall, R. O. Hall Jr, T. A. Kennedy, K. C. Donner, H. A. Wellard Kelly, S. E. Seegert, K. E. Behn, and M. D. Yard. 2013. Food-web dynamics in a large river discontinuum. Ecol. Monogr. 83:311–337.

Cutter, A. D. 2012. The polymorphic prelude to Bateson–Dobzhansky–Muller incompatibilities. Trends in Ecology and Evolution 27:209–218.

Endler, J. A. 1977. Geographic variation, speciation, and clines. Princeton Univ. Press, Princeton, NJ.

Falush, D., M. Stephens, and J. K. Pritchard. 2003. Inference of population structure using multilocus genotype data: linked loci and correlated allele frequencies. Genetics 164:1567–1587.

Fitzpatrick, B. M., J. R. Johnson, D. K. Kump, J. J. Smith, S. R. Voss, and H. B. Shaffer. 2010. Rapid spread of invasive genes into a threatened native species. Proc. Natl. Acad. Sci. 107:3606–3610.

Fitzpatrick, B. M., and H. B. Shaffer. 2007. Hybrid vigor between native and introduced salamanders raises new challenges for conservation. Proc. Natl. Acad. Sci. 104:15793–15798.

Fukami, T. 2015. Historical contingency in community assembly: integrating niches, species pools, and priority effects. Ann. Rev. Ecol. Evol. Syst. 46:1–23.

Gavrilets, S., and J. Gravner. 1997. Percolation on the fitness hypercube and the evolution of reproductive isolation. J. Theoret. Biol. 184:51–64.

Gelwicks, K. R., C. J. Gill, A. Kern, and R. Keith. 2009. Current status of roundtail chub, flannelmouth sucker and bluehead sucker in the Green River Drainage of Wyoming. Wyoming Game and Fish Department, Fish Division.

Gill, C. J., K. R. Gelwicks, and R. M. Keith. 2007. Current distribution of bluehead sucker, flannelmouth sucker, and roundtail chub in seven subdrainages of the Green River, Wyoming. American Fisheries Society Symposium, vol. 53, P. 121. American Fisheries Society.

Gompert, Z., and C. A. Buerkle. 2016. What, if anything, are hybrids: enduring truths and challenges associated with population structure and gene flow. Evol. Appl. 9:909–923.

Gompert, Z., J. A. Fordyce, M. L. Forister, A. M. Shapiro, and C. C. Nice. 2006. Homoploid hybrid speciation in an extreme habitat. Science 314:1923–1925.

Gompert, Z., L. K. Lucas, C. A. Buerkle, M. L. Forister, J. A. Fordyce, and C. C. Nice. 2014. Admixture and the organization of genetic diversity in a butterfly species complex revealed through common and rare genetic variants. Mol. Ecol. 23:4555–4573.

Gompert, Z., T. L. Parchman, and C. A. Buerkle. 2012. Genomics of isolation in hybrids. Philos. Trans. Royal Soc. B Biol. Sci. 367:439–450.

Good, J. M., M. A. Handel, and M. W. Nachman. 2008. Asymmetry and polymorphism of hybrid male sterility during the early stages of speciation in house mice. Evolution 62:50–65.

Haselhorst, M. S. H., and C. A. Buerkle. 2013. Population genetic structure of *Picea engelmannii*, *P. glauca* and their previously unrecognized hybrids in the central Rocky Mountains. Tree Genet. Genomes 9:669–681.

Hewitt, G. M. 1988. Hybrid zones–natural laboratories for evolution studies. Trends Ecol. Evol. 3:158–166.

———. 2011. Quaternary phylogeography: the roots of hybrid zones. Genetica 139:617–638.

Hubbell, S. P. 2001. The unified neutral theory of biodiversity and biogeography (MPB-32), vol. 32. Princeton Univ. Press, Princeton.

Jackson, S. T., J. L. Betancourt, R. K. Booth, and S. T. Gray. 2009. Ecology and the ratchet of events: climate variability, niche dimensions, and species distributions. Proc. Natl. Acad. Sci. 106:19685–19692.

Kaeuffer, R., C. L. Peichel, D. I. Bolnick, and A. P. Hendry. 2012. Parallel and nonparallel aspects of ecological, phenotypic, and genetic divergence across replicate population pairs of lake and stream stickleback. Evolution 66:402–418.

Kopp, A., and A. Frank. 2005. Speciation in progress? A continuum of reproductive isolation in *Drosophila bipectinata*. Genetica 125:55–68.

Kozlowska, J. L., A. R. Ahmad, E. Jahesh, and A. D. Cutter. 2012. Genetic variation for postzygotic reproductive isolation between *Caenorhabditis briggsae* and *Caenorhabditis* Sp. 9. Evolution 66:1180–1195.

Lepais, O., R. Petit, E. Guichoux, J. Lavabre, F. Alberto, A. Kremer, and S. Gerber. 2009. Species relative abundance and direction of introgression in oaks. Mol. Ecol. 18:2228–2242.

Lexer, C., Z. Lai, and L. Rieseberg. 2004. Candidate gene polymorphisms associated with salt tolerance in wild sunflower hybrids: implications for the origin of *Helianthus paradoxus*, a diploid hybrid species. New Phytol. 161:225–233.

Li, H. 2011. A statistical framework for SNP calling, mutation discovery, association mapping and population genetical parameter estimation from sequencing data. Bioinformatics 27:2987–2993.

Li, H., and R. Durbin. 2009. Fast and accurate short read alignment with Burrows–Wheeler transform. Bioinformatics 25:1754–1760.

Li, H., B. Handsaker, A. Wysoker, T. Fennell, J. Ruan, N. Homer, G. Marth, G. Abecasis, R. Durbin, and 1000 Genome Project Data Proc. 2009.

The Sequence Alignment/Map format and SAMtools. Bioinformatics 25:2078–2079.

Lindtke, D., and C. A. Buerkle. 2015. The genetic architecture of hybrid incompatibilities and their effect on barriers to introgression in secondary contact. Evolution 69:1987–2004.

Lindtke, D., Z. Gompert, C. Lexer, and C. A. Buerkle. 2014. Unexpected ancestry of *Populus* seedlings from a hybrid zone implies a large role for postzygotic selection in the maintenance of species. Mol. Ecol. 23:4316–4330.

Losos, J. B. 1998. Contingency and determinism in replicated adaptive radiations of island lizards. Science 279:2115–2118.

———. 2010. Adaptive radiation, ecological opportunity, and evolutionary determinism. Am. Nat. 175:623–639.

Mahler, D. L., T. Ingram, L. J. Revell, and J. B. Losos. 2013. Exceptional convergence on the macroevolutionary landscape in island lizard radiations. Science 341:292–295.

Mandeville, E. G., T. L. Parchman, D. B. McDonald, and C. A. Buerkle. 2015. Highly variable reproductive isolation among pairs of *Catostomus* species. Mol. Ecol. 24:1856–1872.

McDonald, D. B., T. L. Parchman, M. R. Bower, W. A. Hubert, and F. J. Rahel. 2008. An introduced and a native vertebrate hybridize to form a genetic bridge to a second native species. Proc. Natl. Acad. Sci. 105:10837–10842.

Muhlfeld, C. C., S. T. Kalinowski, T. E. McMahon, M. L. Taper, S. Painter, R. F. Leary, and F. W. Allendorf. 2009. Hybridization rapidly reduces fitness of a native trout in the wild. Biol. Lett. 5:328–331.

Muhlfeld, C. C., R. P. Kovach, L. A. Jones, R. Al-Chokhachy, M. C. Boyer, R. F. Leary, W. H. Lowe, G. Luikart, and F. W. Allendorf. 2014. Invasive hybridization in a threatened species is accelerated by climate change. Nat. Climate Change 4:620–624.

Natarajan, C., F. G. Hoffmann, R. E. Weber, A. Fago, C. C. Witt, and J. F. Storz. 2016. Predictable convergence in hemoglobin function has unpredictable molecular underpinnings. Science 354:336–339.

Nolte, A. W., Z. Gompert, and C. A. Buerkle. 2009. Variable patterns of introgression in two sculpin hybrid zones suggest that genomic isolation differs among populations. Mol. Ecol. 18:2615–2627.

Nosil, P., and D. Schluter. 2011. The genes underlying the process of speciation. Trends Ecol. Evol. 26:160–167.

Paine, R. T. 1966. Food web complexity and species diversity. Am. Nat. 100:65–75.

Parchman, T. L., Z. Gompert, J. Mudge, F. Schilkey, C. W. Benkman, and C. A. Buerkle. 2012. Genome-wide association genetics of an adaptive trait in lodgepole pine. Mol. Ecol. 21:2991–3005.

Pritchard, J. K., M. Stephens, and P. Donnelly. 2000. Inference of population structure using multilocus genotype data. Genetics 155:945–959.

R Development Core Team. 2016. R: a language and environment for statistical computing. R foundation for statistical computing, Vienna, Austria. ISBN 3-900051-07-0.

Ramsey, J., H. Bradshaw, and D. Schemske. 2003. Components of reproductive isolation between the monkeyflowers *Mimulus lewisii* and *M. cardinalis* (Phrymaceae). Evolution 57:1520–1534.

Reed, L. K., and T. A. Markow. 2004. Early events in speciation: polymorphism for hybrid male sterility in *Drosophila*. Proc. Natl. Acad. Sci. 101:9009–9012.

Rhymer, J. M., and D. Simberloff. 1996. Extinction by hybridization and introgression. Ann. Rev. Ecol. Syst. 27:83–109.

Riesch, R., M. Muschick, D. Lindtke, R. Villoutreix, A. A. Comeault, T. E. Farkas, K. Lucek, E. Hellen, V. Soria-Carrasco, S. R. Dennis, et al. 2017. Transitions between phases of genomic differentiation during stick-insect speciation 1:0082 EP–.

Rieseberg, L. H. 2000. Crossing relationships among ancient and experimental sunflower hybrid lineages. Evolution 54:859–865.

Rieseberg, L. H., and B. K. Blackman. 2010. Speciation genes in plants. Annal. Bot. 106:439–455.

Rieseberg, L. H., S.-C. Kim, R. A. Randell, K. D. Whitney, B. L. Gross, C. Lexer, and K. Clay. 2007. Hybridization and the colonization of novel habitats by annual sunflowers. Genetica 129:149–165.

Rieseberg, L. H., A. Widmer, A. M. Arntz, and J. M. Burke. 2003. The genetic architecture necessary for transgressive segregation is common in both natural and domesticated populations. Philos. Trans. Royal Soc. Lond. B 358:1141–1147.

Rockman, M. V. 2012. The QTN program and the alleles that matter for evolution: all that's gold does not glitter. Evolution 66:1–17.

Seehausen, O. 2007. Chance, historical contingency and ecological determinism jointly determine the rate of adaptive radiation. Heredity 99:361–363.

Seehausen, O., G. Takimoto, D. Roy, and J. Jokela. 2008. Speciation reversal and biodiversity dynamics with hybridization in changing environments. Mol. Ecol. 17:30–44.

Seehausen, O., J. J. Van Alphen, and F. Witte. 1997. Cichlid fish diversity threatened by eutrophication that curbs sexual selection. Science 277:1808–1811.

Senecal, A. C., K. R. Gelwicks, P. A. Cavalli, and R. M. Keith. 2010. WGFD short-term plan for the three species in the green river drainage of Wyoming; 2009–2014. Wyoming Game and Fish Department, Fish Division.

Shuker, D. M., K. Underwood, T. M. King, and R. K. Butlin. 2005. Patterns of male sterility in a grasshopper hybrid zone imply accumulation of hybrid incompatibilities without selection. Proc. Royal Soc. B Biol. Sci. 272:2491–2497.

Soria-Carrasco, V., Z. Gompert, A. A. Comeault, T. E. Farkas, T. L. Parchman, J. S. Johnston, C. A. Buerkle, J. L. Feder, J. Bast, T. Schwander, et al. 2014. Stick insect genomes reveal natural selection's role in parallel speciation. Science 344:738–742.

Stayton, C. T. 2015. What does convergent evolution mean? The interpretation of convergence and its implications in the search for limits to evolution. Interface Focus 5:20150039.

Stelkens, R., C. Schmid, O. Selz, and O. Seehausen. 2009. Phenotypic novelty in experimental hybrids is predicted by the genetic distance between species of cichlid fish. BMC Evol. Biol. 9:283.

Stelkens, R., and O. Seehausen. 2009. Genetic distance between species predicts novel trait expression in their hybrids. Evolution 63:884–897.

Stuart, Y. E., T. Veen, J. N. Weber, D. Hanson, M. Ravinet, B. K. Lohman, C. J. Thompson, T. Tasneem, A. Doggett, R. Izen, et al. 2017. Contrasting effects of environment and genetics generate a continuum of parallel evolution 1:0158 EP–.

Sweet, D. E., and W. A. Hubert. 2010. Seasonal movements of native and introduced catostomids in the Big Sandy River, Wyoming. Southwestern Nat. 55:382–389.

Sweigart, A. L., A. R. Mason, and J. H. Willis. 2007. Natural variation for a hybrid incompatibility between two species of *Mimulus*. Evolution 61:141–151.

Taylor, E. B., and J. D. McPhail. 2000. Historical contingency and ecological determinism interact to prime speciation in sticklebacks, Gasterosteus. Proc. Royal Soc. London B Biol. Sci. 267:2375–2384.

Teeter, K. C., L. M. Thibodeau, Z. Gompert, C. A. Buerkle, M. W. Nachman, and P. K. Tucker. 2010. The variable genomic architecture of isolation between hybridizing species of house mouse. Evolution 64:472–485.

Underwood, Z. E., C. A. Myrick, and R. I. Compton. 2014. Comparative swimming performance of five Catostomus species and roundtail chub. North Am. J. Fisheries Manag. 34:753–763.

Valone, T. J., and J. H. Brown. 1995. Effects of competition, colonization, and extinction on rodent species diversity. Science 267:880–883.

Vines, T. H., S. C. Kohler, A. Thiel, I. Ghira, T. R. Sands, C. J. MacCallum, N. H. Barton, and B. Nurnberger. 2003. The maintenance of reproductive isolation in a mosaic hybrid zone between the fire-bellied toads *Bombina bombina* and *B. variegata*. Evolution 57:1876–1888.

Vyskočilová, M., Z. Trachtulec, J. Forejt, and J. Piálek. 2005. Does geography matter in hybrid sterility in house mice? Biol. J. Linnean Soc. 84:663–674.

Wade, M. J., N. A. Johnson, R. Jones, V. Siguel, and M. McNaughton. 1997. Genetic variation segregating in natural populations of *Tribolium castaneum* affecting traits observed in hybrids with *T. freemani*. Genetics 147:1235–1247.

Wagner, C. E., L. J. Harmon, and O. Seehausen. 2012. Ecological opportunity and sexual selection together predict adaptive radiation. Nature 487:366–369.

Walsworth, T. E., P. Budy, and G. P. Thiede. 2013. Longer food chains and crowded niche space: effects of multiple invaders on desert stream food web structure. Ecol. Freshwater Fish 22:439–452.

Weiss, K. M. 2008. Tilting at quixotic trait loci (QTL): an evolutionary perspective on genetic causation. Genetics 179:1741–1756.

Williams, D. G., and J. R. Ehleringer. 2000. Carbon isotope discrimination and water relations of oak hybrid populations in southwestern Utah. Western North Am. Nat. 60:121–129.

Wolf, D. E., N. Takebayashi, and L. H. Rieseberg. 2001. Predicting the risk of extinction through hybridization. Conserv. Biol. 15:1039–1053.

Wolf, J. B. W., J. Lindell, and N. Backstrom. 2010. Speciation genetics: current status and evolving approaches. Philos. Trans. Royal Soc. B Biol. Sci. 365:1717–1733.

Yuan, Y.-W., J. M. Sagawa, R. C. Young, B. J. Christensen, and H. D. Bradshaw. 2013. Genetic dissection of a major anthocyanin QTL contributing to pollinator-mediated reproductive isolation between sister species of *Mimulus*. Genetics 194:255–263.

14

Evolution and comparative ecology of parthenogenesis in haplodiploid arthropods

Casper J. van der Kooi,[1,2] (iD) Cyril Matthey-Doret,[1] (iD) and Tanja Schwander[1] (iD)

[1]Department of Ecology and Evolution, University of Lausanne, Lausanne, Switzerland

[2]E-mail: C.J.van.der.Kooi@rug.nl

Changes from sexual reproduction to female-producing parthenogenesis (thelytoky) have great evolutionary and ecological consequences, but how many times parthenogenesis evolved in different animal taxa is unknown. We present the first exhaustive database covering 765 cases of parthenogenesis in haplodiploid (arrhenotokous) arthropods, and estimate frequencies of parthenogenesis in different taxonomic groups. We show that the frequency of parthenogenetic lineages extensively varies among groups (0–38% among genera), that many species have both sexual and parthenogenetic lineages and that polyploidy is very rare. Parthenogens are characterized by broad ecological niches: parasitoid and phytophagous parthenogenetic species consistently use more host species, and have larger, polewards extended geographic distributions than their sexual relatives. These differences did not solely evolve after the transition to parthenogenesis. Extant parthenogens often derive from sexual ancestors with relatively broad ecological niches and distributions. As these ecological attributes are associated with large population sizes, our results strongly suggests that transitions to parthenogenesis are more frequent in large sexual populations and/or that the risk of extinction of parthenogens with large population sizes is reduced. The species database presented here provides insights into the maintenance of sex and parthenogenesis in natural populations that are not taxon specific and opens perspectives for future comparative studies.

KEY WORDS: asexual reproduction, haplodiploidy, Hymenoptera, niche breadth, thelytoky, Thysanoptera, arrhenotoky, polyploidy.

Impact Summary

The animal kingdom exhibits a great diversity in reproductive modes. In addition to the well-known and widespread sexual reproduction, species can reproduce asexually, via parthenogenesis. Males are absent in parthenogenetic species or populations. How this diversity in reproductive systems can be maintained remains a major question in evolutionary biology.

We assembled a database of parthenogenetic arthropod species focusing on groups with haplodiploid sex determination. Haplodiploidy is the sex determination system where males develop from unfertilized eggs and females from fertilized eggs. We use our database to identify ecological traits that contribute to reproductive polymorphisms.

Parthenogenesis evolved many more times than previously thought. We found clear evidence for parthenogenesis in 765 species in many phylogenetically unrelated groups with vastly different ecologies. The frequency of parthenogenesis greatly varies among groups, and many species comprise both sexual and parthenogenetic populations. Overall, the frequency ranges from 0–1.5% between orders, but in species-rich genera, parthenogenesis occurs in up to as much as 38% of the species. Polyploidy is very rare (at most 4%), and endosymbiont-induced parthenogenesis is suggested to occur in approximately 40% of the species. Parthenogens are characterized by larger, polewards extended geographic ranges and utilize more host species than their sexual relatives. Moreover, ecological attributes (i.e. the number of host species and size of geographic distribution) in sexuals favor the transition to and/or the success of derived parthenogenetic lineages. This species database sets the basis for further analyses on sexuals and parthenogens that are not taxon specific.

EVOLUTION AND COMPARATIVE ECOLOGY OF PARTHENOGENESIS

Casper J. van der Kooi; Cyril Matthey-Doret; Tanja Schwander

University of Lausanne, Department of Ecology and Evolution, Lausanne, Switzerland

Correspondence: Casper.vanderKooi@unil.ch

INTRODUCTION

Changing from sexual reproduction to female-producing parthenogenesis (thelytoky), has great evolutionary and ecological consequences, but how many times parthenogenesis evolved in different animal taxa remains unknown.

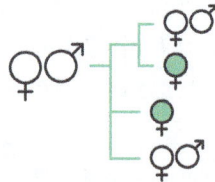

THIS STUDY

We present the first exhaustive database covering 765 cases of obligate parthenogenesis in haplodiploid arthropods. Parthenogenesis can be found in all well-studied haplodiploid taxa, i.e. hymenoptera, thysanoptera, some coleoptera, mites, scale insects and whiteflies.

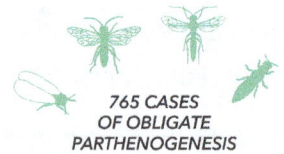

765 CASES OF OBLIGATE PARTHENOGENESIS

RESULTS

1 **The frequency of parthenogenetic lineages extensively varies** among groups.

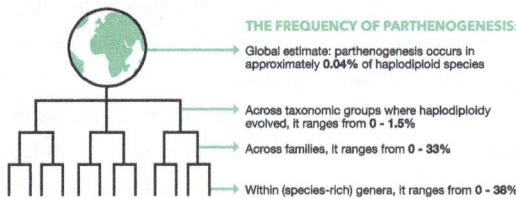

THE FREQUENCY OF PARTHENOGENESIS:

Global estimate: parthenogenesis occurs in approximately **0.04%** of haplodiploid species

Across taxonomic groups where haplodiploidy evolved, it ranges from **0 - 1.5%**

Across families, it ranges from **0 - 33%**

Within (species-rich) genera, it ranges from **0 - 38%**

2 **The incidence of endosymbiont-induced parthenogenesis is** about one third of cases.

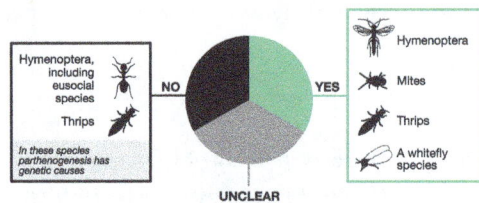

Hymenoptera, including eusocial species

Thrips

In these species parthenogenesis has genetic causes

NO — **YES**

Hymenoptera

Mites

Thrips

A whitefly species

UNCLEAR

3 **Parthenogenetic polyploids are very rare.** Of the 50 parthenogenetic species with available karyotype information, only two (4%) are polyploid.

96% 2n   3n **4%**

4 **Parthenogens are characterized by broad ecological niches.** These differences did not solely evolve after the transition to parthenogenesis.

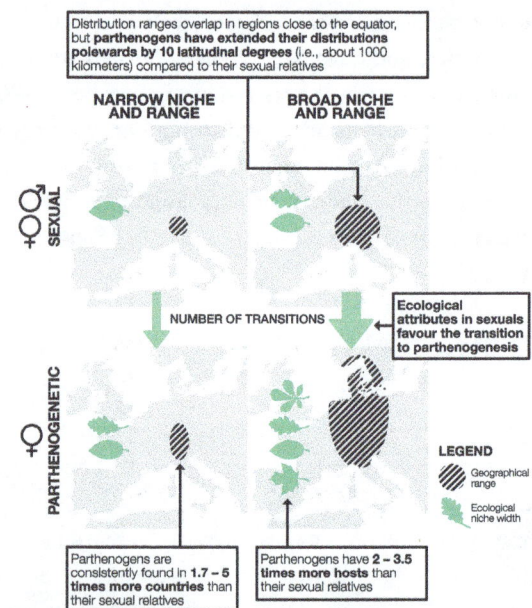

Distribution ranges overlap in regions close to the equator, but **parthenogens have extended their distributions polewards by 10 latitudinal degrees** (i.e., about 1000 kilometers) compared to their sexual relatives

NARROW NICHE AND RANGE   **BROAD NICHE AND RANGE**

SEXUAL

NUMBER OF TRANSITIONS

Ecological attributes in sexuals favour the transition to parthenogenesis

PARTHENOGENETIC

LEGEND

Geographical range

Ecological niche width

Parthenogens are consistently found in **1.7 – 5 times more countries** than their sexual relatives

Parthenogens have **2 – 3.5 times more hosts** than their sexual relatives

CONCLUSION

Parthenogenesis evolved many more times than previously thought, polyploidy is very rare, and parthenogenesis has great ecological consequences.

The species database presented here provides insights into the maintenance of sex and parthenogenesis in natural populations that are not taxon specific and opens perspectives for future comparative studies.

UNIL | Université de Lausanne

Designed by:
significant
science communication
significantcommunication.eu

Changes in reproductive modes, especially from sexual reproduction to female-producing parthenogenesis (also called thelytoky), have great evolutionary and ecological consequences (Bell 1982), but how many times parthenogenesis evolved in different animal taxa is unknown. Whereas parthenogenesis is rare amongst vertebrates and absent in natural bird and mammal populations, it occurs frequently in many species-rich invertebrate groups (Bell 1982; Suomalainen et al. 1987; Normark 2003). Using information mostly from vertebrates, White (1977) estimated that approximately 0.1% of the described animal species reproduce by means of parthenogenesis, a frequency estimate that subsequently has been widely perpetuated (Bell 1982; Schön et al. 2009). This estimate was, however, not based on a species list, and vertebrates represent only a minute and nonrepresentative fraction (at most 1%; (May 2010)) of total animal diversity. More importantly, parthenogenesis in vertebrates is often due to tychoparthenogenesis (i.e., rare, spontaneous hatching of unfertilized eggs in sexual species), rather than facultative or obligate parthenogenesis. Given the inefficiency of tychoparthenogenesis (very low hatching success and low viability of offspring; reviewed by van der Kooi and Schwander 2015), it generally plays no role in the population dynamics of a species.

In contrast to vertebrates, the frequency of obligate parthenogenesis in some species-rich invertebrate groups appears to be much higher than 0.1%. Focusing on hexapods, Normark (2014) recently pointed out that in some insect groups the overall frequency can be orders of magnitudes higher. Indeed, studies that focused on specific invertebrate groups found high frequencies of parthenogenesis, for example it was found in 15% of *Megastigmus* (Boivin et al. 2014) and 30% of *Aphytis* wasp species (DeBach 1969; Rosen and DeBach 1979). Why different taxa vary by several orders of magnitude in the frequency of parthenogenesis remains unknown.

One hypothesis that could explain the extensive variation in parthenogenesis frequency among taxa is that developmental and genetic constraints reduce the transition from sexual reproduction to parthenogenesis in some cases (reviewed by Engelstaedter 2008). For example, the necessity of sperm to initiate embryo development could explain the extreme rareness of parthenogenesis in vertebrates. In many invertebrates such developmental requirements are absent. As an example, in species with haplodiploid sex determination–where males develop from unfertilized eggs and females from fertilized eggs (arrhenotoky)—egg activation and centrosome formation is induced immediately after oviposition, independent of fertilization. However, within taxonomic groups with identical sex determination systems the incidence of parthenogenesis also varies (Normark 2003). This suggests that alternative factors, possibly linked to species ecologies rather than to genetic or developmental factors, also influence the frequency of parthenogenesis. The lack of quantitative frequency estimates of parthenogenetic species however prohibits testing how genetic and ecological effects influence the transition rate to parthenogenesis and the persistence of parthenogenetic lineages across broad taxonomic groups.

Here, we present the first comprehensive survey of female-producing parthenogenesis in haplodiploid arthropods. Haplodiploid sex determination has evolved at least 17 times independently, of which 15 times in arthropods (Otto and Jarne 2001). This allows for comparative analyses across different taxonomic groups but within a single sex-determination system. Furthermore, approximately 12% of all animal species are haplodiploid (Bachtrog et al. 2014), and the ecologies of several haplodiploid taxonomic groups (notably insects) are well studied. In arthropods, haplodiploidy is well known from the insect orders Hymenoptera (ants, bees, sawflies, and wasps) and Thysanoptera (thrips) where all species are haplodiploid, but it also occurs in Hemiptera (whiteflies), a few beetle species and several groups of mites. We focus on obligate female-producing parthenogenesis, thus excluding rare cases of facultative and cyclical parthenogenesis, as these reproductive systems are functionally very similar to sexual reproduction (reviewed by Neiman et al. 2014). We do not include species characterized by paternal genome elimination–where males develop from fertilized eggs, but subsequently eliminate their paternal chromosomes, which has recently been reviewed elsewhere (e.g., Ross et al. 2010). Whenever possible, our database includes information on the causes of parthenogenesis in a lineage, especially whether parthenogenesis is caused by infection with maternally inherited endosymbionts (e.g., *Wolbachia*). Endosymbiont infection is a common cause of parthenogenesis in various haplodiploid groups (Stouthamer et al. 1990; Zchori-Fein et al. 2001). For each taxonomic group, we calculate frequency estimates of parthenogenesis, endosymbiont-induced parthenogenesis, and polyploidy, and we perform comparative tests on ecological characteristics of sexuals and parthenogens.

Materials and Methods
DATA COLLECTION
The species list was compiled via a thorough search through Google Scholar (publications until August 2016) and using previously published reviews on different topics (Table 1). We started with the list provided by Normark (2003) that contained 163 cases of parthenogenesis in haplodiploid taxa, which is about 20% of our database. Other (often overlapping) studies were used to extend our database (Table 1), and various overviews provided a starting point for searches in specific taxa (e.g., Lewis 1973; Cook and Butcher 1999; Wenseleers and Billen 2000; Huigens and Stouthamer 2003; Koivisto and Braig 2003). The recently

Table 1. Important previous overview studies with parthenogenetic haplodiploids.

Taxa studied	Species	Reference
Hemiptera, Hymenoptera, and Thysanoptera	163	(Normark 2003)
Hymenoptera	20	(Flanders 1945; Slobodchikoff and Daly 1971; Gokhman 2009)
Hymenoptera	100	(Stouthamer 2003)
Hymenoptera: Cynipoidea	50	(Askew et al. 2013)
Hymenoptera: *Aphytis*	30	(Rosen and DeBach 1979)
Mites	38	(Norton et al. 1993)
Thysanoptera	46	(Pomeyrol 1929)
Haplodiploid arthropods	765	*This study*

Note that different species lists often overlap.

established Tree of Sex database (Bachtrog et al. 2014) also included parthenogenetic species in various taxa, but did not list any parthenogenetic Hymenoptera or Thysanoptera species. All cases that were obtained from reviews were re-examined.

Reproductive modes in most studies are deduced based on breeding experiments or population sex ratios. Species for which reproductive modes were not assessed via breeding assays were only included if sex ratio estimates were based on large sample sizes, preferably from different locations; hence, we are fairly certain that the species in our list are parthenogenetic. Species not present in the list were assumed to be sexual. However, for the vast majority of these, the reproductive mode has not been studied, and therefore our frequency estimates are underestimates. When available, information on ploidy levels, reproductive polymorphisms, life-history traits as well as the origin and cytological basis of parthenogenesis was included in the database.

Frequencies of parthenogenesis are compared for different taxonomic levels (genera, families, superfamilies, and orders). Total species numbers for each taxonomic level were taken from large-scale overview studies on Hymenoptera in general (Aguiar et al. 2013), Symphyta (Taeger et al. 2010; Taeger and Blank 2011), Chalcidoidea (Noyes 2016), and Thysanoptera (Mound 2013; ThripsWiki 2016).

COMPARATIVE ANALYSES

For our comparative analyses of sexuals and parthenogens, we focused on the mega-diverse Hymenoptera superfamily Chalcidoidea, taking advantage of the many transitions to parthenogenesis within this group (233 parthenogenetic species) as well as the availability of detailed ecological and taxonomic data in the

Universal Chalcidoidea Database (Noyes 2016). Species with reproductive polymorphisms were excluded from these analyses, as separate information for sexual and parthenogenetic lineages within such species was generally not available.

We compared body size, number of host species [a proxy of a species' niche breadth (Jaenike 1990)] and geographic distribution. Whenever possible given the available data, we compared sexuals and parthenogens via two ways: first, parthenogens were compared with their sexual sister-species as deduced from recently published phylogenies (Table 2). In this comparative approach we included 44 parthenogens and 74 sexuals, belonging to eight genera in the families Aphelinidae, Torymidae, and Trichogrammatidae. These species were repartitioned into 32 sexual-parthenogenetic pairs; for clades with multiple species, the mean value was used in the analyses. Second, information on number of host species and geographic distribution was automatically extracted from the Universal Chalcidoidea Database (Noyes 2016) using a custom Python script using BeautifulSoup4 (https://www.crummy.com/software/BeautifulSoup/). This approach allowed us to compare data from sexual and related parthenogens within 52 genera (comprising 134 parthenogens and 8194 sexual species. For the comparisons with species pairs, body size was obtained from taxonomic keys and scientific books or articles as sources. The size was measured in millimeters, as a single value or as a range, and excludes ovipositor length (see Supplementary Material for further details). Information on the number of host species used was mostly obtained from Noyes (2016), except for 13 species for which more recent data was available.

For the per genus analyses, we compared the number of host species and geographic distribution, which were obtained from Noyes (2016). To obtain accurate geographic distribution data in this approach, we replaced redundant location names by currently used names (e.g., Russia instead of USSR) and the large countries Australia, Brazil, Canada, China, India, Russia, and the USA were divided into states or regions. To transform country names to geographical information, we took the geographical center for a country (using OpenStreetMap Contributors, https://www.openstreetmap.org), using the Python package Geocoder (https://github.com/DenisCarriere/geocoder). The extreme latitude values were extracted using a custom python script (available at: https://github.com/cmdoret/chalcid_comparative_analysis).

Frequency of Parthenogenesis
PARTHENOGENESIS EVOLVED IN MANY DIFFERENT TAXONOMIC GROUPS

Cases of parthenogenesis can be found in all well-studied haplodiploid taxa (Table 3). In total, we found evidence for obligate

Table 2. Comparative analyses on morphological and ecological traits.

Approach	Families	Genera	Pairs	Sexual	Parthenogen	Variables
Species pairs	3	8	32	74	44	Body size, host species, geographic distribution
Per genus	11	52	52	8194	134	Host species, geographic distribution

Table 3. Frequency of parthenogenesis in haplodiploid taxa.

Orders	Common name	Parthenogens	Total species	Proportion	Species total reference
Astigmata	Mites	3	5000	0.001	(Norton et al. 1993)
Coleoptera: Micromalthidae	Telephone-pole beetle*	1	1	1	(Normark 2003)
Coleoptera: Scolytinae	Bark beetles	1	4500	0.000	(Farrell et al. 2001)
Hemiptera: Aleyrodoidea	Whiteflies	4	1556	0.003	(Martin and Mound 2007)
Hemiptera: Margarodidae	Scale insects	3	428	0.007	(García Morales et al. 2016)
Hemiptera: Diaspididae	Scale insects	1	2378	0.000	(García Morales et al. 2016)
Hymenoptera	Ants, bees, sawflies and wasps	586	150,000	0.004	(Mayhew 2007)
Mesostigmata	Predatory mites	43	5000	0.009	(Norton et al. 1993)
Prostigmata	Mites	6	14,000	0.000	(Norton et al. 1993)
Thysanoptera	Thrips	91	5938	0.015	(Mayhew 2007)
Trombidiformes	Mites	26	25,821	0.001	(Zhang et al. 2011)

In mites and Scolytinae, the exact origin(s) of haplodiploidy are not known, so the higher taxonomic level was chosen. *The telephone-pole beetle, *Micromalthus debilis*, is the only extant species in this monotypic family, which is considered to have a haplodiploid origin, see Normark (2003). Only taxa with at least one case of parthenogenesis described are shown.

parthenogenesis in 765 species across nine orders, 33 superfamilies, 58 families, and 316 genera, which is about five times the number of species previously reported by Normark (2003) and many times the number of species in other reviews (Table 1). Although information on the phylogenetic relationships among different parthenogens is not available for most taxa, the 765 parthenogenetic species most likely correspond to at least as many independent transitions from sexual reproduction to parthenogenesis. Speciation after the transition to parthenogenesis is considered to be very rare (Bell 1982), and so are reversals to sexuality (van der Kooi and Schwander 2014b). Moreover, what is considered a single parthenogenetic species often corresponds to a pool of parthenogens that derive from multiple, independent transitions to parthenogenesis (e.g., Janko et al. 2008; van der Kooi and Schwander 2014a).

The analysis of parthenogenesis frequencies at different taxonomic levels in haplodiploids indicates that different sex determination systems are not required to generate variation in parthenogenesis frequency among taxa. Indeed, there is extensive variation in parthenogenesis frequency among haplodiploid taxa and no phylogenetic clustering above the genus level. Overall, when excluding the exceptional case of the parthenogenetic *Micromalthus debilis* (the sole extant species in an ancient family), the frequency of parthenogenesis ranges from 0 to 1.5%

across taxonomic groups where haplodiploidy evolved (Table 3). The global estimate of parthenogenesis across these groups is approximately 0.04%. Within the two largest insect clades (Hymenoptera: about 140.000 described species, Thysanoptera: 5000 described species; (Mayhew 2007)) the frequency of parthenogenesis ranges from 0 to 33%, with a mean of 0.9% among families in Hymenoptera and across Thysanoptera families from 0 to 3.7%, with a mean of 0.8%. Parthenogenesis is scattered across the whole phylogeny in hymenopterans and thrips (Figs. 1, S1–S3).

Some groups comprise particularly many parthenogenetic species. For example, Cynipoidea (the superfamily of gall wasps) stand out because they feature a very high frequency of parthenogenetic species. Approximately 4% of species ($n = 122$) are obligately parthenogenetic (Fig. 1). The high frequency of parthenogenesis is robust, because when the cases based on circumstantial evidence for parthenogenesis are excluded, the average frequency of parthenogenesis in gall wasps is still 1.3% ($n = 38$ species), which is markedly higher than in other superfamilies. Sexual gall wasps have a cyclically parthenogenetic life cycle, where a parthenogenetic cycle is generally followed by a sexual cycle (reviewed by Stone et al. 2002). The high number of obligate parthenogens in this system may be explained by simple loss-of-function mutations that suppress the sexual cycle (Neiman et al. 2014). As Cynipoidea is the only known haplodiploid

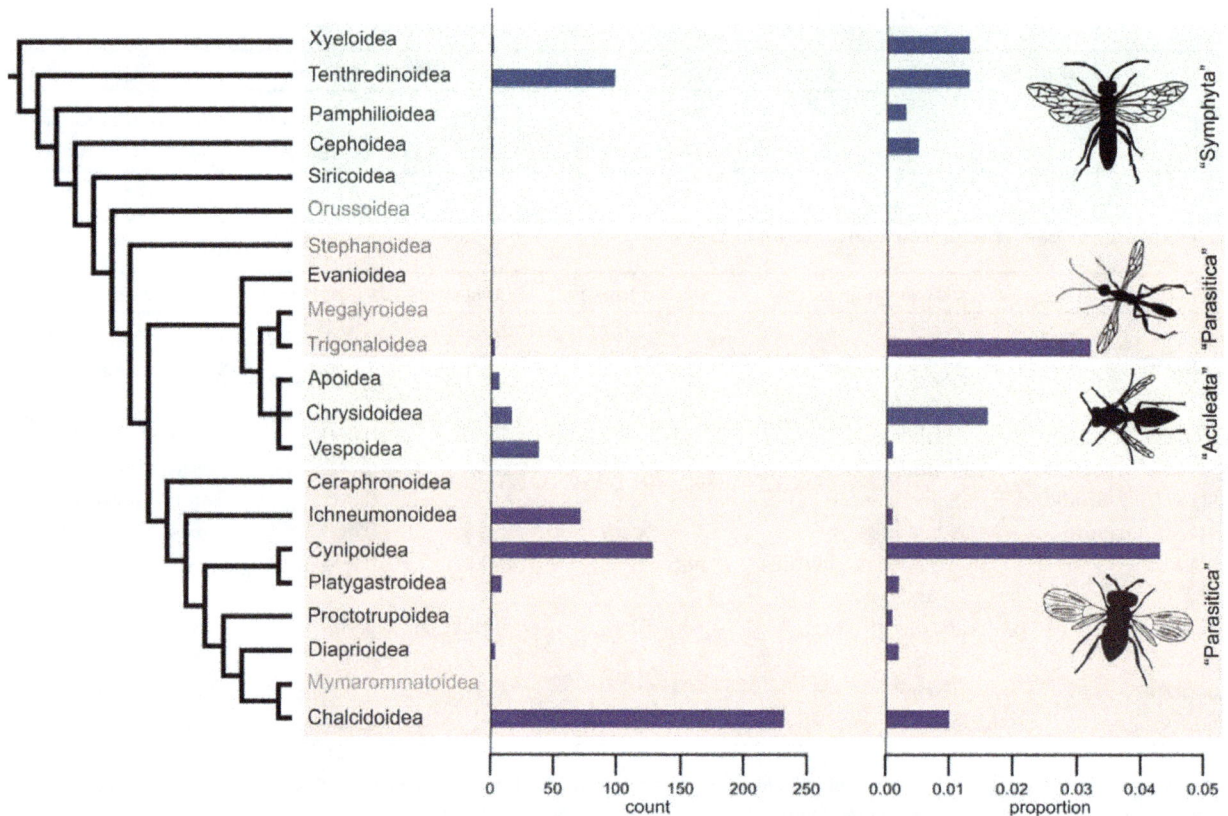

Figure 1. Frequency of parthenogenesis in Hymenoptera superfamilies. The phylogeny is from Klopfstein et al. (2013); taxa with fewer than 100 species described have gray label names. Except for the Ceraphronoidea, Evanioidea, and Siricoidea, all species-rich taxa have parthenogenetic species.

arthropod group with cyclical parthenogenesis, formal tests to infer whether cyclical parthenogenesis favors the transition to obligate parthenogenesis are impossible.

Other groups stand out because of their high frequency of parthenogenesis. When genera with very few species are excluded, the Chalcidoidea genera *Aphytis* and *Trichogramma* feature a high frequency of parthenogenesis. In line with earlier estimates (DeBach 1969; Rosen and DeBach 1979), the highest proportion of parthenogens was found in the parasitoid wasp genus *Aphytis* s.l. where parthenogenesis occurs in 42 out of 110 (38%) species. In *Trichogramma*, 27 out of 239 species (11%) have parthenogenetic lineages. The reason for the high incidence of parthenogenesis in *Aphytis* is unknown, but for *Trichogramma* it is presumably due to ascertainment bias; in this genus the first cases of endosymbiont-induced parthenogenesis were described (Stouthamer et al. 1990). This may have stimulated researchers to study reproductive systems in this genus. Similar frequencies were found among thrips genera (Supporting file 5). In summary, parthenogenesis is found in all major haplodiploid groups and the frequency greatly varies between taxonomic groups (Table 3; Figs. 1, S1–S3).

REPRODUCTIVE POLYMORPHISMS ARE WIDESPREAD

In many plant and animal species both sexual and parthenogenetic lineages can be found, either sympatrically or in different geographic areas (Bell 1982; Lynch 1984). Such reproductive polymorphisms are interesting, as they can be used to study possible costs and benefits of sex under natural conditions. In our survey, we found that for 143 parthenogenetic haplodiploid species (19% of the parthenogenetic species in our study) there is clear evidence for existence of sexual lineages as well. Reproductive polymorphisms occur frequently across many taxonomic levels and at comparable frequencies in species with parthenogenesis caused by genetic factors or endosymbiont infection (respectively 10/27 species 37%, and 26/58 species 45%; $\chi^2 = 0.21$, $df = 1$, $P = 0.65$). This suggests that the ecological and/or evolutionary factors that maintain reproductive polymorphisms seem similar for species with different causes of parthenogenesis. How frequently reproductive polymorphisms occur in other (nonhaplodiploid) taxonomic groups remains unknown, as there currently are no estimates for other taxa.

The frequency of reproductive polymorphisms is most likely underestimated. Species that do not occur in our database are assumed to be sexual, and parthenogenetic species for which

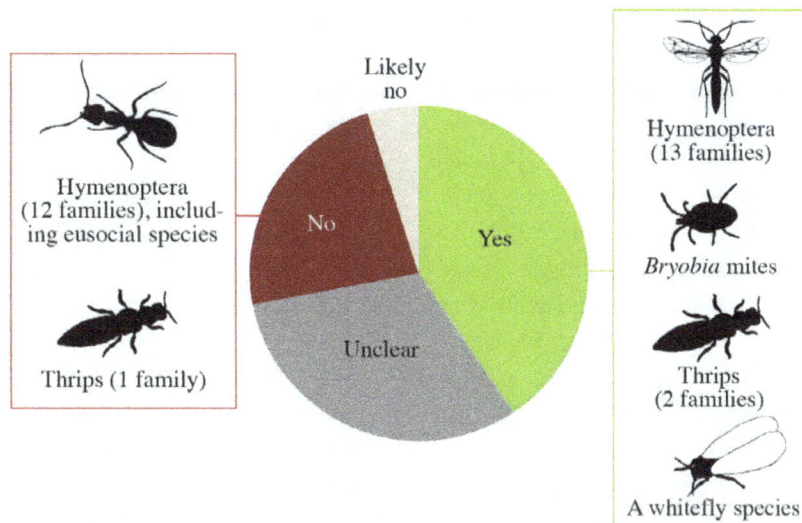

Figure 2. Frequency of endosymbiont-induced parthenogenesis. Yes: sexual reproduction can be restored via antibiotic treatment and/or after exposure to heat. No: sexual reproduction cannot be restored and is not caused by endosymbionts. Unclear: only circumstantial evidence points to endosymbionts (e.g. only PCR screening and no antibiotic treatment; see main text). Likely no: no indication of parthenogenesis-inducing endosymbionts, but reversibility to sexuality has not been tested. Based on 143 cases.

no sexual lineages are known are considered to be parthenogenetic. Considerable research effort (sampling of populations across the species distribution range and breeding experiments) is required in order to detect reproductive polymorphisms. That increasing research effort may increase the chance of detecting reproductive polymorphism becomes clear when the number of citations for species with reproductive polymorphisms is compared with that for obligate parthenogens. Species with reproductive polymorphisms have more than twice as many citations as obligate parthenogens (Supplementary Material). An alternative, nonmutually exclusive explanation is that reproductive polymorphic species are more studied because of their polymorphism. We currently cannot formally test the two hypotheses; it nonetheless is likely that many sexual lineages in putatively parthenogenetic species as well as parthenogenetic lineages in putatively sexual species remain undetected.

Origins of Parthenogenesis

ENDOSYMBIONT-INDUCED PARTHENOGENESIS IS PROVEN TO OCCUR IN 42% OF SPECIES

Transitions from sex to parthenogenesis can have different causes. Parthenogenesis can have a genetic basis, such as a hybridization event between related species or a mutation in sex-specific genes, which may result in the origin of a parthenogenetic lineage (Normark 2003). Transitions to parthenogenesis can also be caused by infection with maternally inherited endosymbionts (Stouthamer et al. 1990). In haplodiploids, at least three taxa include parthenogenesis-inducing endosymbionts

(PI-endosymbionts): *Wolbachia, Cardinium,* and *Rickettsia* (e.g., Stouthamer et al. 1993; Zchori-Fein et al. 2001; Hagimori et al. 2006). Presence of PI-endosymbionts can be tested by removing endosymbionts from parthenogenetic females, either by exposing them to high temperatures during early development (Flanders 1945) or by treating them with antibiotics (Stouthamer et al. 1990). This "cures" parthenogenetic females from their endosymbionts, and causes them to produce sons.

The frequency of endosymbiont-induced parthenogenesis in haplodiploids is surprisingly high. Out of 139 species for which we obtained information on the causes of the transition to parthenogenesis, for 105 species it was suggested that parthenogenesis was caused by endosymbionts. However, in only 58 cases (42% of the investigated species) this was convincingly shown (Fig. 2). Clear cases of endosymbiont-induced parthenogenesis are currently known in three insect orders (Hemiptera, Hymenoptera, and Thysanoptera) and in *Bryobia* mites (Table S3) (Weeks and Breeuwer 2001).

Three factors may generate an overestimate of the frequency of endosymbiont-induced parthenogenesis. First, publication bias will skew the frequency of endosymbiont-induced parthenogenesis toward positive results, as studies showing PI-endosymbionts to be present are much more likely to be published than studies that show that PI-endosymbionts are absent in a species ["negative results"; see (Monti et al. 2016) for a notable exception]. Second, infection with *Wolbachia* is often interpreted as evidence for an endosymbiont inducing parthenogenesis in a lineage (e.g., Weeks and Breeuwer 2001; Huigens and Stouthamer 2003; Boivin et al. 2014). These interpretations should be considered with caution, however, because *Wolbachia*-infection is very widespread, including in many sexual species (Zug and Hammerstein 2012). Third, when parthenogenesis-inducing endosymbionts have been found in a species, they are often assumed to occur in related parthenogenetic species, which is not necessarily true. Mechanisms underlying parthenogenesis can greatly differ between highly related species, as shown in *Trichogramma* (Vavre et al. 2004) and *Encarsia* parasitoids wasps (Gokhman 2009) as well as *Aptinothrips* grass thrips (van der Kooi and Schwander 2014a), where parthenogenesis is caused by endosymbionts in some species, but via another mechanism in others. In conclusion, in 42% of the investigated haplodiploid species parthenogenesis is due to endosymbionts, but the actual frequency presumably is much lower.

In many cases the endosymbiont species causing parthenogenesis remains unknown. *Cardinium* and *Rickettsia* have been shown to cause parthenogenesis in Chalcidoidea (Hymenoptera) only. Strong correlations with *Rickettsia* and parthenogenesis are found in three Eulophidae species, and *Cardinium* is suggested to cause parthenogenesis in several species of Aphelinidae and Encyrtidae. In the majority of systems *Wolbachia* is considered to be the causal endosymbiont, but this view is most likely biased because the other endosymbiont species were described more than a decade later. Consequently, numerous studies screened parthenogens for presence of *Wolbachia*, but did not screen for the other PI-endosymbionts *Cardinium* and *Rickettsia*. This lack of complete screens, together with the fact that some parthenogenetic lineages are known to harbor more than one endosymbiont species (e.g., in *Aphytis* parasitoid wasps with PI-endosymbionts, both *Cardinium* and *Wolbachia* are found (Zchori-Fein and Perlman 2004)), highlight that we should be cautious in ascribing parthenogenesis to a specific endosymbiont species.

POLYPLOIDY IS EXCEEDINGLY RARE

In vertebrates and plants, polyploidy occurs more frequently in parthenogens than sexuals (Suomalainen et al. 1987; Otto and Whitton 2000); as a consequence, polyploidy is sometimes considered the norm in parthenogens (e.g., Kearney et al. 2009). At least in haplodiploids this is not the case. Of the 50 parthenogenetic species with available karyotype information, only two (4%) are polyploid (Fig. S4), namely the gall wasp *Diplolepis eglanteriae* and the sawfly *Pachyprotasis youngiae* (both triploids) (Sanderson 1988; Naito and Inomata 2006). The frequency of polyploids drops even further when species with parthenogenesis-inducing endosymbionts are included. Parthenogenesis-inducing endosymbionts are unlikely to occur in polyploids, because endosymbiont-induced parthenogenesis generally involves meiotic parthenogenesis with secondary restoration of diploidy. Including the 48 species with parthenogenesis-inducing endosymbionts (but for which no kary-

otypes are known), the frequency of polyploid parthenogens drops to 2%. Objective frequency estimates of polyploidy in sexuals are lacking for any haplodiploid group, though it is thought to be rare in hymenopterans (Gokhman 2009). Nonetheless, given there are many parthenogens in haplodiploid groups, and very few of these are polyploid, polyploidy is certainly more rare among parthenogens than commonly thought.

Ecological Differences between Sexuals and Parthenogens

Several hypotheses predict that sexual and parthenogenetic species differ in ecological generalism and the size of distribution ranges, with opposite predictions depending on the hypothesis (Vrijenhoek 1979; Bell 1982; Lynch 1984). For example, parthenogens might have broader ecological niches, because there may be lineage-level selection for general-purpose genotypes in parthenogens (Lynch 1984). Alternatively, parthenogens might have narrower niches, because when a parthenogen derives from a sexual ancestor it inherits a single genotype from a genetically heterogeneous sexual group (the "frozen niche variation hypothesis," sensu Vrijenhoek 1979; see also Bell 1982).

We compared the number of host species and distribution ranges for obligate sexual and obligate parthenogenetic wasps in parasitoid and phytophagous wasps in the mega-diverse Hymenoptera superfamily Chalcidoidea. Chalcidoidea was chosen because of the many independent transitions to parthenogenesis in this group ($n = 233$; Fig. 1) and because their ecology and geographic distribution are well-documented (see Noyes 2016). We consider the number of host species as a proxy of a species' niche width, that is, generalists will have more hosts than specialists (Jaenike 1990). Sexuals and parthenogens were compared in two different datasets. First using sexual-parthenogenetic sister-species pairs as inferred from phylogenetic trees (32 sexual-parthenogen species pairs) and second via a coarser approach where data from all parthenogens and sexuals was compared per genus (8328 species in 52 genera). The data subsets used for comparisons of specific ecological traits depended on the availability of ecological and phylogenetic information (Materials and Methods, Supplementary Information).

Parthenogens consistently parasitize more host species, indicating they have wider ecological niches than sexuals. A significant 2–3.5-fold increase in number of host species used was found via both the species-pair and per genus approach (Fig. 3), and these results were robust with respect to publication bias (i.e., more hosts known for more intensively studies species; see Supplementary Information). Broad niches in parthenogens can stem from two mechanisms: few successful genotypes that have a broad ecological niche (i.e., "general purpose genotypes") or a large mixture of genetically different clones

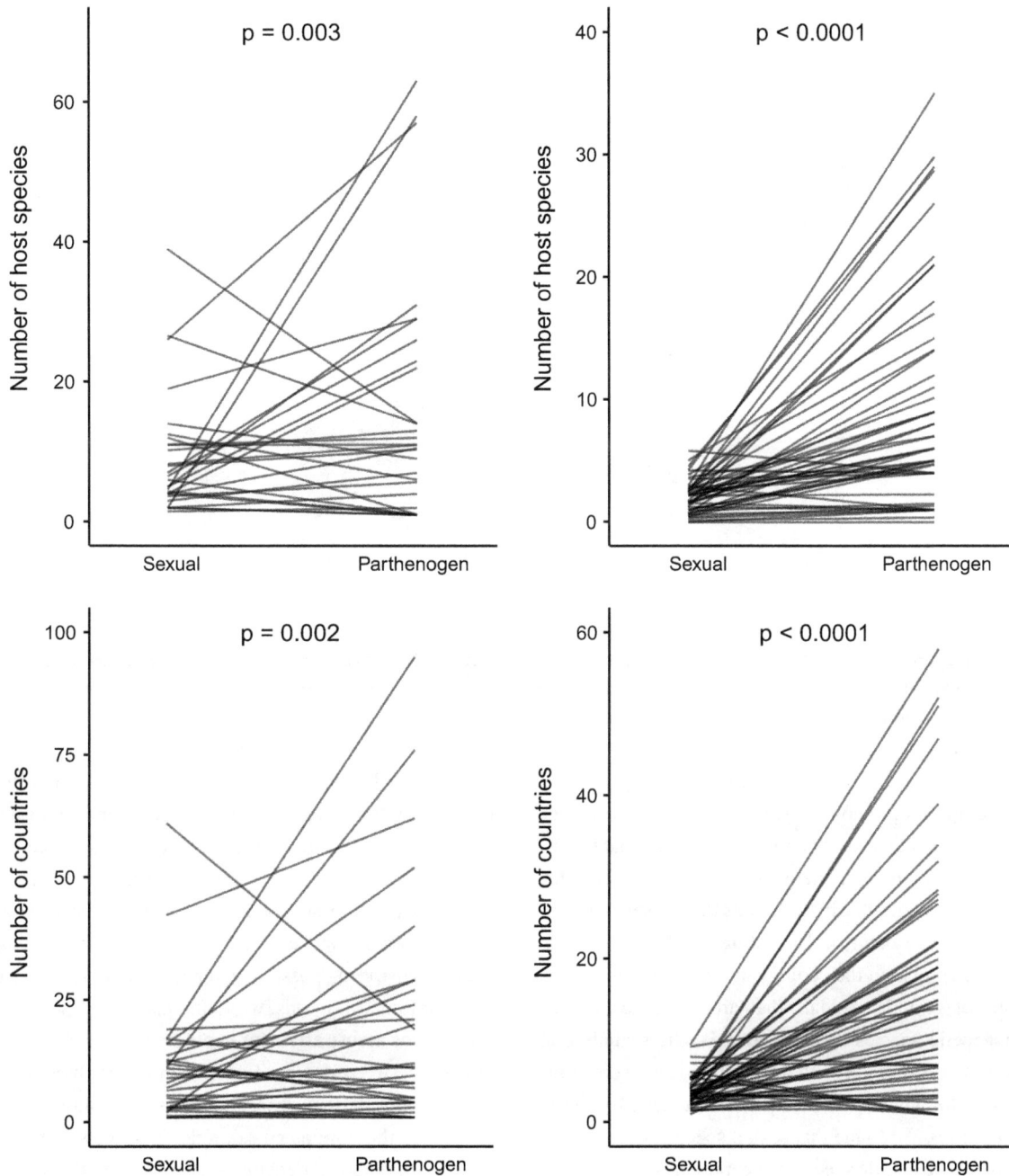

Figure 3. Number of host species for sexual and parthenogenetic Chalcidoidea parasitoid and phytophagous wasps. Left panels: pairwise analysis, right panels: analysis incorporating all data from the database; note the different y-axis ranges. For visualization purposes, one pair with a parthenogen with extremely many host species is not shown in the upper panels.

each with a distinct and narrow ecological niche (Bell 1982; Lynch 1984). To disentangle these hypotheses, detailed studies on the genetic vs. niche diversity of parthenogenetic lineages are required.

Animals with a wider ecological niche are more likely to expand their geographic range (Normark and Johnson 2011). In line with this theory, we found that parthenogens occur in larger geographic regions than their related sexual species. Parthenogens are consistently found in 1.7–5 times more countries than their sexual

relatives (Fig. 3). Distribution ranges of sexuals and parthenogens largely overlap in regions close to the equator, but parthenogens have extended their distributions polewards by 10 latitudinal degrees (i.e., about 1000 kilometers) compared to their sexual relatives (Fig. S5). These effects are robust with respect to publication bias (more occurrences known for more intensively studied species; see Supplementary Material).

The observed ecological and geographical differences between sexuals and parthenogens can stem from two

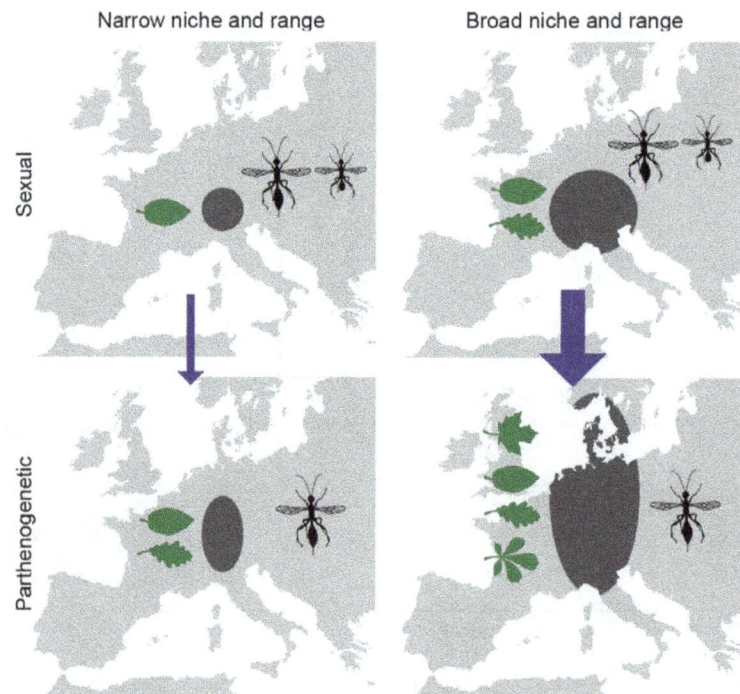

Figure 4. Parthenogens have a wider niche and polewards extended distribution as compared to sexuals, and parthenogenesis is more likely to evolve in sexuals with relatively wide niches and distribution ranges. Width of the arrow indicates the number of transitions, leaves represent the ecological niche width and dark shading represents geographical range.

nonmutually exclusive mechanisms. First, selection for a broad ecological niche and large geographical range can follow after the transition to parthenogenesis. Second, it is possible that transitions to parthenogenesis are more likely to occur in sexual species with broad niches or geographic distributions than in sexuals with narrow niches and distributions. To distinguish between these two scenarios, we compared the ecology and distribution ranges for two groups of sexuals: sexuals that are the sister-species for a parthenogenetic species or clade, that is, the sexuals that likely share a common ancestor with the parthenogen, were compared with sexuals for which no related parthenogenetic species is known (the outgroup). We found that sexual species that share an ancestor with parthenogenetic species have more host species and occur in larger geographic areas than sexuals from clades where no parthenogenetic species are known (Fig. S6). This means that the increased niche width and enlarged geographic distribution (partly) arose before the transition to parthenogenesis and that (successful) parthenogenetic species arise more frequently from sexuals with wide ecological niches and geographic distributions, than from sexuals with narrow ecologies and geographic distributions (Fig. 4).

General Discussion

Many isolated studies examined the frequency, mechanism, ploidy, and/or ecology of parthenogenesis in one or a few species

(e.g., DeBach 1969; Vrijenhoek et al. 1989; Stouthamer et al. 1990; Boivin et al. 2014; van der Kooi and Schwander 2014a; Monti et al. 2016), but large-scale, quantitative studies that provide insights into general patterns of parthenogens are lacking. We here present the first broad comparison on the ecology and evolution of parthenogenesis that is based on a species database that is as exhaustive as possible given the available data. Focusing on haplodiploid arthropods, we found parthenogenesis in more than 750 species across all major haplodiploid groups (ranging from 0 to 1.5% between orders; Table 3). In many phylogenetically different groups parthenogenesis occurs much more frequently than previously thought; as an example, in species-rich Hymenoptera and Thysanoptera genera the frequency of parthenogenesis ranges from 0 to 38%. The absence of phylogenetic clustering above the genus level as well as the observation that parthenogenetic species are found in 2–3% of the genera (Supplementary Information) are very similar to frequency estimates of (facultative) asexual seed production in plants, which was found to occur in about 2.2% of phylogenetically distinct plant genera (Hojsgaard et al. 2014). However, the relative frequency of facultative versus obligate parthenogenesis in plants remains unknown. Our results are of great importance to our overall understanding of reproductive system evolution, as haplodiploidy is the sex determination system of many animal taxa (Bachtrog et al. 2014), including Hymenoptera, which is one of the most species-rich insect groups (Mayhew 2007).

The number of obligate parthenogens in our database and the frequency estimates calculated here are certainly underestimates, as many cases of parthenogenesis remain undetected. For instance, in poorly studied species, female-biased sex ratios will often remain unnoticed. More importantly, we assumed that species that do not occur in our list are by default sexual, while the reproductive mode of these species has not been directly investigated. Many species contain both sexual and parthenogenetic lineages (see section on reproductive polymorphisms), which may easily co-occur within populations. Unless studied in detail via breeding assays, such mixed populations will generally be considered sexual, because they comprise males.

It is unlikely that the parthenogenesis frequency we report here is specific for haplodiploids, although addressing this question requires development of detailed species lists for other groups. In organisms with other sex-determination systems, developmental processes, such as egg activation and centrosome formation, are supposed to impose constraints on the transition to parthenogenesis (Engelstaedter 2008). In haplodiploids, where males always develop parthenogenetically, several of these developmental constraints are overcome. For example, in haplodiploids egg activation and centrosome formation is induced immediately after oviposition–independent of fertilization. Hence, a transition to parthenogenesis may involve a relatively small change— and thus occur frequently–in haplodiploids (Engelstaedter 2008). Nevertheless, the extensive variation in the frequency of parthenogenesis among haplodiploids (Figs. 1, S1–S3) clearly shows that different sex determination systems are not necessary to explain the variation in the frequency of parthenogenesis among major taxa.

The incidence of polyploidy and endosymbiont-induced parthenogenesis is much lower than commonly thought. We find that polyploidy is extremely rare in parthenogenetic haplodiploids. This shows that—at least in haplodiploids— polyploidy is not a common consequence of parthenogenesis. Parthenogenesis-inducing endosymbionts are found in 42% of the cases, but this is almost certainly an overestimate due to publication bias. Convincing evidence supporting endosymbiont-induced parthenogenesis is reported for several taxonomic groups, albeit in the vast majority of cases the causal endosymbiont remains unknown. There is a clear need for more studies that provide balanced evidence supporting or rejecting endosymbionts as causal agents for parthenogenesis (e.g., van der Kooi and Schwander 2014a; Monti et al. 2016), as well as detailed studies that characterize the endosymbiont species in systems where endosymbiont-induced parthenogenesis is suggested.

An in-depth comparative analysis of parasitoid and phytophagous wasps showed that parthenogens have more host species and wider geographical distributions than sexuals (Figs. 3, 4). These differences mimic the large distribution ranges and/or poleward expansions in self-fertilizing (Grossenbacher et al. 2015) and asexual plants (Johnson et al. 2010). The extended distribution of parthenogenetic lineages toward the poles could stem from enhanced colonization abilities in these lineages, for example because parthenogenesis confers reproductive assurance ["Baker's law"; (reviewed by Pannell et al. 2015)] and/or because parthenogens have an advantage over sexual relatives in colder climates. For example, cold climates can generate colonization-extinction cycles (i.e., strong reductions in population size during extreme winters), which can lead to mate limitation and inbreeding in sexuals (Haag and Ebert 2004). Given that mate limitation and inbreeding will never be a problem for parthenogenetic females, parthenogenetic lineages may be more successful in (re)colonizing habitats after extreme winters. Ecological niche width and geographic distribution range are likely to be interrelated, but the relative importance of either remains currently unknown.

The wider ecological niches and geographical ranges found in parthenogens only partially evolved after the transition to asexuality. Sexual sister-species–that have a shared ancestor with the parthenogenetic lineage–exhibit relatively large ecological niches and distribution ranges, as compared to sexuals for which no related parthenogenetic species is known (Figs. 4, S6). A similar pattern was suggested to occur in (diploid) parthenogenetic scale insects (Ross et al. 2013), and emphasizes that large population sizes of sexuals are paramount in the origin and/or evolutionary success of derived parthenogenetic lineages. Large population sizes should favor the evolution of parthenogenesis because more mutants capable of parthenogenesis are expected in species with larger population size. Parthenogens with large population and range sizes are also less prone to extinction than those with small populations (e.g., Otto and Barton 2001). We also found weak evidence that parthenogenesis emerges more frequently in species with small body sizes, which generally also have larger populations (Supplementary Information).

Finally, the development of this database opens perspectives on future comparative studies on the evolution of sex. Of particular interest would be to develop parthenogenetic species lists for groups with other sex determination systems.

AUTHOR CONTRIBUTIONS

C.J.v.d.K. and T.S. designed the study; C.J.v.d.K. developed the species database with input from all authors; C.M.D. developed the tool to automatically obtain data from the Universal Chalcidoidea Database; C.J.v.d.K. and C.M.D. performed the comparative analyses; C.J.v.d.K. and T.S. wrote the manuscript with input from all authors.

ACKNOWLEDGMENTS

We thank three reviewers for useful comments on an early version of the manuscript, and W. Reen for figure editing. Drs. J. Noyes and A. Polaszek are acknowledged for construction and maintaining the Universal Chalcidoidea Database, and useful comments.

LITERATURE CITED

Aguiar, A. P., A. R. Deans, M. S. Engel, M. Forshage, J. T. Huber, J. T. Jennings, N. F. Johnson, A. S. Lelej, J. T. Longino, and V. Lohrmann. 2013. Order hymenoptera. Animal biodiversity: An outline of higher-level classification and survey of taxonomic richness (addenda 2013). Zootaxa 3703:51–62.

Askew, R. R., G. Melika, J. Pujade-Villar, K. Schoenrogge, G. N. Stone, and J. L. Nieves-Aldrey. 2013. Catalogue of parasitoids and inquilines in cynipid oak galls in the West Palaearctic. Zootaxa 3643:001–133.

Bachtrog, D., J. E. Mank, C. L. Peichel, M. Kirkpatrick, S. P. Otto, T.-L. Ashman, M. W. Hahn, J. Kitano, I. Mayrose, R. Ming, et al. 2014. Sex determination: why so many ways of doing it? PLoS Biol. 12:e1001899.

Bell, G. 1982. The masterpiece of nature: The evolution and genetics of sexuality. California Univ. Press, Berkeley, CA.

Boivin, T., H. Henri, F. Vavre, C. Gidoin, P. Veber, J. N. Candau, E. Magnoux, A. Roques, and M. A. Auger-Rozenberg. 2014. Epidemiology of asexuality induced by the endosymbiotic Wolbachia across phytophagous wasp species: host plant specialization matters. Mol. Ecol. 23:2362–2375.

Cook, J. M., and R. D. Butcher. 1999. The transmission and effects of Wolbachia bacteria in parasitoids. Res. Popul. Ecol. 41:15–28.

DeBach, P. 1969. Uniparental, sibling and semi-species in relation to taxonomy and biological control. Israelian J. Entomol. 4:11–28.

Engelstaedter, J. 2008. Constraints on the evolution of asexual reproduction. BioEssays 30:1138–1150.

Farrell, B. D., A. S. Sequeira, B. C. O'Meara, B. B. Normark, J. H. Chung, and B. H. Jordal. 2001. The evolution of agriculture in beetles (Curculionidae: Scolytinae and Platypodinae). Evolution 55:2011–2027.

Flanders, S. E. 1945. The bisexuality of uniparental Hymenoptera, a function of the environment. Am. Nat. 79:122–141.

García Morales, M., B. Denno, D. Miller, G. Miller, Y. Ben-Dov, and N. Hardy. 2016. ScaleNet: a literature-based model of scale insect biology and systematics. Database accessable via http://scalenet.info.

Gokhman, V. E. 2009. Karyotypes of parasitic Hymenoptera. Springer Science & Business Media, The Netherlands.

Grossenbacher, D., R. Briscoe Runquist, E. E. Goldberg, and Y. Brandvain. 2015. Geographic range size is predicted by plant mating system. Ecol. Lett. 18:706–713.

Haag, C. R., and D. Ebert. 2004. A new hypothesis to explain geographic parthenogenesis. Annales Zoologici Fennici 41:539–544.

Hagimori, T., Y. Abe, S. Date, and K. Miura. 2006. The first finding of a Rickettsia bacterium associated with parthenogenesis induction among insects. Curr. Microbiol. 52:97–101.

Hojsgaard, D., S. Klatt, R. Baier, J. G. Carman, and E. Hörandl. 2014. Taxonomy and biogeography of apomixis in angiosperms and associated biodiversity characteristics. Crit. Rev. Plant Sci. 33:414–427.

Huigens, M. E., R. Stouthamer. 2003. Parthenogenesis associated with Wolbachia. Pp. 247–266 in K. Bourtzis, and T. A. Miller, eds. Insect symbiosis. CRC Press, Boca Raton.

Jaenike, J. 1990. Host specialization in phytophagous insects. Ann. Rev. Ecol. Syst. 21:243–273.

Janko, K., P. Drozd, J. Flegr, and J. R. Pannell. 2008. Clonal turnover versus clonal decay: a null model for observed patterns of asexual longevity, diversity and distribution. Evolution 62:1264–1270.

Johnson, M. T. J., S. D. Smith, and M. D. Rausher. 2010. Effects of plant sex on range distributions and allocation to reproduction. New Phytol. 186:769–779.

Kearney, M., M. Fujita, J. Ridenour, I. Schön, K. Martens, and P. van Dijk. 2009. Lost sex in the reptiles: constraints and correlations. In I. Schön, P. van Dijk, and K. Martens, eds. Lost sex: The evolutionary biology of parthenogenesis, Springer, Netherlands.

Klopfstein, S., L. Vilhelmsen, J. M. Heraty, M. Sharkey, and F. Ronquist. 2013. The hymenopteran tree of life: evidence from protein-coding genes and objectively aligned ribosomal data. PloS One 8:e69344.

Koivisto, R. K., and H. R. Braig. 2003. Microorganisms and parthenogenesis. Biol. J. Linnean Soc. 79:43–58.

Lewis, T. 1973. Thrips, their biology, ecology and economic importance. Academic Press, London.

Lynch, M. 1984. Destabilizing hybridization, general-purpose genotypes and geographic parthenogenesis. Quart. Rev. Biol. 59:257–290.

Martin, J., and L. Mound. 2007. An annotated check list of the world's whiteflies (Insecta: Hemiptera: Aleyrodidae). Zootaxa 1:1–84.

May, R. M. 2010. Tropical arthropod species, more or less? Science 329:41–42.

Mayhew, P. J. 2007. Why are there so many insect species? Perspectives from fossils and phylogenies. Biol. Rev. 82:425–454.

Monti, M. M., F. Nugnes, L. Gualtieri, M. Gebiola, and U. Bernardo. 2016. No evidence of parthenogenesis-inducing bacteria involved in Thripoctenus javae thelytoky: an unusual finding in Chalcidoidea. Entomol. Exp. Appl. 160:292–301.

Mound, L. A. 2013. Order thysanoptera haliday, 1836. Animal biodiversity: An outline of higher-level classification and survey of taxonomic richness (Addenda 2013). Zootaxa 3703:49–50.

Naito, T., and R. Inomata. 2006. A new triploid thelytokous species of the genus Pachyprotasis Hartig, 1837 (Hymenoptera: Tenthredinidae) from Japan and Korea. Pp. 279–283 in S. Blank, S. Schmidt, and A. Taeger, eds. Recent sawfly research: Synthesis and prospects. Goecke & Evers, Keltern.

Neiman, M., T. F. Sharbel, and T. Schwander. 2014. Genetic causes of transitions from sexual reproduction to asexuality in plants and animals. J. Evol. Biol. 27:1346–1359.

Normark, B. B. 2003. The evolution of alternative genetic systems in insects. Ann. Rev. Entomol. 48:397–423.

———. 2014. Modes of reproduction. Oxford Univ. Press, Oxford.

Normark, B. B., and N. A. Johnson. 2011. Niche explosion. Genetica 139:551–564.

Norton, R. A., J. B. Kethley, D. E. Johnston, B. M. O'Connor, D. Wrensch, and M. Ebbert. 1993. Phylogenetic perspectives on genetic systems and reproductive modes of mites. Pp.8–99 in D. Wrensch, M. Ebbert, eds. Evolution and diversity of sex ratio in insects and mites. Chapmann & Hall Publishers, New York.

Noyes, J. S. 2016. Universal Chalcidoidea Database. World Wide Web electronic publication. http://www.nhm.ac.uk/chalcidoids.

Otto, S. P., and N. H. Barton. 2001. Selection for recombination in small populations. Evolution 55:1921–1931.

Otto, S. P., and P. Jarne. 2001. Haploids–Hapless or happening? Science 292:2441–2443.

Otto, S. P., and J. Whitton. 2000. Polyploid incidence and evolution. Ann. Rev. Genet. 34:401–437.

Pannell, J. R., J. R. Auld, Y. Brandvain, M. Burd, J. W. Busch, P. O. Cheptou, J. K. Conner, E. E. Goldberg, A. G. Grant, and D. L. Grossenbacher. 2015. The scope of Baker's law. New Phytol. 208:656–667.

Pomeyrol, R. 1929. La parthenogenese des thysanopteres. Bull. Biol. France Belgium 62:3–12.

Rosen, D., and P. DeBach. 1979. Species of aphytis of the world: Hymenoptera: aphelinidae. Springer Science & Business Media, Dr W. Junk Publishers, The Hague.

Ross, L., N. B. Hardy, A. Okusu, and B. B. Normark. 2013. Large population size predicts the distribution of asexuality in scale insects. Evolution 67:196–206.

Ross, L., I. Pen, and D. M. Shuker. 2010. Genomic conflict in scale insects: the causes and consequences of bizarre genetic systems. Biol. Rev. 85:807–828.

Sanderson, A. R. 1988. Cytological investigations of parthenogenesis in gall wasps (Cynipidae, Hymenoptera). Genetica 77:189–216.

Schön, I., K. Martens, and P. Van Dijk. 2009. Lost sex: The evolutionary biology of parthenogenesis. Springer Science & Business Media, Springer, Netherlands.

Slobodchikoff, C., and H. V. Daly. 1971. Systematic and evolutionary implications of parthenogenesis in the Hymenoptera. Am. Zool. 11:273–282.

Stone, G. N., K. Schönrogge, R. J. Atkinson, D. Bellido, and J. Pujade-Villar. 2002. The population biology of oak gall wasps (Hymenoptera: Cynipidae). Ann. Rev. Entomol. 47:633–668.

Stouthamer, R. 2003. The use of unisexual wasps in biological control. Pp. 93–113 in Quality control and production of biological control agents theory and testing procedures. CABI Publishing, Wallingford.

Stouthamer, R., J. Breeuwer, R. Luck, and J. Werren. 1993. Molecular identification of microorganisms associated with parthenogenesis. Nature 361:66–68.

Stouthamer, R., R. F. Luck, and W. Hamilton. 1990. Antibiotics cause parthenogenetic Trichogramma (Hymenoptera/Trichogrammatidae) to revert to sex. Proc. Natl. Acad. Sci. 87:2424–2427.

Suomalainen, E., A. Saura, and J. Lokki. 1987. Cytology and evolution in parthenogenesis. CRC Press, Boka Raton.

Taeger, A., and S. Blank. 2011. ECatSym—Electronic World Catalog of Symphyta (Insecta, Hymenoptera). Program version 3.10, data version 38. Digital Entomological Information, http://www.sdei.de/ecatsym/index.html. Müncheberg.

Taeger, A., S. M. Blank, and A. D. Liston. 2010. World catalog of symphyta (Hymenoptera). Zootaxa 2580:1–1064.

ThripsWiki. 2016. ThripsWiki—providing information on the World's thrips. http://thrips.info/wiki.

van der Kooi, C. J., and T. Schwander. 2014a. Evolution of asexuality via different mechanisms in grass thrips (Thysanoptera: Aptinothrips). Evolution 68:1883–1893.

van der Kooi, C. J., and T. Schwander. 2014b. On the fate of sexual traits under asexuality. Biol. Rev. 89:805–819.

van der Kooi, C. J., and T. Schwander. 2015. Parthenogenesis: birth of a new lineage or reproductive accident? Curr. Biol. 25:R659–R661.

Vavre, F., J. De Jong, R. Stouthamer. 2004. Cytogenetic mechanism and genetic consequences of thelytoky in the wasp Trichogramma cacoeciae. Heredity 93:592–596.

Vrijenhoek, R. C. 1979. Factors affecting clonal diversity and coexistence. Am. Zool. 19:787–797.

Vrijenhoek, R. C., R. M. Dawley, C. J. Cole, and J. P. Bogart. 1989. A list of the known unisequal vertebrates. In R. M. Dawley, and J. P. Bogart, eds. Evolution and cytology of unisexual vertebrates. The University of the State of New York, New York.

Weeks, A., and J. Breeuwer. 2001. Wolbachia-induced parthenogenesis in a genus of phytophagous mites. Proc. R Soc. Lond. B 268:2245–2251.

Wenseleers, T., and J. Billen. 2000. No evidence for Wolbachia-induced parthenogenesis in the social Hymenoptera. J. Evol. Biol. 13:277–280.

White, M. J. D. 1977. Animal cytology and evolution: CUP Archive.

Zchori-Fein, E., Y. Gottlieb, S. Kelly, J. Brown, J. Wilson, T. Karr, and M. Hunter. 2001. A newly discovered bacterium associated with parthenogenesis and a change in host selection behavior in parasitoid wasps. Proc. Natl. Acad. Sci. 98:12555–12560.

Zchori-Fein, E., and S. J. Perlman. 2004. Distribution of the bacterial symbiont Cardinium in arthropods. Mol. Ecol. 13:2009–2016.

Zhang, Z., Q. Fan, V. Pesic, H. Smit, A. Bochkov, A. Khaustov, A. Baker, A. Wohltmann, T. Wen, and J. Amrine. 2011. Order Trombidiformes Reuter, 1909. Zootaxa 3148:129–138.

Zug, R., and P. Hammerstein. 2012. Still a host of hosts for Wolbachia: analysis of recent data suggests that 40% of terrestrial arthropod species are infected. PloS One 7:e38544.

15

Single-gene speciation: Mating and gene flow between mirror-image snails

Paul M. Richards,[1] Yuta Morii,[2] Kazuki Kimura,[2] Takahiro Hirano,[2] Satoshi Chiba,[2] and Angus Davison[1,3]

[1]*School of Life Sciences, University of Nottingham, Nottingham NG7 2RD, United Kingdom*

[2]*Division of Ecology and Evolutionary Biology, Graduate School of Life Sciences, Tohoku University, Aobayama, Sendai 980–8578, Japan*

[3]*E-mail: angus.davison@nottingham.ac.uk*

Variation in the shell coiling, or chirality, of land snails provides an opportunity to investigate the potential for "single-gene" speciation, because mating between individuals of opposite chirality is believed not possible if the snails mate in a face-to-face position. However, the evidence in support of single-gene speciation is sparse, mostly based upon single-gene mitochondrial studies and patterns of chiral variation between species. Previously, we used a theoretical model to show that as the chiral phenotype of offspring is determined by the maternal genotype, occasional chiral reversals may take place and enable gene flow between mirror image morphs, preventing speciation. Here, we show empirically that there is recent or ongoing gene flow between the different chiral types of Japanese *Euhadra* species. We also report evidence of mating between mirror-image morphs, directly showing the potential for gene flow. Thus, theoretical models are suggestive of gene flow between oppositely coiled snails, and our empirical study shows that they can mate and that there is gene flow in *Euhadra*. More than a single gene is required before chiral variation in shell coiling can be considered to have created a new species.

KEY WORDS: Behavioral genetics, evolutionary genomics.

Impact Summary

Although most snails have a right-handed spiraling shell, rare "mirror-image" individuals have a shell that coils to the left. This curious inherited condition has attracted attention because the genitals of mirror image snails are on different sides of the head, and so mating is difficult or impossible. If they are unable to mate, then does a change in the direction of the shell coil make a new species? In investigating a Japanese snail genus, *Euhadra*, we were surprised to find that different-coiling individuals can sometimes mate, against expectations, and that there is evidence for this in their genetic make-up. It turns out that the mating problem is mainly behavioral, rather than a physical incompatibility. This new work therefore suggests that the two types of Japanese snail should be considered a single species, and has implications for the classification of other snail species. As it is has previously been shown that the same sets of genes that make mirror image snails are also involved in making mirror image bodies in other animals– including humans–then further research using the natural variation snails could offer the chance to develop an understanding of how organs are placed in the body and why this process can sometimes go wrong.

Understanding the extent and underlying causes of speciation under gene flow is a longstanding challenge in evolutionary biology. Strong reproductive isolation usually depends upon the evolution and maintenance of associations between multiple traits contributing to different reproductive barriers (Coyne and Orr 2004). However, a problem is that gene flow is fundamentally antagonistic to this process because it is expected to homogenize divergence at individual loci, and through recombination, randomize associations between the different loci contributing to reproductive isolation (Felsenstein 1981; Coyne and Orr 2004; Gavrilets 2004; Servedio 2009). Consequently, the complete cessation of gene flow by means of geographic isolation has traditionally been viewed as necessary for reproductive isolation to evolve. Unopposed by recombination, processes such as mutation, selection, and genetic drift can drive genome-wide divergence between allopatric populations, leading to the build-up of linkage disequilibrium between loci contributing to reproductive barriers (Felsenstein 1981; Coyne and Orr 2004).

Despite the theoretical difficulties, it is now clear that speciation with gene flow may be relatively common in nature (Servedio and Noor 2003; Gavrilets 2004; Bolnick and Fitzpatrick 2007; Nosil 2008; Smadja and Butlin 2011). For instance, *de novo* divergence in sympatry may occur through assortative mating resulting from associations between loci subject to divergent ecological selection (e.g., differential local adaptation to habitat, or predation) and loci underlying mating traits (Rundle and Nosil 2005). Alternatively, geographic isolation may be important for initiating speciation, with divergent ecological selection, or reinforcement strengthening reproductive barriers following secondary contact (Servedio and Noor 2003; Rundle and Nosil 2005). The challenge is in determining the relative contributions of spatial isolation and gene flow to the evolution of reproductive isolation (Smadja and Butlin 2011; Martin et al. 2013), and elucidating mechanisms that act in lieu of spatial isolation to prevent recombination from disrupting associations between the different components of reproductive isolation (Smadja and Butlin 2011).

One exceptional means by which speciation with gene flow could be facilitated is through occasional reversals of left-right asymmetry, or *chirality*, in snails. Due to pleiotropic effects of the maternal effect locus that determines snail chirality (Boycott and Diver 1923; Sturtevant 1923; Schilthuizen and Davison 2005), mating is believed not possible between mirror-image individuals with low spired shells. Switches in chirality may therefore be a driver of so-called "single-gene" speciation (Gittenberger 1988; Asami et al. 1998; Coyne and Orr 2004; Schilthuizen and Davison 2005; Hoso et al. 2010). However, the likelihood of single-gene speciation in snails, and the mechanisms by which it could occur have been the subject of much debate because it is both theoretically challenging (Johnson et al. 1990; Orr 1991; Davison et al. 2005) and the empirical evidence is extremely limited.

First, theoretical models have shown that while individual snails of opposite coil may be unable to mate, gene flow could be substantial between morphs. As the chiral phenotype of offspring is determined by the maternal genotype, occasional chiral reversals will take place and enable gene flow, unless there is complete reciprocal fixation of chirality-determining alleles (Davison et al. 2005).

Second, as predicted by classic two-locus models of speciation (Orr 1996), fixation of a novel chiral allele is unlikely because the new chiral morph might lack potential intrachiral mating partners (Johnson 1982; Orr 1991, 1996). Consequently, several studies have investigated the conditions under which this mating disadvantage could be overcome, including founder ef-

fects, population size, and density, as well as selection, such as reproduction character displacement or predation (Johnson 1982; Orr 1991; van Batenburg and Gittenberger 1996; Davison et al. 2005; Yamamichi and Sasaki 2013).

Third, sparse empirical data mean that putative instances of single-gene speciation have been inferred from single-gene mitochondrial phylogenies (Ueshima and Asami 2003; Davison et al. 2005; Uit de Weerd et al. 2006; Feher et al. 2013; Modica et al. 2016), or by combining single mitochondrial genes with relatively invariable ribosomal RNA sequences (Hoso et al. 2010; Kornilios et al. 2015). From this data alone, it is impossible to definitively distinguish between low levels of gene flow, introgressive hybridization, or speciation.

Finally, the other main approach has been to investigate and compare patterns of chiral variation between species, across wide geographical scales (Hoso et al. 2010; Gittenberger et al. 2012). While this is useful in understanding broad patterns, especially in explaining the high frequency of sinistrals in South East Asia (Hoso et al. 2010), the phylogenetic relationship between the species is often not clear, and beset by the taxonomic problem that species are sometimes defined on the basis of chirality alone.

Chiral reversal in the Japanese snail genus *Euhadra* perhaps presents one of the best candidate systems for investigating the potential for single-gene speciation, but also illustrates the lack of empirical data. Two independent studies (Ueshima and Asami 2003; Davison et al. 2005) have used mitochondrial DNA sequences to investigate the phylogenetic relationships between the five sinistral *Euhadra* species and the other dextral species. Both phylogenies supported a single origin of the sinistral species from a dextral ancestor, but also found evidence supporting recent evolution of dextral *E. aomoriensis* from sinistral *E. quaesita*. Specifically, three lineages of the dextral species were polyphyletically distributed within *E. quaesita*, leading to the suggestion that this is due repeated single-gene speciation of the dextral from the sinistral (Ueshima and Asami 2003). We therefore set out to test the evidence for single-gene speciation in *Euhadra*, by combining a fine-scale RAD-seq phylogeographic study with behavioral observations of snail mating.

Methods
SAMPLING

There are 22 taxonomically defined species/subspecies of *Euhadra* (Bradybaenidae) distributed throughout Japan and the neighboring Korean island of Jeju (Davison et al. 2005). For this study, three of the five sinistral species, *E. decorata*, *E. murayamai*, and *E. quaesita*, the dextral species *E. senckenbergiana*, and the nominal dextral species *E. "aomoriensis"* were sampled.

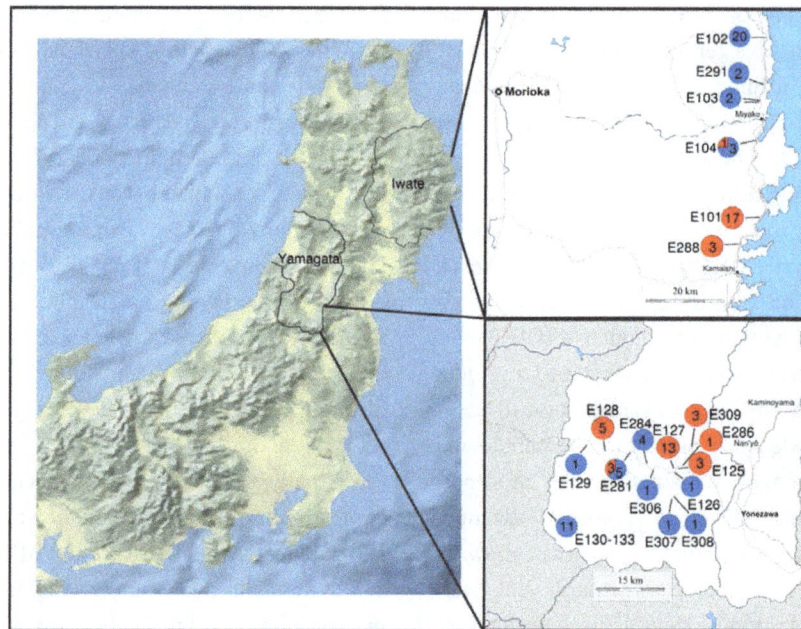

Figure 1. Map showing topography of northern and central Honshu, Japan. Insets: two newly identified contact zones between sinistral (red) *E. quaesita* and dextral (blue) *E. aomoriensis*, showing sample size and site ID.

Sinistral *E. quaesita* were collected from across the Tohoku (northern Honshu) region of Japan. *E. aomoriensis* has a distribution that is largely allopatric with *E. quaesita*, being more frequently found in sympatry with sinistral *E. decorata*, especially in the northern part of Tohoku. Like *E. quaesita*, *E. aomoriensis* was also sampled opportunistically across Tohoku, but with a concentrated effort on two dextral/sinistral contact zones that we identified, one in Iwate prefecture (NE Tohoku) and another in Yamagata (SW Tohoku), approximately 250 km apart (Fig. 1). Further samples were obtained of sinistral *E. murayamai*, a species that is endemic to a small limestone outcrop and is also polyphyletic within *E. quaesita*, based on mtDNA (Davison et al. 2005). Finally, *E. senkenbergiana* and *E. decorata* were included because they are sometimes sympatric with dextral *E. aomoriensis* and sinistral *E. quaesita*. Thus, the collection contained samples of sympatric and parapatric *E. quaesita*/*E. aomoriensis,* and for comparison, sympatric *E. quaesita*/*E. senckenbergiana*, and *E. aomoriensis*/*E. decorata*.

BEHAVIORAL OBSERVATIONS

It is commonly assumed that dextral and sinistral low-spired snails are either unable to mate, or can only mate very rarely (Asami et al. 1998; Davison and Mordan 2007). There are no known reports of mating, and we are not aware of any systematic studies. We used a network of malacological contacts, and a knowledge of Japanese language sources to investigate evidence of possible matings between dextral and sinistral *Euhadra*.

DE NOVO GENERATION OF RAD-SEQ SNP MARKERS

RAD-seq was used to generate SNP markers for 16 individuals representing four species. The samples included two sinistral *E. quaesita* populations ($n = 6$) that are largely parapatric with two dextral *E. aomoriensis* populations ($n = 6$) in East Iwate and South Yamagata, where geographic and mtDNA data suggest interchiral contact may have been recent or ongoing (see Results). For comparison, one population of dextral *E. senkenbergiana* ($n = 3$) and one individual of sinistral *E. decorata* were used.

From the final filtered set of SNPs (see Supplementary Methods for further details) a number of datasets were generated, allowing for varying degrees of missing data. After quality filtering, 13,167 biallelic loci were found in eight or more individuals, which reduced to 7871 loci once singleton SNPs were removed. There were still a substantial number of missing genotypes in this dataset, so to refine the loci used further, only one null was allowed in each of the four main population samples of interest (sinistral and dextral snails from Iwate and Yamagata), leaving 4598 loci. This reduced dataset was used for all subsequent analyses. Although not shown, other datasets produced similar outputs in terms of subsequent analyses.

RAD-SEQ PHYLOGENOMIC ANALYSES

We conducted four separate phylogenomic analyses, based on the RAD-seq dataset. First, the concatenated SNPs were used to build a maximum likelihood phylogeny, using the same methods as for the mitochondrial data. However, phylogenies are not useful in understanding conflicting signals in the underlying data, as might be produced by varying degrees of linkage between markers, recombination, and introgression. Therefore, we also constructed a network, using the neighbor-net method in SplitsTree 4 (Huson and Bryant 2006), based on a matrix of uncorrected p-distances, and using the equal-angle split transformation and ignoring ambiguous states. Also, the relationship between the individuals was investigated using principal components analysis (PCA), conducted using ADEGENET (Jombart 2008), and ADE4 (Dray and Dufour 2007) in R 3.2.3.

To test for signals of admixture between population samples, and correspondingly, whether any inferred tree is truly bifurcating, we used Treemix (Pickrell and Pritchard 2012). This software uses allele frequencies within groups to relate a sample of populations to their common ancestor, including as output a maximum-likelihood (ML) tree of estimated migration events, including the direction. Five populations used were Yamagata sinistral, Yamagata dextral, Iwate sinistral, Iwate dextral, and *E. senckenbergiana* outgroup. F_{ST} between sites was also estimated, using Genepop and the same populations (Rousset 2008). To complement these analyses, population structure and admixture was estimated using individual genotypes with STRUCTURE v2.3.4 (Falush et al. 2003; Evanno et al. 2005).

TESTING OF SCENARIOS VIA APPROXIMATE BAYESIAN COMPUTATION METHOD

We used an approximate Bayesian computation (ABC) approach, implemented in the software DIYABC v 2.1.0 (Cornuet et al. 2014) to compare hypotheses. In brief, simulated datasets were produced for five scenarios, by sampling parameter values in defined prior distributions. Three scenarios were similar in that the populations showed a bifurcating topology, only differing in divergence order. Two other models included ancestral admixture, because the shared chirality between dextral *E. aomoriensis* from Yamagata and Iwate might be because of shared ancestry. The analysis was restricted to the four population samples of *E. quaesita* and *E. aomoriensis*, primarily because the large genetic distance between *E. senckenbergiana* and the other samples meant that it was difficult to find a suitable range of parameter values. See Supplementary Methods for further detail.

MITOCHONDRIAL PHYLOGENETIC AND POPULATION ANALYSES

The number of individuals in the phylogenomic analysis was necessarily limited by resources.1 To sample more individuals and over a greater geographic area, ~800 bp fragments of 16S rRNA were amplified and sequenced using standard conditions and buffers (see Supplementary Methods). For maximum likelihood phylogenies, an appropriate model of evolution was selected using jModelTest and the Akaike Information Criterion (Darriba et al. 2012), followed by tree construction and visualization using

Table 1. Reports of mating between dextral and sinistral *Euhadra* species.

Dextral	Sinistral		Year	Observer	Notes	
E. senckenbergiana Nada, Osaka	*E. quaesita* Nada, Osaka	Wild	1980	Hiroyuki Nishitani	Pers. comm.	Reciprocal dart shooting observed
E. senckenbergiana	*E. quaesita*	Wild	1979	Kazuhisa Shinagawa	Lecture, Hanshin Shell Club	
E. peliomphala Setagaya, Tokyo	*E. quaesita* Setagaya, Tokyo	Wild	1999	Seiichi Takase, Kentaro Nakano	Report[1]	see Figure 1
E. amaliae Kobe	*E. grata* Katsuki, Niigata	Laboratory	1982	Ryoji Takada	Pers. comm.	Prolonged mating
E. sandai Mita, Hyogo	*E. quaesita* Miyakejima	Laboratory	1981	Ryoji Takada	Pers. comm.	Mating on branch in aquarium

[1]Heterospecific mating between sinistral and dextral species in *Euhadra*. Kainakama 33(4): 1, Hanshin Molluscan Research Group, Nishinomiya.

Figure 2. Reciprocal mating between sinistral *E. quaesita* and dextral *E. peliomphala*. Mating between these distantly related species may not produce viable offspring, but illustrates the general point that dextrals and sinistrals are able to mate, if only rarely. Photo: Kentaro Nakao and Seiichi Takase, reproduced with permission.

PhyML (Guindon and Gascuel 2003) and TreeExplorer, including bootstrap support, with the tree rooted on *E. senkenbergiana* (Ueshima and Asami 2003; Davison et al. 2005).

Results

CONTACT ZONES BETWEEN SINISTRAL AND DEXTRAL SNAILS

In both East Iwate and in South Yamagata prefecture we found one site (E104 and E281, respectively) that contained both chiral morphs of the two species (Fig. 1). Sympatric sites were also recorded between sinistral *E. quaesita* and dextral *E. senckenbergiana* in Chubu (e.g., Noto peninsula, site E261, Anamizu), between sinistral *E. quaesita* and dextral *E. senckenbergiana* in Chubu (D24, Mt. Myojo), and between dextral *E. aomoriensis* and sinistral *E. decorata* in Tohoku (E106, Kabayama; E227, Nohira; E299, Wakasennin; D31, Tamayama).

MATING BETWEEN DEXTRAL AND SINISTRAL SNAILS

We found five records of mating between dextral and sinistral *Euhadra* (Table 1; Fig. 2). These observations included matings between sympatric sinistral *E. quaesita* and dextral *E. senckenbergiana*.

PHYLOGENOMICS

Phylogenies (Fig. 3A–B) based on whole genome RAD-seq data (see Table S1 for read depths) clearly showed that dextral *E. aomoriensis* and sinistral *E. quaesita* group together and are distinct from sinistral *E. decorata* and dextral *E. senkenbergiana*. Within the *E. quaesita*/*E. aomoriensis* groups, dextrals and sinistrals grouped together by geographic region, Yamagata or Iwate, rather than by chirality, with strong bootstrap support. Within regions, individuals clustered with other individuals from the same sampling location, with the exception of dextral individual E102-4, which clustered with sinistral individuals from the nearby site, E101. This general result was confirmed using a principal components analysis (Fig. 3C). In the latter, when the analysis was restricted to just *E. quaesita* and *E. aomoriensis* samples, the first three axes explained 48.4% of the variation, respectively separating individuals by region (Iwate or Yamagata, 23.0%), then dextral and sinistrals within Yamagata (13.8%) and dextral and sinistrals within Iwate (11.6%). As above, the position of dextral individual E102-4 was different to the other two individuals from the same site.

Using population allele frequencies, a TreeMix phylogeny (Fig. 3D) showed the same overall topology, except also containing two putative migration events, from sinistral *E. quaesita* in Iwate into sinistral *E. quaesita* in Yamagata (14%) and from the dextral or sinistral ancestor of the Yamagata snails into sinistral Iwate *E. quaesita* (8%). Using individual genotypes, STRUCTURE identified evidence for more recent gene flow. When the analysis was confined to dextral *E. aomoriensis* and sinistral *E. quaesita* from Iwate, the "optimal" K was 2, but with the dextral E102-4 clustering with the other sinistrals. When dextral *E. aomoriensis* and sinistral *E. quaesita* from Yamagata were analyzed together, the optimal number of clusters was three, but with sinistrals and dextrals showing mixed ancestry. Estimates of F_{ST} (Table S2) showed that divergence within regions is low to moderate (e.g., F_{ST} ~0.2 between sinistral and dextral sites in Yamagata), higher between different regions (e.g., Yamagata – Iwate, F_{ST} ~0.4) and very high when comparing with *E. senckenbergiana* (F_{ST} ~0.6).

TESTING OF SCENARIOS

Approximate Bayesian computation (Cornuet et al. 2014) was used to compare five different models of cladogenesis, two including admixture (Fig. 4). A scenario involving admixture (0–10%) from dextral Yamagata snails into Iwate snails was the optimal model (model #4 in Table 2; see also Fig. 4), significantly better using logistic regression. To further evaluate confidence in the models, test datasets were simulated. The true scenario had

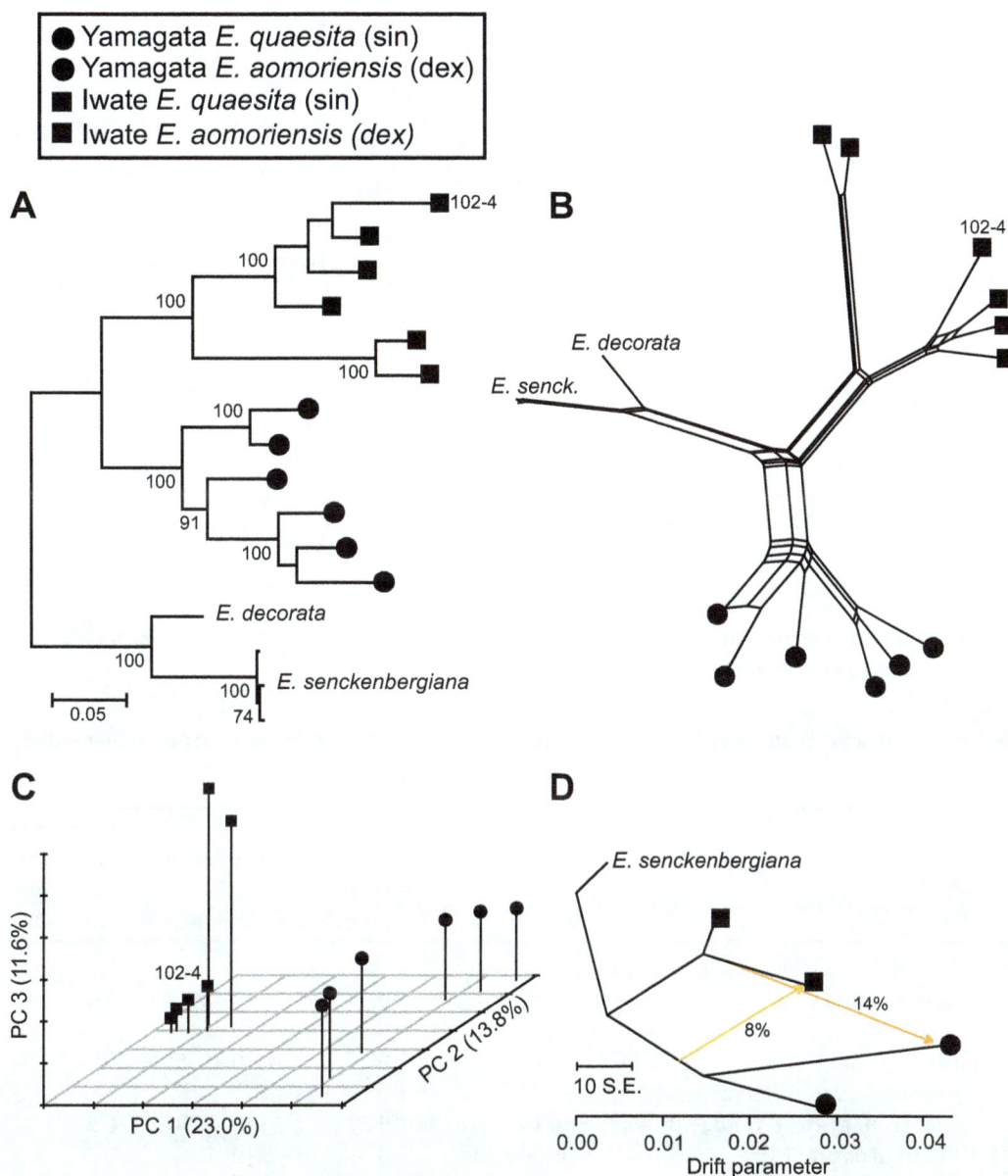

Figure 3. Phylogenomic analysis of 4598 RAD-seq derived biallelic loci. (A) Maximum likelihood phylogeny using the GTR model, with bootstrap support. (B) Neighbor-net broadly shows the same relationship between individuals. (C) Principal components analysis, carried out on only *E. quaesita* and *E. aomoriensis*, separates individuals by region (Iwate or Yamagata), then by sites within regions. (D) Treemix analysis of allele frequencies within populations indicates evidence of ancestral migration between populations.

the highest posterior probability for 0.72/0.73 (direct/logistic) test datasets, giving a posterior error rate of 0.27/0.28 (Table 2). Similarly, test datasets were simulated and a prior based error analysis conducted to understand the probability with which true models might be rejected. The proportion of wrongly identified scenarios was 0.33/0.25 (direct/logistic). Finally, scenario specific prior error rate was estimated, by drawing test datasets from the parameter prior distribution under a given scenario. By drawing pods against model #4, and comparing to the next best model (#1), the type I error was 0.10/0.09; by drawing pods against model #1, and comparing to model #4, the type II error for model #1

was 0.29/013. In both cases, similar values were obtained when comparing against other models.

MITOCHONDRIAL PHYLOGENIES

In the Iwate contact zone, it was found that most sinistral and dextral *E. quaesita* and *E. amoriensis* snails were of the same haplogroup, QUA2a, including three haplotypes were shared between sinistral and dextral coiling snails; in the Yamagata contact zone, the same two species were mostly of haplogroup QUA1 (Fig. S1). In comparison, in sites containing *E. quaesita/E. senckenbergiana* or *E. aomoriensis/E. decorata*, different species

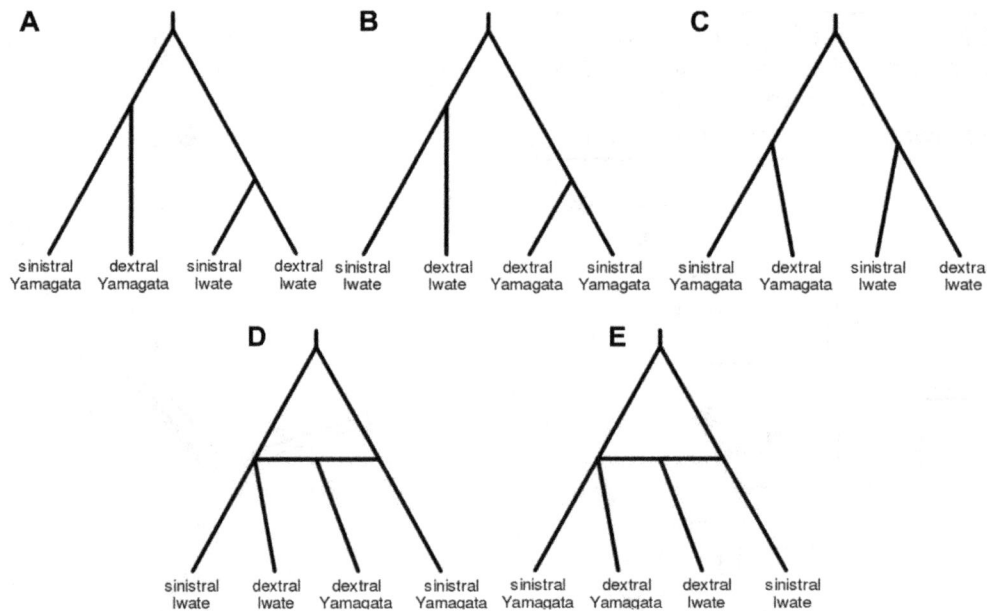

Figure 4. The five scenarios tested in the ABC analysis, three with bifurcating topologies (A, B, C), which only differ in the relative timing of events, and two with admixture between dextral populations (D, E).

Table 2. Scenarios for the repeated evolution of dextral populations of snails within a sinistral species, either independently (1–3), or involving admixture.

		Direct		Logistic regression	
	Scenarios	Posterior probability	95% confidence	Posterior probability	95% confidence
	Independent origin of dextral/sinistrals in Yamagata versus Iwate				
1.	Yamagata snails diverged first	0.27	[0–0.6]	0.03	[0.03]
2.	Iwate snails diverged first	0.12	[0–0.41]	0	[0]
3.	Diverged at same time	0.26	[0–0.65]	0	[0]
	Shared ancestry between dextrals (admixture)				
4.	**Admixture (<10%) from dextral Yamagata snails into Iwate snails**	**0.37**	**[0–0.79]**	**0.97**	**[0.96–0.97]**
5.	Admixture (<10%) from dextral Iwate snails into Yamagata snails	0	[0]	0	[0]

The best model is highlighted in bold.

contained divergent mitochondrial haplogroups (Figs. S1 and S2; Table S3).

Discussion

Previously, we used a theoretical model to show that occasional chiral reversals may take place and enable gene flow between mirror image snail-shell morphs. This is because the maternal inheritance of snail chirality means that there is sometimes a discord between phenotype and genotype, leading to gene flow between different types (Davison et al. 2005). By collecting together reports and observations from Japanese naturalists, we found direct evidence that dextral and sinistral *Euhadra* are sometimes able to mate. Moreover, the genomic data indirectly suggests recent

or ongoing gene flow between sinistral *E. quaesita* and dextral *E. aomoriensis*. Thus, overall, the theoretical model (Davison et al. 2005), behavioral observations, genomic data, and biogeographic context do not support a model of single-gene speciation. More than a single gene is required before chiral variation can be considered to have created a new species, at least in *Euhadra*.

A fundamental new insight from this work is that analyses of genome-wide SNP markers show that dextral and sinistral *Euhadra* in Iwate, Japan are distinct from dextral *Euhadra* in Yamagata. Moreover, genomic and mtDNA divergence was low in both locations where dextral and sinistral morphs were found together (Figs. 3, S1; Table S3), especially in comparison to other snails that show greater differentiation over shorter geographic distances (Davison and Clarke 2000). Some dextral and sinistral snails shared identical mtDNA haplotypes, and individual snails showed traces of mixed ancestry (Fig. 3A–C; also STRUCTURE analyses). There are only two explanations for this pattern—either there is ongoing gene flow between dextral *E. aomoriensis* and sinistral *E. quaesita* in two separate locations, or else there was gene flow, but this has recently ceased.

A second important insight from this work is that face-to-face mating between low-spired snails sometimes take place, albeit at an unknown frequency. Previously, the assumption has been that the genitals are on the "wrong" side of the head in mating between sinistral and dextral snails, and so intromission is not possible. However, the data here suggest that the problem is mainly behavioral. As in *Amphidromus*, the only snail genus that routinely has interchiral mating (Schilthuizen et al. 2007; Schilthuizen and Looijestijn 2009), it is likely that the long, thin, flexible genital organs are able to twist to match the partners chirality. Moreover, in high-spired snails, it is already known that interchiral mating (by "shell-mounting") is less of an issue—and AFLP markers have recently been used to show that gene flow is extensive between the two types, as expected (Koch et al. 2017).

Putting our findings together, we are able to reinterpret previous studies (Ueshima and Asami 2003; Davison et al. 2005). The fact that both the mtDNA and nuclear trees presented here are concordant for the polyphyletic pattern puts beyond doubt the hypothesis that dextral *E. aomoriensis* and sinistral *E. quaesita* are coderived, and should probably be treated as forms of the same species, *E. quaesita*. Although it is possible—or even likely—that the dextral chirality determining allele is ultimately derived from another dextral species, *E. aomoriensis* must have mainly shared ancestry with *E. quaesita*, because otherwise we would have instead expected to observe a signal of admixture with dextral *E. senckenbergiana* in the genomic RAD-seq data. The two chiral types cannot be defined as separate species, given that they are sometimes found in sympatry, they sometimes mate, the genomic evidence for gene flow between the two types, and the underlying theory that reproductive isolation is unstable (Davison et al. 2005). Altogether, these observations critically weaken the argument for chirality directly leading to single-gene speciation, at least without implicating other factors such as ecology or predation.

A challenging question to consider how dextral *E. aomoriensis* evolved from sinistral *E. quaesita*, or indeed, whether sinistrals evolved from dextrals? Do the geographically separate regions represent independent transition events from sinistral to dextral, or did dextrality evolve once, then introgressing with a local sinistral in secondary contact? Although further investigations are required and the precise details differ, both the Treemix analysis and the ABC results are consistent with past admixture, suggesting that a common origin is likely. Specifically, the ABC analysis is suggestive of admixture from dextral Yamagata snails

into Iwate (Table 1), and thus, that the dextrals in both locations may share the same dextral-determining allele. Alternatively, the Treemix analysis (Fig. 3D) suggests admixture in the opposite direction, as well as from an ancestor of unknown chirality.

This gene flow need not have been direct. As has been suggested in parallel incipient speciation by local adaptation in other species like intertidal *Littorina* snails (Butlin et al. 2008) and sticklebacks (Colosimo et al. 2005), a feasible scenario is that chirality-determining alleles exists at low frequencies, especially the recessive version, but under certain selective and/or chance biogeographic conditions (see below) reach fixation, establishing a new chiral morph (Johnson 1982; Orr 1991; van Batenburg and Gittenberger 1996; Davison et al. 2005; Hoso et al. 2010; Yamamichi and Sasaki 2013). Unfortunately, it is unknown which allele is dominant in *Euhadra* and this may vary, depending upon genomic context (Clarke and Murray 1969; Schilthuizen and Davison 2005).

A similar interpretation may also be applied to a recent AFLP study on *Alopia* door snails (Koch et al. 2017) and another mitochondrial rRNA study between sinistral and dextral *Satsuma* species (Hoso et al. 2010), a genus that is relatively closely related to *Euhadra,* and which might even share the same chiral-varying alleles. In the latter, the authors concluded that the presence of multiple sinistral lineages cannot be explained by introgression via hybridization, because *Satsuma* snails mate face-to-face. Our data instead suggest that both gene flow via maternal inheritance and direct face-to-face mating should be considered in interpreting whether the patterns are really caused by a "speciation gene."

Evidently, further work is needed to disentangle the geographic context of the chiral reversion event (s) that lead to the evolution of new chiral types in snails. Such a challenge is fundamental to the field of speciation in general (Coyne and Orr 2004; Smadja and Butlin 2011; Martin et al. 2013). More broadly, part of the interest in snail chirality arises from attempts to understand chiral invariance across the metazoans (Grande and Patel 2008; Okumura et al. 2008; Davison et al. 2009; Utsuno et al. 2011; Davison et al. 2016). Thus, while chiral variation in snails is perhaps a small step toward speciation, this chiral variation may be an invaluable genetic resource in helping reveal the earliest steps of symmetry breaking across the Bilateria (Davison et al. 2016).

AUTHOR CONTRIBUTIONS

A.D. and S.C. conceived the study. All authors conducted field work. A.D., T.H., and P.M.R. conducted lab work. P.M.R. and A.D. analyzed the genetic data. A.D., S.C., and P.M.R. wrote the manuscript.

ACKNOWLEDGMENTS

The main part of this study was funded by a BBSRC studentship to Paul Richards, with additional funding from BBSRC grant BB/F018940/1 to A.D., the Japanese Society for the Promotion of Science, the Genetics Society, and the Daiwa Foundation. We thank Karim Gharbi, Marian

Thompson, and colleagues at The GenePool Genomics Facility, University of Edinburgh (now Edinburgh Genomics) for general advice and DNA sequencing, Maureen Liu for advice on RAD-seq, and Johan Michaux for advice on ABC analyses. Thanks also to Morito Hayashi and Naoyuki Takahashi for help with some of the field work, and to Ryoji Takada, Jamen Uiriamu Otani, Hiroshi Minato, Misao Kawana, Akira Tada, Shoji Suzuki, and Kentaro Nakao for providing information regarding heterochiral mating. Insight on an early version by Axios review, and comments from two referees and a subject editor substantially improved the manuscript.

LITERATURE CITED

Asami, T., R. H. Cowie, and K. Ohbayashi. 1998. Evolution of mirror images by sexually asymmetric mating behavior in hermaphroditic snails. Am. Nat. 152:225–236.

Bolnick, D. I., and B. M. Fitzpatrick. 2007. Sympatric speciation: models and empirical evidence. Pp. 459–487 in Annual review of ecology evolution and systematics. Annual Reviews, Palo Alto, California. http://www.annualreviews.org/toc/ecolsys/38/1.

Boycott, A. E., and C. Diver. 1923. On the inheritance of sinistrality in *Limnaea peregra*. Proc. R Soc. Biol. Sci. Ser. B 95:207–213.

Butlin, R. K., J. Galindo, and J. W. Grahame. 2008. Sympatric, parapatric or allopatric: the most important way to classify speciation? Philos. Trans. R Soc. B Biol. Sci. 363:2997–3007.

Clarke, B., and J. Murray. 1969. Ecological genetics and speciation in land snails of the genus *Partula*. Biol. J. Linn. Soc. 1:31–42.

Colosimo, P. F., K. E. Hosemann, S. Balabhadra, G. Villarreal, M. Dickson, J. Grimwood, J. Schmutz, R. M. Myers, D. Schluter, D. M. Kingsley, et al. 2005. Widespread parallel evolution in sticklebacks by repeated fixation of ectodysplasin alleles. Science 307:1928–1933.

Cornuet, J.-M., P. Pudlo, J. Veyssier, A. Dehne-Garcia, M. Gautier, R. Leblois, et al. 2014. DIYABC v2.0: a software to make approximate Bayesian computation inferences about population history using single nucleotide polymorphism, DNA sequence and microsatellite data. Bioinformatics 30:1187–1189.

Coyne, J. A., and H. A. Orr. 2004. Speciation. Sinauer Associates, Sunderland, Massachusetts, USA.

Darriba, D., G. L. Taboada, R. Doallo, and D. Posada. 2012. jModelTest 2: more models, new heuristics and parallel computing. Nat. Methods 9:772–772.

Davison, A., N. H. Barton, and B. Clarke. 2009. The effect of coil phenotypes and genotypes on the fecundity and viability of *Partula suturalis* and *Lymnaea stagnalis*: implications for the evolution of sinistral snails. J. Evol. Biol. 22:1624–1635.

Davison, A., S. Chiba, N. H. Barton, and B. C. Clarke. 2005. Speciation and gene flow between snails of opposite chirality. PLoS Biol. 3:e282.

Davison, A., and B. Clarke. 2000. History or current selection? A molecular analysis of 'area effects' in the land snail *Cepaea nemoralis*. Proc. R Soc. Lond. B Biol. Sci. 267:1399–1405.

Davison, A., G. S. McDowell, J. M. Holden, H. F. Johnson, G. D. Koutsovoulos, M. M. Liu, et al. 2016. Formin is associated with left-right asymmetry in the pond snail and the frog. Curr. Biol. 26:654–660.

Davison, A., and P. Mordan. 2007. A literature database on the mating behavior of stylommatophoran land snails and slugs. Am. Malacol. Bull. 23:411–415.

Dray, S., and A.-B. Dufour. 2007. The ade4 package: implementing the duality diagram for ecologists. J. Stat. Software 22, 1–20.

Evanno, G., S. Regnaut, and J. Goudet. 2005. Detecting the number of clusters of individuals using the software structure: a simulation study. Mol. Ecol. 14, 2611–2620.

Falush, D., M. Stephens, and J. K. Pritchard. 2003. Inference of population structure using multilocus genotype data: linked loci and correlated allele frequencies. Genetics 164:1567–1587.

Feher, Z., L. Nemeth, A. Nicoara, and M. Szekeres. 2013. Molecular phylogeny of the land snail genus Alopia (Gastropoda: Clausiliidae) reveals multiple inversions of chirality. Zool. J. Linn. Soc. 167:259–272.

Felsenstein, J. 1981. Skepticism towards Santa Rosalia, or why are there so few kinds of animals. Evolution 35:124–138.

Gavrilets, S. 2004. Fitness landscapes and the origin of species. Princeton Univ. Press, Princeton.

Gittenberger, E. 1988. Sympatric speciation in snails—a largely neglected model. Evolution 42:826–828.

Gittenberger, E., T. D. Hamann, and T. Asami. 2012. Chiral speciation in terrestrial pulmonate snails. PLOS One 7:1–5. http://journals.plos.org/plosone/article?id=10.1371/journal.pone.0034005.

Grande, C., and N. H. Patel. 2008. Nodal signalling is involved in left-right asymmetry in snails. Nature 405:1007–1011.

Guindon, S., and O. Gascuel. 2003. A simple, fast, and accurate algorithm to estimate large phylogenies by maximum likelihood. Syst. Biol. 52:696–704.

Hoso, M., Y. Kameda, S.-P. Wu, T. Asami, M. Kato, and M. Hori. 2010. A speciation gene for left-right reversal in snails results in anti-predator adaptation. Nat. Commun. 1:1–7. https://www.nature.com/articles/ncomms1133.

Huson, D. H., and D. Bryant. 2006. Application of phylogenetic networks in evolutionary studies. Mol. Biol. Evol. 23:254–267.

Johnson, M. S. 1982. Polymorphism for direction of coil in *Partula suturalis*—behavioral isolation and positive frequency-dependent selection. Heredity 49:145–151.

Johnson, M. S., B. Clarke, and J. Murray. 1990. The coil polymorphism in *Partula suturalis* does not favor sympatric speciation. Evolution 44:459–464.

Jombart, T. 2008. adegenet: a R package for the multivariate analysis of genetic markers. Bioinformatics 24:1403–1405.

Koch, E. L., M. T. Neiber, F. Walther, and B. Hausdorf. 2017. High gene flow despite opposite chirality in hybrid zones between enantiomorphic door-snails. Mol. Ecol 15:3998–4012.

Kornilios, P., E. Stamataki, and S. Giokas. 2015. Multiple reversals of chirality in the land snail genus Albinaria (Gastropoda, Clausiliidae). Zool. Scr. 44:603–611.

Martin, S. H., K. K. Dasmahapatra, N. J. Nadeau, C. Salazar, J. R. Walters, F. Simpson, et al. 2013. Genome-wide evidence for speciation with gene flow in *Heliconius* butterflies. Genome Res. 23:1817–1828.

Modica, M. V., P. Colangelo, A. Hallgass, A. Barco, and M. Oliverio. 2016. Cryptic diversity in a chirally variable land snail. Italian J. Zool. 83:351–363.

Nosil, P. 2008. Speciation with gene flow could be common. Mol. Ecol. 17:2103–2106.

Okumura, T., H. Utsuno, J. Kuroda, E. Gittenberger, T. Asami, and K. Matsuno. 2008. The development and evolution of left-right asymmetry in invertebrates: lessons from *Drosophila* and snails. Dev. Dyn. 237:3497–3515.

Orr, H. A. 1991. Is single-gene speciation possible? Evolution 45:764–769.

———. 1996. Dobzhansky, Bateson, and the genetics of speciation. Genetics 144:1331–1335.

Pickrell, J. K., and J. K. Pritchard. 2012. Inference of population splits and mixtures from genome-wide allele frequency data. Plos Genet. 8:1–17.

Rousset, F. 2008. GENEPOP' 007: a complete re-implementation of the GENEPOP software for Windows and Linux. Mol. Ecol. Res. 8:103–106.

Rundle, H. D., P. Nosil. 2005. Ecological speciation. Ecol. Lett. 8:336–352.

Schilthuizen, M., P. G. Craze, A. S. Cabanban, A. Davison, J. Stone, E. Gittenberger, et al. 2007. Sexual selection maintains whole-body chiral dimorphism in snails. J. Evol. Biol. 20:1941–1949.

Schilthuizen, M., and A. Davison. 2005. The convoluted evolution of snail chirality. Naturwissenschaften 92:504–515.

Schilthuizen, M., and S. Looijestijn. 2009. The sexology of the chirally dimophic snail species *Amphidromus inversus* (Gastropoda: Camaenidae). Malacologia 51:379–387.

Servedio, M. R. 2009. The role of linkage disequilibrium in the evolution of premating isolation. Heredity 102:51–56.

Servedio, M. R., and M. A. F. Noor. 2003. The role of reinforcement in speciation: theory and data. Ann. Rev. Ecol. Evol. Syst. 34:339–364.

Smadja, C. M., and R. K. Butlin. 2011. A framework for comparing processes of speciation in the presence of gene flow. Mol. Ecol. 20:5123–5140.

Sturtevant, A. H. 1923. Inheritance of direction of coiling in *Limnaea*. Science 58:269–270.

Ueshima, R., and T. Asami. 2003. Single-gene speciation by left-right reversal—a land-snail species of polyphyletic origin results from chirality constraints on mating. Nature 425:679–679.

Uit de Weerd, D. R., D. S. J. Groenenberg, M. Schilthuizen, and E. Gittenberger. 2006. Reproductive character displacement by inversion of coiling in clausiliid snails (Gastropoda, Pulmonata). Biol. J. Linn. Soc. 88:155–164.

Utsuno, H., T. Asami, T. J. M. Van Dooren, and E. Gittenberger. 2011. Internal selection against the evolution of left-right reversal. Evolution 65:2399–2411.

van Batenburg, F. H. D., and E. Gittenberger. 1996. Ease of fixation of a change in coiling: computer experiments on chirality in snails. Heredity 76:278–286.

Yamamichi, M., and A. Sasaki. 2013. Single-gene speciation with pleiotropy: effects of allele dominance, population size, and delayed inheritance. Evolution 67:2011–2023.

16

Mitogenome evolution in the last surviving woolly mammoth population reveals neutral and functional consequences of small population size

Patrícia Pečnerová,[1,2,3] (iD) Eleftheria Palkopoulou,[1,2,4] Christopher W. Wheat,[2] Pontus Skoglund,[4,5] Sergey Vartanyan,[6] Alexei Tikhonov,[7,8] Pavel Nikolskiy,[9] Johannes van der Plicht,[10,11] David Díez-del-Molino,[1] and Love Dalén[1,12]

[1] *Department of Bioinformatics and Genetics, Swedish Museum of Natural History, Stockholm, Sweden*

[2] *Department of Zoology, Stockholm University, Stockholm, Sweden*

[3] *E-mail: pata.pecnerova@gmail.com*

[4] *Department of Genetics, Harvard Medical School, Boston, Massachusetts 02115*

[5] *Broad Institute of Harvard and MIT, Cambridge, Massachusetts 02142*

[6] *North-East Interdisciplinary Scientific Research Institute N.A.N.A. Shilo, Far East Branch, Russian Academy of Sciences (NEISRI FEB RAS), Magadan, Russia*

[7] *Zoological Institute of Russian Academy of Sciences, Saint-Petersburg, Russia*

[8] *Institute of Applied Ecology of the North, North-Eastern Federal University, Yakutsk, Russia*

[9] *Geological Institute of the Russian Academy of Sciences, Moscow, Russia*

[10] *Centre for Isotope Research, Groningen University, Groningen, The Netherlands*

[11] *Faculty of Archaeology, Leiden University, Leiden, The Netherlands*

[12] *E-mail: Love.Dalen@nrm.se*

The onset of the Holocene was associated with a global temperature increase, which led to a rise in sea levels and isolation of the last surviving population of woolly mammoths on Wrangel Island. Understanding what happened with the population's genetic diversity at the time of the isolation and during the ensuing 6000 years can help clarify the effects of bottlenecks and subsequent limited population sizes in species approaching extinction. Previous genetic studies have highlighted questions about how the Holocene Wrangel population was established and how the isolation event affected genetic diversity. Here, we generated high-quality mitogenomes from 21 radiocarbon-dated woolly mammoths to compare the ancestral large and genetically diverse Late Pleistocene Siberian population and the small Holocene Wrangel population. Our results indicate that mitogenome diversity was reduced to one single haplotype at the time of the isolation, and thus that the Holocene Wrangel Island population was established by a single maternal lineage. Moreover, we show that the ensuing small effective population size coincided with fixation of a nonsynonymous mutation, and a comparative analysis of mutation rates suggests that the evolutionary rate was accelerated in the Holocene population. These results suggest that isolation on Wrangel Island led to an increase in the frequency of deleterious genetic variation, and thus are consistent with the hypothesis that strong genetic drift in small populations leads to purifying selection being less effective in removing deleterious mutations.

KEY WORDS: *Mammuthus primigenius*, mitochondrial genomes, woolly mammoth, Wrangel Island.

Impact Summary

While most of the Pleistocene megafauna species became extinct at the end of the last ice age, the woolly mammoth survived in small insular populations, most notably on Wrangel Island where it survived until 4000 years before present. Genetic data suggest that compared to the large and diverse Pleistocene population, Holocene mammoths on Wrangel Island had low genetic diversity. However, it is still unclear to what extent genetic diversity was lost as a consequence of a founder effect when rising sea levels led to the formation of the island, compared to the subsequent effect of small effective population size during the ensuing 6000 years. To examine this, we sequenced mammoth mitogenomes from before and after the isolation on Wrangel Island. Our results show a severe loss in genetic diversity and fixation of a mutation with potential functional consequences at the time the population was established, supporting the hypothesis of a founder effect. However, the observation of an increase in the evolutionary rate following isolation on the island is consistent with an elevated impact of genetic drift leading to purifying selection becoming less efficient. Our findings add some details into the mosaic of complex processes that preceded the woolly mammoth's extinction and serve as a rare example of testing basic population genetic concepts in a wild population.

The fate of taxa during periods of climate change can be simplified into three processes: adaptation, migration, or extinction (Davis and Shaw 2001; Aitken et al. 2008). Changes in environmental conditions can potentially lead to adaptation through selection on standing genetic variation. However, if these changes are too rapid, or the amount of genetic variation is insufficient, species may need to track geographical changes in habitat availability to persist. The woolly mammoth (*Mammuthus primigenius*) inhabited the Northern Hemisphere for ~800 thousand years (kyr), from the late Middle Pleistocene to early Holocene (Lister and Sher 2001). Mammoths thus survived a number of alternating glacials and interglacials, but went extinct during the Holocene interglacial period ~4 thousand years before present (kyr calBP) (Vartanyan et al. 1995). Even though mammoths were originally thought to have become extinct at the Pleistocene/Holocene boundary (~11,700 yr calBP), along with other representatives of the Pleistocene megafauna in a phenomenon known as the *Quaternary extinction event* (Stuart 1991), mammoth remains with radiocarbon ages as young as 4 kyr calBP have been found on Wrangel Island (Vartanyan et al. 1993; Stuart et al. 2002). Wrangel Island thus represents the refugium of the last surviving population of the species.

Wrangel Island became separated from continental Siberia ~10 kyr calBP following the rise in sea levels after the Last Glacial Maximum (LGM) (Vartanyan et al. 2008; Arppe et al. 2009). During the Late Pleistocene, Wrangel formed a mountainous area in the otherwise flat Beringian landscape, with parts covered by an inactive rock glacier and perennial snowfields (Vartanyan et al. 2008). The fossil record and strontium isotopes in bones indicate that mammoths were not permanent residents in the region during the Late Pleistocene, but rather visited (what later became) Wrangel Island during seasonal migrations (Vartanyan et al. 2008; Arppe et al. 2009).

After the isolation, however, mammoths were confined to and survived on Wrangel Island for an additional ~6 kyr. The process leading up to the mammoth's extinction has been studied using both nuclear and mitochondrial DNA. Initial studies on short mitochondrial sequences (Nyström et al. 2010) and microsatellite loci (Nyström et al. 2012) suggested that, after an abrupt loss of genetic variation related to the isolation event, genetic diversity was retained or even slightly increased during the subsequent ~6 kyr. More recent analyses of two complete nuclear genomes revealed that the Wrangel mammoth genome – compared to a Pleistocene mainland mammoth genome – had lower observed heterozygosity and a higher proportion of the genome allocated in runs of homozygosity (ROH), which could have been either the result of a founder effect during the isolation of the island, and/or repeated breeding between distant relatives due to the small Holocene effective population size (Palkopoulou et al. 2015).

To further examine the genetic changes that took place during and after the isolation of Wrangel Island, we generated high-quality mitogenomes from 21 woolly mammoths, including 14 Holocene Wrangel individuals (Fig. 1), and compared these with 21 previously published sequences. We used this data to test the hypothesis that the Wrangel Island population was established by a small number of individuals (i.e., a founder effect), as well as to characterize in situ evolution of neutral and functional mitochondrial variation in the small isolated Holocene population.

Methods

Fifty-one woolly mammoth specimens collected on Wrangel Island ($n = 35$) and the Siberian mainland ($n = 16$) were analyzed in this study. The samples consisted of bones, tusks, and teeth, and those with unknown age were radiocarbon dated using Acceleration Mass Spectrometry (AMS) in Oxford and Groningen (Table 1). The results are reported in conventional radiocarbon years (BP), which includes correction for isotopic fractionation and usage of the conventional half-life (Mook and van der Plicht 1999). The ^{14}C dates were calibrated into calendar ages using the recommended calibration curve IntCal13 (Reimer et al. 2013)

Table 1. Characteristics of mammoth samples used in this study, including the estimated haplotypes (W = Wrangel, S = Siberia).

Lab ID	¹⁴C Lab no.	¹⁴C BP date ± error	Median CalBP	Material	Location	Endogenous DNA (%)	Average fragment length (bp)	Accession no.	Reference	Average Coverage	Haplotype
E469D	Ua-13366	3685 ± 60	4024	Tooth	Wrangel Island	29.59	64.6	MG334270	This study	9.5	W7
E468	LU-2741	3730 ± 40	4079	Tusk	Wrangel Island	16.23	101.5	MG334269	This study	31.9	W1
E467	AA40665	3905 ± 47	4336	Tooth	Wrangel Island	80.35	81.9	MG334268	This study	11.7	W1
E466	GIN-6985	3920 ± 40	4354	Tusk	Wrangel Island	17.68	88.9	MG334267	This study	8.8	W6
E465	LU-4448	4120 ± 110	4643	Tusk	Wrangel Island	12.23	98.8	MG334266	This study	38.2	W5
E464	Ua-13375	4210 ± 70	4726	Tooth	Wrangel Island	70.55	84.2	MG334265	This study	21.8	W1
E460	LU-2756	4400 ± 40	4969	Tusk	Wrangel Island	87.13	90.1	MG334264	This study	39.0	W4
M28	GIN-6988	5610 ± 40	6380	Tusk	Wrangel Island	56.59	151.6	MG334281	This study	276.7	W4
L459	OxA-30117*	6148 ± 32	7060	Tusk	Wrangel Island	0.47	77.8	MG334276	This study	5.5	W4
M26	LU-2799	6260 ± 50	7194	Tooth	Wrangel Island	27.69	71.2	MG334280	This study	25.8	W3
L386	Ua-13374	6410 ± 90	7336	Tooth	Wrangel Island	82.37	73.0	MG334274	This study	12.0	W4
M23	LU-4449	6560 ± 60	7470	Tusk	Wrangel Island	75.85	111.5	MG334279	This study	170.9	W2
M17	Ua-13372	7510 ± 80	8318	Tooth	Wrangel Island	68.16	71.2	MG334278	This study	27.2	W1
L468	OxA-30122*	7711 ± 36	8491	Bone	Wrangel Island	15.30	70.8	MG334277	This study	7.8	W1
P011	GrA-65691*	10,240 ± 50	11972	Tusk	Taimyr Peninsula	2.11	66.6	MG334285	This study	5.5	S5
P005	GrA-65686*	10,920 ± 50	12775	Tusk	New Siberian Islands	91.70	75.7	MG334283	This study	10.8	S3
Ber28	UCIAMS38670	12,125 ± 30	14011	n.a.	Berelekh	n.a.	n.a.	KX027495	Enk et al. 2016		S24
Krause	KIA-25289	12,170 ± 50	14056	Bone	Yakutia	n.a.	n.a.	DQ188829	Krause et al. 2006		S22
L410	OxA-31180*	12,370 ± 55	14408	Tooth	Wrangel Island	29.31	80.0	MG334275	This study	6.8	W9
L158	OxA-20046	12,380 ± 45	14431	Humerus	Pioneyveem River, Chukotka	47.91	76.7	MG334272	This study	17.7	S1
P009	GrA-65689*	13,030 ± 60	15602	Tusk	New Siberian Islands	82.40	67	MG334284	This study	6.8	S4
L164	OxA-20048	13,935 ± 50	16901	tusk	Pioneyveem River, Chukotka	85.92	83.9	MG334273	This study	9.8	S2
GilbertM15	OxA-19605	13,995 ± 55	16996	Hair	Ayon Island, Chukotka	n.a.	n.a.	EU153446	Gilbert et al. 2008		S13
GilbertM18	OxA-17116	17,125 ± 70	20655	Hair	Gydan Peninsula	n.a.	n.a.	EU153447	Gilbert et al. 2007		S14
GilbertM4	OxA-17098	18,545 ± 70	22422	Hair	n.a.	n.a.	n.a.	EU153456	Gilbert et al. 2007		S9
GilbertM19	GrN-28258	18,560 ± 50	22434	Hair	Yakutia	n.a.	n.a.	EU153448	Gilbert et al. 2008		S9
GilbertM2	UtC-8138	20,380 ± 140	24507	Hair	Taimyr Peninsula	n.a.	n.a.	EU153449	Gilbert et al. 2007		S7
GilbertM3	Beta-148647	20,620 ± 70	24833	Hair	Taimyr Peninsula	n.a.	n.a.	EU153455	Gilbert et al. 2007		S8
GilbertM26	Beta-17114	24,740 ± 110	28769	Hair	Indigirka	n.a.	n.a.	EU153454	Gilbert et al. 2007		S19
Poinar	Beta-210777	27,740 ± 220	31501	Bone	Taimyr Peninsula	n.a.	n.a.	EU155210	Poinar et al. 2006		S20
GilbertM13	T-171	35,800 ± 1200	40418	Hair	Lena River	n.a.	n.a.	EU153445	Gilbert et al. 2007		S12
E470	LU-3511	37,080 ± 1650	41632	Bone	Wrangel Island	31.72	73.4	MG334271	This study	11.5	W8
Oimyakon	GrA-30727	41,300 ± 900	44828	Skin	Yakutia	n.a.	n.a.	MG334282	Palkopoulou et al. 2015		S23
GilbertM8	OxA-17102	46,900 ± 700	46962	Hair	Magadan	n.a.	n.a.	EU153458	Gilbert et al. 2007		S11
GilbertM22	OxA-17111	50,200 ± 900	50304	Hair	New Siberian Islands	n.a.	n.a.	EU153452	Gilbert et al. 2007		S17
GilbertM25	OxA-19610	59,300 ± 2700	60495	Hair	Yakutia	n.a.	n.a.	EU153453	Gilbert et al. 2008		S18
GilbertM1	n.a.	n.a.	n.a.	Hair	n.a.	n.a.	n.a.	EU153444	Gilbert et al. 2007		S6
GilbertM5	n.a.	n.a.	n.a.	Hair	n.a.	n.a.	n.a.	EU153457	Gilbert et al. 2007		S10
GilbertM20	OxA-19608	>63,500	n.a.	Hair	New Siberian Islands	n.a.	n.a.	EU153450	Gilbert et al. 2008		S15
GilbertM21	OxA-19609	>58,000	n.a.	Hair	New Siberian Islands	n.a.	n.a.	EU153451	Gilbert et al. 2008		S16
Rogaev	MAG-1000	33,750—31,950	n.a.	Muscle	Enmynveem River, Chukotka	n.a.	n.a.	DQ316067	Rogaev et al. 2006		S21
2002/472	UCIAMS38677	>48,800	n.a.	n.a.	Taimyr Peninsula	n.a.	n.a.	KX027489	Enk et al. 2016		S25

Asterisks indicate new radiocarbon dates; n.a. = not available.

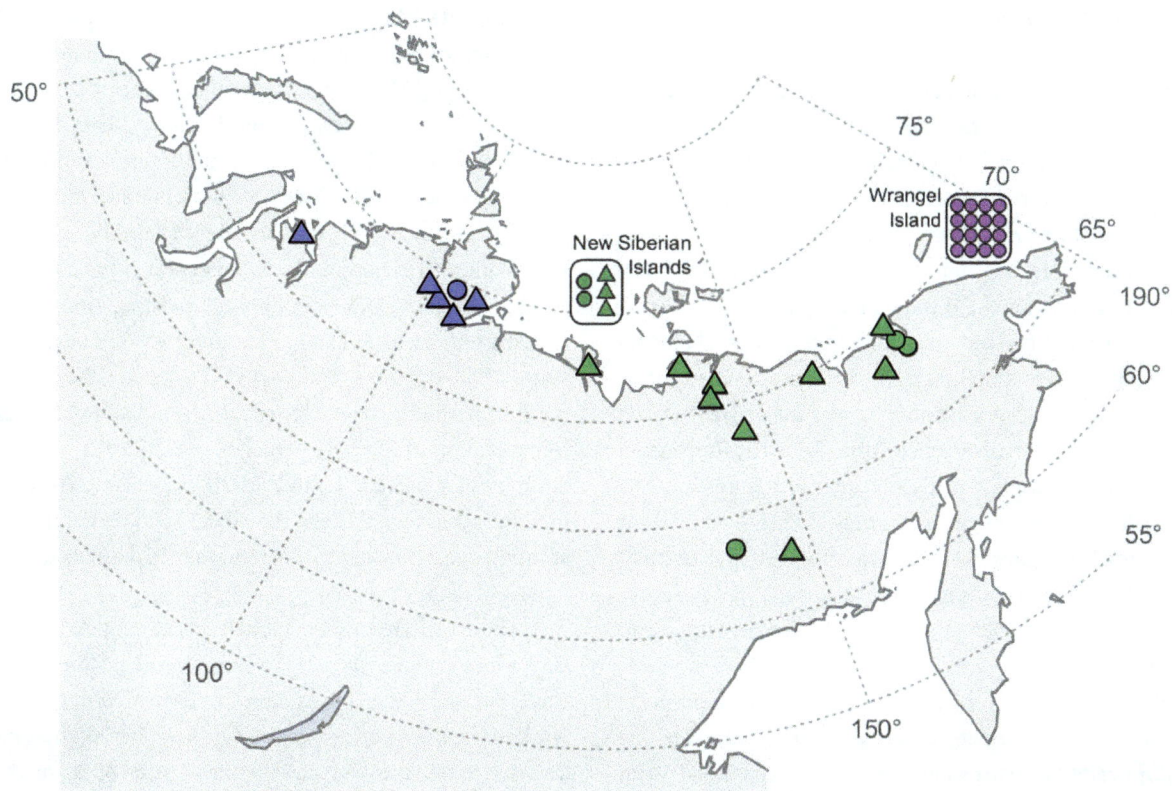

Figure 1. Map depicting the geographic origin of the samples, including 16 mitogenomes from Wrangel Island. Samples analyzed in this study are depicted as circles, while previously published samples are shown as triangles. Colors show the geographical classification used in this study: blue – Western Siberia, green – Central Siberia, and purple – Wrangel Island. The map was created using R (R Development Core Team 2016; available from https://www.R-project.org/).

using the program OxCal 4.2 (Ramsey 2009). Medians of the calibrated dates are reported in calBP, that is calendar years relative to 1950 AD.

DATA PREPARATION

Sample E469D was extracted using a method optimized for highly degraded samples (Dabney et al. 2013), while sequence data for samples labeled "E" come from (Palkopoulou et al. 2015), but with new consensus sequences generated in this study. Samples labeled "L," "M," and "P" were extracted according to protocol C in Yang et al. (1998) as modified in Brace et al. (2012). Double stranded Illumina libraries were prepared from 20 μL of DNA extract according to Meyer and Kircher (2010), using uracil-treatment with the USER enzyme (New England Biolabs; Briggs et al. 2010). During the blunt-end repair, USER enzyme was added so that the final concentration was 0.15 U/μL in the reaction mix described in "Step 4" of Meyer and Kircher (2010). T4 DNA polymerase was added to the reaction mix following a three-hour incubation at 37°C. Subsequently, blunt-end repair incubation and all following steps were performed according to the protocol by Meyer and Kircher (2010). Indexing amplifications were prepared with AccuPrime™ Pfx DNA Polymerase (Life

Technologies) using one indexing primer per library and the following amplification conditions: 95°C for 2 minutes and between 8 and 14 cycles of: 95°C for 15 seconds, 60°C for 30 seconds, 68°C for 30 seconds. Libraries were purified along with size selection using Agencourt AMPure XP beads (Beckman Coulter) targeting fragments between 100 and 500 base pairs to remove unligated adapters, primer dimers, and long contaminant sequences. Library concentrations were measured with a high-sensitivity DNA chip on a Bioanalyzer 2100 (Agilent). Multiplexed libraries were pooled in two separate pools in equimolar concentrations and shotgun-sequenced on two lanes of Illumina HiSeq2500 with a 2 × 125 bp setup in the HighOutput mode.

DATA PROCESSING

Bcl to Fastq conversion was performed using bcl2Fastq 1.8.3 from the CASAVA software suite. SeqPrep 1.1 (https://github.com/jstjohn/SeqPrep) was used to trim adapters and merge paired-end reads, using default settings and a minor modification to the source code, allowing us to choose the best quality scores of bases in the merged region instead of aggregating the scores (Palkopoulou et al. 2015). The modified file is available for download at the webpage www.palaeogenetics.com/adna/data.

Sequencing reads were processed with BWA 0.7.8 (Li and Durbin 2010) and SAMtools 0.1.19 (Li et al. 2009). Following Prufer et al. (2014), we modified the woolly mammoth mitochondrial genome (GenBank accession no. DQ188829) by copying the first 240 bp to the end of the sequence to facilitate mapping and to avoid lower coverage in the marginal parts of the sequence. This modified mitochondrial genome was merged with the African savanna elephant nuclear genome (LoxAfr4) generated by the Broad Institute, and the merged sequence was used as a reference to avoid mapping nuclear copies of mitochondrial DNA (numts) to the mitochondrial DNA reference genome (note, however, that this would not identify numts that have evolved in mammoths since their divergence from the African savanna elephant). Merged sequencing reads were mapped against the reference using the BWA aln algorithm with parameters adapted for ancient DNA reads that deactivate seeding (-l 16500), allow more substitutions (-n 0.01) and up to two gaps (-o 2). BWA samse command was used to generate alignments. Reads mapping to the mitochondrial genome were extracted and processed in SAMtools 0.1.19 (Li et al. 2009), including converting the alignments in SAM format to BAM format, coordinate sorting, indexing, and removing duplicates (with the single-end option "-s"). Reads with mapping qualities below 20 were filtered out.

BAM files generated using SAMtools were uploaded to Geneious® 7.0.3 (Kearse et al. 2012) and consensus sequences were called for positions with at least 3X coverage using the majority rule, with ambiguous and low-coverage positions called as undetermined.

To avoid incorrect consensus calling due to DNA damage characteristic for ancient samples, two steps were taken: (a) USER (Uracil-Specific Excision Reagent) Enzyme was used to remove damaged sites, and (b) only positions covered by at least three bases, that is three individual replicates, were called.

Twenty-one samples with consensus sequences resolved for at least 80% of positions of the mitogenome were aligned to 21 previously published mammoth mitogenomes (Table 1) in MAFFT 7.245 (Katoh and Standley 2013). The variable number tandem repeats (VNTR) section of the alignment was removed since it had not been assembled in the previously published data (Gilbert et al. 2007; Gilbert et al. 2008).

To verify that there was no bias introduced by using a clade I mammoth (Krause; DQ188829) as a reference, the data was also mapped against a clade II mammoth mitogenome (GilbertM25; EU153453). Despite only using a mitochondrial reference (rather than a merged nuclear-mitochondrial reference as in the original processing) and omitting the "E" samples (Palkopoulou et al. 2015) from the analyses, the results were consistent and the haplotype network maintained the same structure, including the star-like pattern of the Holocene haplotypes (Fig. S1).

DEMOGRAPHIC AND PHYLOGENETIC ANALYSES

A median-joining haplotype network of all mitogenomes was created in PopART (available at http://popart.otago.ac.nz). Sequences of mammoth mitochondrial clade I with finite radiocarbon dates were analyzed by Bayesian Inference in BEAST 1.8.0 (Drummond et al. 2012) using tip-dating and the HKY+I substitution model, which was selected according to the Bayesian Information Criterion in ModelGenerator (Keane et al. 2006). A strict molecular clock was applied and mutation rate was estimated from the data, using a starting rate of 8.07×10^{-8} site^{-1} year^{-1} (Palkopoulou et al. 2013). To test for the extent of temporal signal in the data, we performed a date-randomization test (Palkopoulou et al. 2013; Duchene et al. 2015). We used Site-Sampler v1.1 (Ho and Lanfear 2010) to generate 20 data sets with randomly reassigned labels (i.e., dates) and we compared the substitution rate estimated by the randomized datasets to the estimate from real data.

Three different tree models were tested: constant size, Bayesian Skyline, and Bayesian Skyride. Constant size and Bayesian Skyride analyses were performed with default settings while the number of groups in the Bayesian Skyline model was adjusted to five to avoid overparametrization of the model (Drummond et al. 2005). To decide which model provides the best fit, we calculated the marginal likelihoods using path and stepping-stone sampling as implemented in BEAST 1.8.0 (Baele et al. 2012; Baele et al. 2013). Bayes Factors were estimated using the marginal likelihoods, and we used the approach by Kass and Raftery (1995) to select the most appropriate model for further analyses (Table S1).

For all models, the Markov Chain Monte Carlo was set to run for 50 million generations, sampling every 5000th generation. Information from the sampled trees was summarized in TreeAnnotator. Tracer 1.6 (Rambaut et al. 2014) was used to compare the tested tree models, to verify convergence of the runs and to perform Bayesian Skyride reconstruction (using the default settings) estimating the female effective population size (N_{ef}). The output tree was visualized in FigTree 1.4.2 (available at http://tree.bio.ed.ac.uk/software/figtree/). Both the haplotype network and the phylogenetic tree were graphically edited in Inkscape 0.91 (available at https://inkscape.org/en/).

MUTATION RATE

We took advantage of the known evolutionary history and the star-like pattern of haplotypes to estimate the mutation rate in the Wrangel Island samples. Simulations were performed using *fastsimcoal ver. 2.5.2.2* (Excoffier et al. 2013) and the number of haplotypes was estimated with *arlsumstats ver. 3.5.2* (Excoffier and Lischer 2010), both controlled by custom R scripts (R Development Core Team 2013). We assumed that the population on

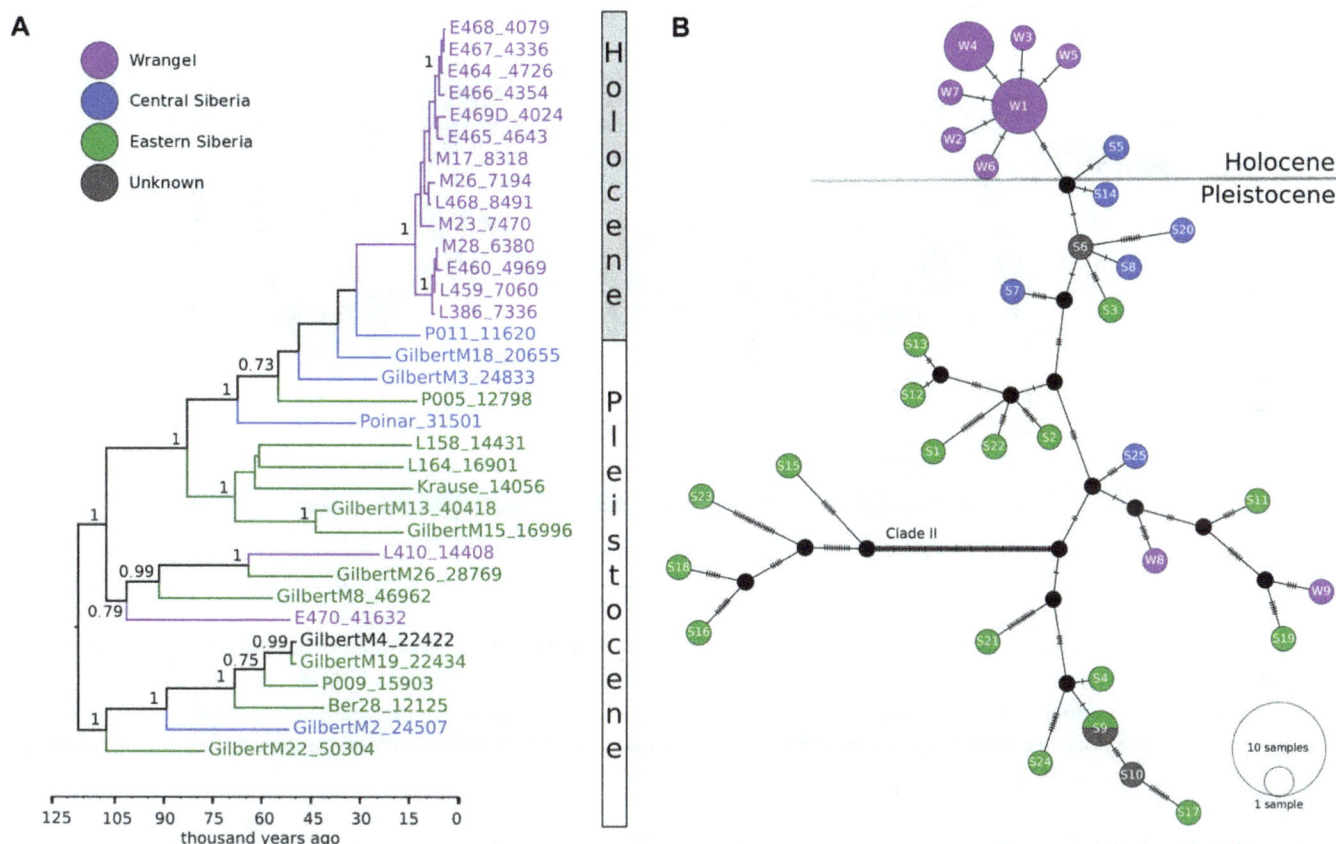

Figure 2. A Bayesian Phylogeny of clade I mammoths with finite radiocarbon dates (A) and median-joining haplotype network of all 42 Pleistocene and Holocene mammoths (B; W = Wrangel, S = Siberia). In the phylogeny, nodes with posterior probabilities above 0.7 are shown and numbers in the sample names indicate age.

Wrangel Island can be modeled as an isolated and continuous population from the time of the bottleneck (~12 kyr calBP) to the time of extinction (~4 kyr calBP). We performed coalescent simulations on a 100×100 grid composed of values from a range of constant female effective population sizes (N_{ef}: $1 - 10,000,000$ individuals) and mutation rates (μ: $0.07 - 667 \times 10^{-8}$ site^{-1} year^{-1}, corresponding to a range $0.1 - 1000 \times 10^{-7}$ site^{-1} generation^{-1}) both on an equally spaced log-scale. For each combination of parameters we performed 1000 simulations of a DNA fragment of the same size as our alignment (16,506 bp), forcing all lineages to coalesce into one at the time of isolation so that only one haplotype would be present in the founder population of the island. In each simulation, we sampled individuals at the same time as the mean calibrated age of our samples (Table 1) and the probability of observing exactly seven haplotypes (as inferred from the haplotype network; see the Results) in the simulated Wrangel samples was reported. The female generation time was set to 15 years (Palkopoulou et al. 2013). Coalescent simulations were run both assuming no transition bias (0.33), and a high transition bias (0.98, as in Palkopoulou et al. 2013) with no marked effect in the estimated mutation rates.

Results

MITOCHONDRIAL HAPLOTYPES

Across the 16,506 base pairs (bp) of the mitochondrial genome, mammoths in our study were assigned to 34 unique haplotypes. The haplotype median-joining network (Fig. 2B) clearly differentiated mammoths belonging to clades I and II, with Wrangel Island mammoths nested within clade I as shown previously (Nyström et al. 2010; Palkopoulou et al. 2013).

In the Holocene Wrangel population, we found seven mitochondrial haplotypes that formed a unique subgroup shaped in a star-like pattern with multiple haplotypes surrounding what was presumably a single ancestral haplotype (W1, Fig. 2B). These seven haplotypes were not closely related to the haplotypes observed in Wrangel mammoths radiocarbon dated to the Late Pleistocene (E470, L410), nor did they show a close affinity to the end-Pleistocene haplotypes from Chukotka, which is the geographically closest mainland region. Instead, the Holocene Wrangel Island haplotypes were most closely related to haplotypes observed in mammoths from Central Siberia (Fig. 2B; Table 1). The latter were also basal to the Holocene Wrangel mammoths in the phylogeny (Fig. 2A).

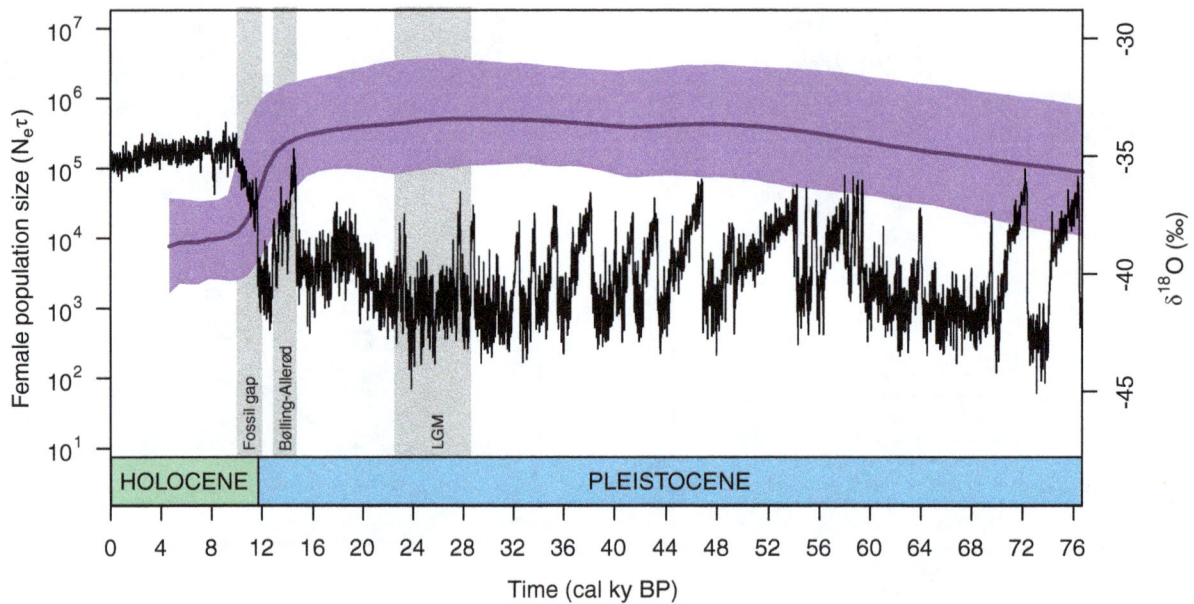

Figure 3. Bayesian Skyride plot of the female effective population size (N_{ef}) based on 34 clade I mammoth samples with finite radiocarbon dates. The left *y*-axis and the solid purple line represent median values of N_{ef} with the purple area indicating the 95% highest posterior density. The *x*-axis is in calendar years before present. N_{ef} is scaled by a factor of generation time assumed to be 15 years. The right *y*-axis and the black line show the climate record from the North Greenland ice core (North Greenland Ice Core Project members 2004).

The Holocene Wrangel mammoths differed from all other mammoths by three unique mutations: one mutation in a region coding for transfer RNA (tRNA) valine and two mutations in protein-coding genes, a synonymous mutation in *ND6* and a non-synonymous mutation in *ATP6* (G457A).

Bayesian phylogenies and effective population size

The phylogeny inferred using Bayesian Skyride model (Fig. 2A) indicated a differentiation among Late Pleistocene and Holocene mammoths. While the Late Pleistocene mammoths generally formed long and well-supported branches (posterior probability ≥0.95), Holocene Wrangel mammoths clustered into a monophyletic group with unresolved internal relationships and short branches, indicative of a rapid diversification.

The Bayesian Skyride plot (Fig. 3) revealed a sharp decrease in effective population size starting at about 15 kyr calBP, which coincides with the beginning of the Bølling-Allerød interstadial (14.7–12.9 kyr calBP). The steep decline continued through the Pleistocene/Holocene boundary until approximately 9 kyr calBP. Assuming an average generation time of 15 years (as in Palkopoulou et al. 2013), N_{ef} dropped 15-fold from ca 26,000 individuals (95% highest posterior density (HPD): 105,000–3500) prior to Bølling-Allerød (14.7 kyr calBP) to ca 1700 individuals (95% HPD: 9000–160) in the early Holocene (10 kyr calBP).

Mutation rate

The mutation rate of the mitogenomes estimated by tip calibration (Table S2) with a Skyride tree prior yielded a mean rate of 1.31×10^{-8} site^{-1} year^{-1} (95% HPD: 0.79–1.84×10^{-8} site^{-1} year^{-1}), whereas using a constant size tree prior yielded a mean rate of 1.36×10^{-8} site^{-1} year^{-1} (95% HPD: 0.83 – 1.93×10^{-8} site^{-1} year^{-1}). The rate estimated by Bayesian Skyline analyses was 1.22×10^{-8} site^{-1} year^{-1} (95% HPD: 0.66 – 1.75×10^{-8}). The extent of temporal signal in the data was validated by the date-randomization test, which showed that the 95% HPD of the substitution rate estimated from 20 randomized datasets (2.41×10^{-14} – 0.5×10^{-8}) is outside the 95% HPD of the estimate from real data (0.79 – 1.84×10^{-8}).

Coalescent simulations of the mutation rate of the Holocene Wrangel population suggested that prior knowledge about the population size is paramount for the estimation of the mutation rate. As depicted in Fig. 4A, the mutation rate estimate is stable and a little lower than the estimates obtained from BEAST for effective population sizes over 5000 females ($\sim$0.9 $\times 10^{-8}$ site^{-1} year^{-1}; Table S2), but it changes rapidly when lower effective population sizes are assumed. However, an effective size larger than 5000 females is rather unlikely for the Wrangel Island population, which was recently estimated to around 328 individuals for both sexes using genomic data (Palkopoulou et al. 2015). Moreover, it is unlikely that the effective population size exceed the carrying capacity of Wrangel Island, which has been estimated

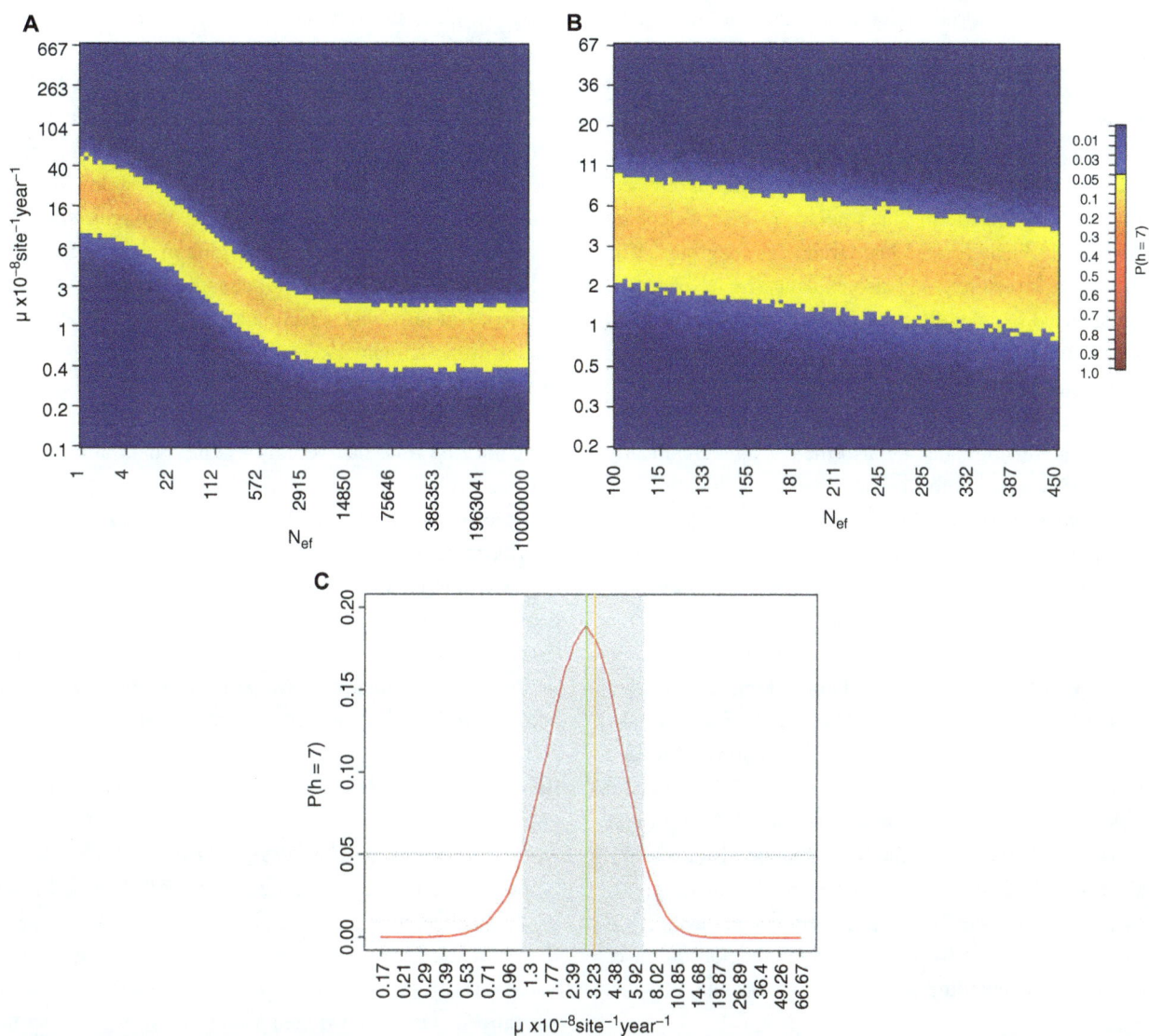

Figure 4. Probability of observing seven mitochondrial haplotypes in 14 Holocene Wrangel samples estimated from coalescent simulations using the mutation rate (μ) and the female effective population size (N_{ef}) as exploratory parameters: (A) using wide priors, μ: $0.07–667 \times 10^{-8}$ site^{-1} year^{-1} and N_{ef}: 1–10,000,000 individuals; (B) using narrower priors better fitting the scenario of low effective population size on Wrangel Island (N_{ef}: 100–450 individuals, and μ: $0.17–66.7 \times 10^{-8}$ site^{-1} year^{-1}); (C) summary of probabilities from simulations with narrower priors and incorporating all values of N_{ef} to account for uncertainty. The green line indicates the mutation rate corresponding with the highest probability. The orange line represents the estimated mutation rate using 328 individuals as effective population size as in (Palkopoulou et al. 2015) and assuming a 1:1 sex ratio (Nyström et al. 2010).

to between 149 and 819 individuals using Damuth's equation (Nyström et al. 2010). To more precisely estimate the mutation rate, assuming more realistic Wrangel population sizes and an 1:1 sex ratio (Nyström et al. 2010), we performed a second set of simulations with much narrower priors (N_{ef}: 100–450 individuals, and μ: $0.17 – 66.7 \times 10^{-8}$ site^{-1} year^{-1}, Fig. 4B). The results from this second set of simulations indicated that the probability of observing seven Holocene Wrangel haplotypes was maximized when the mutation rate was 3.05×10^{-8} site^{-1} year^{-1} (95% HPD: $1.31 – 6.69 \times 10^{-8}$ site^{-1} year^{-1}; Fig. 4C).

Discussion

THE ORIGINS AND DEMOGRAPHY OF WRANGEL MAMMOTHS

The haplotype network showed signs of a highly diverse mammoth population throughout the Pleistocene. However, our results suggest that immediately after the isolation on Wrangel Island only one haplotype was present, based on the star-like pattern of the haplotype network where all haplotypes are only one or two mutational steps from the modal haplotype. This conclusion is further supported by the results reported by Nyström et al. (2010)

where only one mitochondrial haplotype was observed during the first 1500 years following the isolation of Wrangel Island. Reduction to a single maternal lineage is in agreement with mammoths going through a bottleneck and founder effect as they became isolated on Wrangel Island. The Wrangel Island's mammoth population was thus likely founded by a very limited number of females, presumably one herd considering that similarly to present elephant species mammoths likely formed matriarchal family groups. The matriarchal social structure could also explain why we observed a more pronounced founder effect using mitogenomes compared to previous results from nuclear data (Palkopoulou et al. 2015; Rogers and Slatkin 2017).

Interestingly, the samples genetically most similar to the Holocene Wrangel mammoths are also the ones geographically most distant, originating from Central Siberia, specifically from the Gydan Peninsula (GilbertM18), Taimyr Peninsula (P011, GilbertM3, Poinar), and New Siberian Islands (P005). The area comprising the Taimyr and Gydan Peninsulas is possibly the only region in Siberia with a continuous fossil record throughout the warm Bølling-Allerød Interstadial when mammoths disappeared from other parts of Eurasia (Stuart 2005). During the Younger Dryas (YD; 12.9–11.7 kyr calBP), a colder period following the Bølling-Allerød Interstadial, cold climate conditions allowed a reexpansion of the open steppe tundra habitat, which is thought to have enabled a reexpansion of the woolly mammoth from Taimyr to northwestern Siberia and northeastern Europe (Stuart 2005). Although speculative, these are the first genetic results supporting the hypothesis that the Taimyr Peninsula could have served as a source population for a Younger Dryas recolonization of northeastern Siberia, including what was later to become Wrangel Island.

Post-LGM population size reduction and accelerated rate of evolution

The estimated female effective population size was rather stable during the Pleistocene and only started to decrease about 15 kyr calBP, at roughly the same time as the onset of the Bølling-Allerød warming period and the time when mammoths disappeared from most of the Siberian mainland (Fig. 3). This decline in N_{ef} was approximately 15-fold, which is comparable to the reduction observed in a previous study based on short mitochondrial sequences (Palkopoulou et al. 2013). The Bayesian Skyride analysis also indicated that the decline in effective population size continued until the final extinction, but at a considerably lower rate.

The coalescent simulations suggested a two- to three-fold higher mutation rate in Holocene Wrangel Island samples (3.05×10^{-8} site^{-1} year^{-1}; 95% HPD: 1.31–6.69×10^{-8} site^{-1} year^{-1}), as compared to the estimates for all clade I mammoths from BEAST (1.31×10^{-8} site^{-1} year^{-1}; 95% HPD: 0.79–1.84×10^{-8} site^{-1} year^{-1}; Table S2). Although the HPDs

are wide, these substitution rate estimates are lower than previously published rates based on a short hypervariable fragment of 741 bp (Barnes et al. 2007; Debruyne et al. 2008; Palkopoulou et al. 2013), but clearly higher than any other mutation rate published for mammoth or other proboscidean complete mitogenomes (Rohland et al. 2007). One possible explanation is that the higher substitution rate in the Wrangel Island population is associated with reduced purging of deleterious or slightly deleterious variants in the mitochondrial genome due to lower efficiency of purifying selection at long-term low effective population sizes (Kimura 1957). This process potentially results in a higher than expected amount of polymorphism, and consequently an increased measurable rate of evolution. This is to our knowledge the first time that serially sampled data are used to show that the evolutionary rate changes in a species through time, as the species' population size decreases. These results consequently provide tentative support for expectations from the nearly neutral theory of molecular evolution (Ohta 1992). It should, however, be noted that a somewhat higher mutation rate could also result from a reduction in generation times in Wrangel Island mammoths as a consequence of insularity, as hypothesized by Rogers and Slatkin (2017). However, we find it unlikely that shorter generations would lead to an increase in evolutionary rate as high as two- to threefold (Nabholz et al. 2008; Bromham 2009).

Fixed mutations in the Wrangel population

Although the Holocene Wrangel mammoths formed a separate cluster genetically differentiated from the Pleistocene specimens, the Wrangel population was defined by only three synapomorphic mutations across the whole mitochondrial genome. Interestingly, the third mutation constituted a nonsynonymous substitution in the gene *ATP6* (G457A) encoding for subunit *a* of the ATP synthase enzyme, resulting in an alanine to threonine substitution at amino acid 157 (A157T), which had a derived state in all 14 Holocene Wrangel mitogenomes (Fig. S2). Mitochondrial ATP synthase is a key enzyme of the oxidative phosphorylation pathway and is responsible for ATP production in all living beings except for archaea (Vantourout et al. 2010). To assess genetic variation at the *ATP6* gene, we aligned the mammoth *ATP6* sequence with those of other taxa, randomly choosing one GenBank (Benson et al. 2013) sequence per species (Fig. S3) to stochastically capture variability in the dataset. We observed that the G457A mutation was located in a conserved part of the *ATP6* sequence and that no other species had a nonsynonymous mutation at that site (Fig. S3).

Assessing the potential fitness consequences of the A157T amino acid fixation is aided by the highly conserved structure of ATP6 across taxa. First, the location of this substitution is near an active site residue at position 159. Second, in humans, two different substitutions at the neighboring site 156 give rise to disease phenotypes. Both the L156R and L156P substitutions affect ATP synthase, resulting in a decreased proton flux and thereby decreased ATP synthase (Jonckheere et al. 2012). If we assume that these neighboring substitutions are a good proxy for the phenotypic effect of A157T, a reduced ATP synthase phenotype is possible (Cortes-Hernandez et al. 2007).

One of the key predictions from the mutational meltdown theory (Lynch et al. 1995) is that deleterious mutations can become fixed in small populations due to strong genetic drift. Although an excess of deleterious mutations has previously been described in genome from a 4300-year-old Wrangel mammoth (Rogers and Slatkin 2017), our results from ATP6 might represent the first identification of a deleterious mutation that appears to have become fixed as a consequence of genetic drift during the establishment of the Wrangel Island population.

Rather than being the consequence of small effective population size for hundreds of generations (Rogers and Slatkin 2017), the excess of deleterious mutations in the Wrangel Island mammoth genome could be due to a severe bottleneck and founder effect at the time of the isolation, which may have also contributed to an increase in detrimental mutations immediately after the establishment of the population, as suggested by our results. Consequently, analyses of serially sampled genomic data from additional Wrangel Island mammoths are needed to resolve to what extent genetic drift during the Holocene led to a gradual accumulation of deleterious genetic variation.

AUTHOR CONTRIBUTIONS

L.D. and P.P. designed the study; S.V., A.T., P.N., and J.vd.P. provided samples and conducted radiocarbon dating; P.P. and E.P. performed lab work; P.P., D.D., E.P., and P.S. performed data analysis; P.P., D.D., C.W.W., and L.D. drafted the manuscript.

ACKNOWLEDGMENTS

The authors are grateful to Veronica Nyström Edmark for assistance with sampling and to Swapan Mallick for the modification of the SeqPrep source code. The authors would like to acknowledge support from Science for Life Laboratory, the National Genomics Infrastructure, and UPPMAX (project number: b2015028) for providing assistance in massive parallel sequencing and computational infrastructure. The authors would also like to thank Estelle Proux-Wéra from the Science for Life Laboratory for bioinformatics advice. The genetic analyses were funded through a grant from the Swedish Research Council (VR grant 2012–3869). P.S. was supported by the Swedish Research Council (VR grant 2014–453).

LITERATURE CITED

Aitken, S. N., S. Yeaman, J. A. Holliday, T. L. Wang, and S. Curtis-McLane. 2008. Adaptation, migration or extirpation: climate change outcomes for tree populations. Evol. Appl. 1:95–111.

Arppe, L., J. A. Karhu, and S. L. Vartanyan. 2009. Bioapatite Sr-87/Sr-86 of the last woolly mammoths—implications for the isolation of Wrangel Island. Geology 37:347–350.

Baele, G., P. Lemey, T. Bedford, A. Rambaut, M. A. Suchard, and A. V. Alekseyenko. 2012. Improving the accuracy of demographic and molecular clock model comparison while accommodating phylogenetic uncertainty. Mol. Biol. Evol. 29:2157–2167.

Baele, G., W. L. S. Li, A. J. Drummond, M. A. Suchard, and P. Lemey. 2013. Accurate model selection of relaxed molecular clocks in Bayesian phylogenetics. Mol. Biol. Evol. 30:239–243.

Barnes, I., B. Shapiro, A. Lister, T. Kuznetsova, A. Sher, D. Guthrie, and M. G. Thomas. 2007. Genetic structure and extinction of the woolly mammoth, *Mammuthus primigenius*. Curr. Biol. 17:1072–1075.

Benson, D. A., K. Clark, I. Karsch-Mizrachi, D. J. Lipman, J. Ostell, and E. W. Sayers. 2013. GenBank. Nucleic Acids Res. 41(Database issue):D36–D42.

Brace, S., E. Palkopoulou, L. Dalén, A. M. Lister, R. Miller, M. Otte, M. Germonpré, S. P. E. Blockley, J. R. Stewart, and I. Barnes. 2012. Serial population extinctions in a small mammal indicate Late Pleistocene ecosystem instability. Proc. Natl. Acad. Sci. USA 109:20532–20536.

Briggs, A. W., U. Stenzel, M. Meyer, J. Krause, M. Kircher, and S. Pääbo. 2010. Removal of deaminated cytosines and detection of in vivo methylation in ancient DNA. Nucleic Acids Res 38:e87.

Bromham, L. 2009. Why do species vary in their rate of molecular evolution? Biol. Lett. 5:401–404.

Cortes-Hernandez, P., M. E. Vazquez-Memije, and J. J. Garcia. 2007. ATP6 homoplasmic mutations inhibit and destabilize the human F1F0-ATP synthase without preventing enzyme assembly and oligomerization. J. Biol. Chem. 282:1051–1058.

Dabney, J., M. Knapp, I. Glocke, M. T. Gansauge, A. Weihmann, B. Nickel, C. Valdiosera, N. Garcia, S. Pääbo, J. L. Arsuaga, et al. 2013. Complete mitochondrial genome sequence of a Middle Pleistocene cave bear reconstructed from ultrashort DNA fragments. Proc. Natl. Acad. Sci. USA 110:15758–15763.

Davis, M. B., and R. G. Shaw. 2001. Range shifts and adaptive responses to Quaternary climate change. Science 292:673–679.

Debruyne, R., G. Chu, C. E. King, K. Bos, M. Kuch, C. Schwarz, P. Szpak, D. R. Gröcke, P. Matheus, G. Zazula, et al. 2008. Out of America: ancient DNA evidence for a new world origin of late quaternary woolly mammoths. Curr. Biol. 18:1320–1326.

Drummond, A. J., A. Rambaut, B. Shapiro, and O. G. Pybus. 2005. Bayesian coalescent inference of past population dynamics from molecular sequences. Mol. Biol. Evol. 22:1185–1192.

Drummond, A. J., M. A. Suchard, D. Xie, and A. Rambaut. 2012. Bayesian phylogenetics with BEAUti and the BEAST 1.7. Mol. Biol. Evol. 29:1969–1973.

Duchene, S., D. Duchene, E. C. Holmes, and S. Y. W. Ho. 2015. The performance of the date-randomization test in phylogenetic analyses of time-structured virus data. Mol. Biol. Evol. 32:1895–1906.

Enk, J., A. Devault, C. Widga, J. Saunders, P. Szpak, J. Southon, J.-M. Rouillard, B. Shapiro, G. B. Golding, G. Zazula, et al. 2016. Mammuthus population dynamics in Late Pleistocene North America: divergence, phylogeography, and introgression (original research). Frontiers Ecol. Evol. 4:42.

Excoffier, L., I. Dupanloup, E. Huerta-Sánchez, V. C. Sousa, and M. Foll. 2013. Robust demographic inference from genomic and SNP data. PLoS Genet. 9:e1003905.

Excoffier, L., and H. E. L. Lischer. 2010. Arlequin suite ver 3.5: a new series of programs to perform population genetics analyses under Linux and Windows. Mol. Ecol. Resour. 10:564–567.

Gilbert, M. T. P., D. I. Drautz, A. M. Lesk, S. Y. W. Ho, J. Qi, A. Ratan, C. H. Hsu, A. Sher, L. Dalén, A. Götherström, et al. 2008. Intraspecific phylogenetic analysis of Siberian woolly mammoths using complete mitochondrial genomes. Proc. Natl. Acad. Sci. USA 105:8327–8332.

Gilbert, M. T. P., L. P. Tomsho, S. Rendulic, M. Packard, D. I. Drautz, A. Sher, A. Tikhonov, L. Dalen, T. Kuznetsova, P. Kosintsev, et al. 2007. Whole-genome shotgun sequencing of mitochondria from ancient hair shafts. Science 317:1927–1930.

Ho, S. Y. W., and R. Lanfear. 2010. Improved characterisation of among-lineage rate variation in cetacean mitogenomes using codon-partitioned relaxed clocks. Mitochondr. DNA 21:138–146.

Jonckheere, A. I., J. A. M. Smeitink, and R. J. T. Rodenburg. 2012. Mitochondrial ATP synthase: architecture, function and pathology (journal article). J. Inherited Metab. Dis. 35:211–225.

Kass, R. E., and A. E. Raftery. 1995. Bayes factors. J. Am. Stat. Assoc. 90:773–795.

Katoh, K., and D. M. Standley. 2013. MAFFT multiple sequence alignment software version 7: improvements in performance and usability. Mol Biol Evol 30:772–780.

Keane, T. M., C. J. Creevey, M. M. Pentony, T. J. Naughton, and J. O. McInerney. 2006. Assessment of methods for amino acid matrix selection and their use on empirical data shows that ad hoc assumptions for choice of matrix are not justified. BMC Evol. Biol. 6:29.

Kearse, M., R. Moir, A. Wilson, S. Stones-Havas, M. Cheung, S. Sturrock, S. Buxton, A. Cooper, S. Markowitz, C. Duran, et al. 2012. Geneious basic: an integrated and extendable desktop software platform for the organization and analysis of sequence data. Bioinformatics 28:1647–1649.

Kimura, M. 1957. Some problems of stochastic-processes in genetics. Ann. Math. Stat. 28:882–901.

Krause, J., P. H. Dear, J. L. Pollack, M. Slatkin, H. Spriggs, I. Barnes, A. M. Lister, I. Ebersberger, S. Pääbo, and M. Hofreiter 2006. Multiplex amplification of the mammoth mitochondrial genome and the evolution of Elephantidae. Nature 439:724–727.

Li, H., and R. Durbin. 2010. Fast and accurate long-read alignment with Burrows-Wheeler transform. Bioinformatics 26:589–595.

Li, H., B. Handsaker, A. Wysoker, T. Fennell, J. Ruan, N. Homer, G. Marth, G. Abecasis, R. Durbin, and G. P. D. Proc. 2009. The sequence alignment/map format and SAMtools. Bioinformatics 25:2078–2079.

Lister, A. M., and A. V. Sher. 2001. The origin and evolution of the woolly mammoth. Science 294:1094–1097.

Lynch, M., J. Conery, and R. Burger. 1995. Mutation accumulation and the extinction of small populations. Am. Nat. 146:489–518.

Meyer, M., and M. Kircher. 2010. Illumina sequencing library preparation for highly multiplexed target capture and sequencing (Research Support, Non-U.S. Gov't). Cold Spring Harbor Protocols 2010: pdb prot5448.

Mook, W. G., and J. van der Plicht. 1999. Reporting C-14 activities and concentrations. Radiocarbon 41:227–239.

Nabholz, B., S. Glemin, and N. Galtier. 2008. Strong variations of mitochondrial mutation rate across mammals—the longevity hypothesis. Mol. Biol. Evol. 25:120–130.

Nyström, V., L. Dalén, S. Vartanyan, K. Lidén, N. Ryman, and A. Angerbjörn. 2010. Temporal genetic change in the last remaining population of woolly mammoth. P Roy Soc. B Biol. Sci. 277:2331–2337.

Nyström, V., J. Humphrey, P. Skoglund, N. J. McKeown, S. Vartanyan, P. W. Shaw, K. Lidén, M. Jakobsson, I. Barnes, A. Angerbjörn, et al. 2012. Microsatellite genotyping reveals end-Pleistocene decline in mammoth autosomal genetic variation. Mol. Ecol. 21:3391–3402.

Ohta, T. 1992. The nearly neutral theory of molecular evolution. Annu. Rev. Ecol. Syst. 23:263–286.

Palkopoulou, E., L. Dalén, A. M. Lister, S. Vartanyan, M. Sablin, A. Sher, V. N. Edmark, M. D. Brandstrom, M. Germonpre, I. Barnes, et al. 2013. Holarctic genetic structure and range dynamics in the woolly mammoth. P Roy Soc. B Biol. Sci. 280:20131910.

Palkopoulou, E., S. Mallick, P. Skoglund, J. Enk, N. Rohland, H. Li, A. Omrak, S. Vartanyan, H. Poinar, A. Götherstrom, et al. 2015. Complete genomes reveal signatures of demographic and genetic declines in the woolly mammoth. Curr. Biol. 25:1395–1400.

Poinar, H. N., C. Schwarz, J. Qi, B. Shapiro, R. D. E. MacPhee, B. Buigues, A. Tikhonov, D. H. Huson, L. P. Tomsho, A. Auch, et al. 2006. Metagenomics to paleogenomics: large-scale sequencing of mammoth DNA. Science 311:392–394

Prufer, K., F. Racimo, N. Patterson, F. Jay, S. Sankararaman, S. Sawyer, A. Heinze, G. Renaud, P. H. Sudmant, C. de Filippo, et al. 2014. The complete genome sequence of a Neanderthal from the Altai Mountains. Nature 505:43-+.

Rambaut, A., M. A. Suchard, D. Xie, and A. J. Drummond. 2014 'Tracer' 1.6. Available at: http://beast.bio.ed.ac.uk/Tracer.

Ramsey, C. B. 2009. Bayesian Analysis of Radiocarbon Dates. Radiocarbon 51:337–360.

Reimer, P. J., E. Bard, A. Bayliss, J. W. Beck, P. G. Blackwell, C. B. Ramsey, C. E. Buck, H. Cheng, R. L. Edwards, M. Friedrich, et al. 2013. Intcal13 and marine13 radiocarbon age calibration curves 0–50,000 years Cal Bp. Radiocarbon 55:1869–1887.

Rogaev, E. I., Y. K. Moliaka, B. A. Malyarchuk, F. A. Kondrashov, M. V. Derenko, I. Chumakov, and A. P. Grigorenko. 2006. Complete mitochondrial genome and phylogeny of Pleistocene mammoth Mammuthus primigenius. PLoS Biol. 4:403–410.

Rogers, R. L., and M. Slatkin. 2017. Excess of genomic defects in a woolly mammoth on Wrangel island. PLoS Genet. 13:e1006601.

Rohland, N., A. S. Malaspinas, J. L. Pollack, M. Slatkin, P. Matheus, and M. Hofreiter. 2007. Proboscidean mitogenomics: chronology and mode of elephant evolution using mastodon as outgroup. PLoS Biol. 5:1663–1671.

Stuart, A. J. 1991. Mammalian extinctions in the Late Pleistocene of Northern Eurasia and North-America. Biol. Rev. Camb. Philos. Soc. 66:453–562.

———. 2005. The extinction of woolly mammoth (Mammuthus primigenius) and straight-tusked elephant (Palaeoloxodon antiquus) in Europe. Quatern. Int. 126:171–177.

Stuart, A. J., L. D. Sulerzhitsky, L. A. Orlova, Y. V. Kuzmin, and A. M. Lister. 2002. The latest woolly mammoths (Mammuthus primigenius Blumenbach) in Europe and Asia: a review of the current evidence. Quatern. Sci. Rev. 21:1559–1569.

Vantourout, P., C. Radojkovic, L. Lichtenstein, V. Pons, E. Champagne, and L. O. Martinez. 2010. Ecto-F-1-ATPase: a moonlighting protein complex and an unexpected apoA-I receptor. World J. Gastroentero. 16:5925–5935.

Vartanyan, S. L., K. A. Arslanov, J. A. Karhu, G. Possnert, and L. D. Sulerzhitsky. 2008. Collection of radiocarbon dates on the mammoths (Mammuthus primigenius) and other genera of Wrangel Island, northeast Siberia, Russia. Quatern. Res. 70:51–59.

Vartanyan, S. L., K. A. Arslanov, T. V. Tertychnaya, and S. B. Chernov. 1995. Radiocarbon dating evidence for mammoths on Wrangel Island, Arctic-Ocean, until 2000-Bc. Radiocarbon 37:1–6.

Vartanyan, S. L., V. E. Garutt, and A. V. Sher. 1993. Holocene dwarf mammoths from Wrangel-Island in the Siberian Arctic. Nature 362:337–340.

Yang, D. Y., B. Eng, J. S. Waye, J. C. Dudar, and S. R. Saunders. 1998. Technical note: improved DNA extraction from ancient bones using silica-based spin columns. Am. J. Phys. Anthropol. 105:539–543.

17

Egg chemoattractants moderate intraspecific sperm competition

Rowan A. Lymbery,[1,2] W. Jason Kennington,[1] and Jonathan P. Evans[1]

[1]School of Biological Sciences, The University of Western Australia, Crawley, WA 6009, Australia

[2]E-mail: rowan.lymbery@research.uwa.edu.au

Interactions among eggs and sperm are often assumed to generate intraspecific variation in reproductive fitness, but the specific gamete-level mechanisms underlying competitive fertilization success remain elusive in most species. Sperm chemotaxis–the attraction of sperm by egg-derived chemicals—is a ubiquitous form of gamete signaling, occurring throughout the animal and plant kingdoms. The chemical cues released by eggs are known to act at the interspecific level (e.g., facilitating species recognition), but recent studies have suggested that they could have roles at the intraspecific level by moderating sperm competition. Here, we exploit the experimental tractability of a broadcast spawning marine invertebrate to test this putative mechanism of gamete-level sexual selection. We use a fluorescently labeled mitochondrial dye in mussels to track the real-time success of sperm as they compete to fertilize eggs, and provide the first direct evidence in any species that competitive fertilization success is moderated by differential sperm chemotaxis. Furthermore, our data are consistent with the idea that egg chemoattractants selectively attract ejaculates from genetically compatible males, based on relationships inferred from both nuclear and mitochondrial genetic markers. These findings for a species that exhibits the ancestral reproductive strategy of broadcast spawning have important implications for the numerous species that also rely on egg chemoattractants to attract sperm, including humans, and have potentially important implications for our understanding of the evolutionary cascade of sexual selection.

KEY WORDS: Gamete interactions, genetic compatibility, sexual selection, sperm chemotaxis, sperm competition.

Impact Summary

Gamete interactions are a critical component of competitive reproductive fitness. In many organisms, multiple mating (for internal fertilizers) or multi-individual spawning (for external fertilizers) lead to competition among ejaculates for fertilization and the opportunity for females (or eggs) to promote the success of preferred sperm. However, despite the pervasiveness of these forms of sexual selection, we know very little about the specific mechanisms of interaction among eggs and sperm that underlie such processes. One emerging putative mechanism is sperm chemotaxis, a taxonomically widespread phenomenon involving the attraction of sperm toward eggs by egg-derived chemicals. Here, we exploit the experimental versatility of a broadcast spawning mussel to provide the first empirical evidence that differential sperm chemotaxis allows females to bias the outcomes of intraspecific sperm competition toward sperm from "preferred" males. Additionally, patterns of genetic relatedness at both nuclear and microsatellite markers suggest that females base these chemoattractant-induced preferences on complex patterns of genetic compatibility. Together, our results provide rare mechanistic insight into the interactions underlying gamete-level sexual selection. Moreover, this mechanism (sperm chemotaxis) has the potential to play similar roles across many taxa, given the ubiquity of egg chemoattractants. Indeed, as broadcast spawning was the ancestral mode of reproduction, gamete-level mechanisms that mediate competitive fertilizations likely played an important role in the evolution of sexual reproduction. The identification of such mechanisms, therefore, represents a crucial step forward in our understanding of sexual selection.

Sexual selection, which acts on variation in traits that influence reproductive success, almost certainly began in the sea with externally fertilizing organisms (Levitan 2010; Parker 2014). In these systems, before the evolution of advanced mobility and sensory structures, there would have been limited opportunity for mating competition or mate choice prior to gamete release. Instead, synchronous broadcast spawning (where gametes from both sexes are expelled externally) and the co-occurrence of gametes from multiple individuals likely fuelled sexual selection in the form of sperm competition (competition for fertilization among ejaculates from multiple males; Parker 1970) and cryptic female choice (biasing of fertilization by females or their eggs toward particular ejaculates; Thornhill 1983; Eberhard 1996). Recent theory suggests that these ancestral processes of sexual selection instigated the evolutionary cascade toward many derived features of animal reproductive systems, including sexual dimorphism, internal fertilization, and precopulatory sexual selection (Parker 2014). However, sperm competition and cryptic female choice have themselves remained pervasive forms of sexual selection in most sexually reproducing taxa (Pitnick and Hosken 2010). There is, therefore, considerable empirical value in studying gamete-level interactions in extant broadcast spawners as they may provide clues into the mechanisms underlying sperm-egg interactions in a broad range of taxonomic groups (Levitan 2010; Evans and Sherman 2013).

A key goal in reproductive and evolutionary biology is to seek mechanistic insights into the processes that generate fertilization biases during sperm competition, and in particular into the role that females play in moderating this competition (Pitnick et al. 2009; Pitnick and Hosken 2010; Firman et al. 2017). While evidence for female control over fertilization is now compelling in many systems (e.g., Clark et al. 1999; Nilsson et al. 2003; Pilastro et al. 2004; Lovlie et al. 2013; Young et al. 2013; Firman and Simmons 2015), direct demonstrations of the underlying mechanisms remain largely elusive (but see Gasparini and Pilastro 2011; Alonzo et al. 2016). Broadcast spawning taxa offer particularly amenable and experimentally tractable systems with which to identify such mechanisms (Evans and Sherman 2013). Unlike internal fertilizers, in broadcast spawners the interactions between gametes are not hidden from view within the female reproductive tract, making it possible to visualize processes (e.g., gamete selection) that would otherwise have to be inferred indirectly. For example, eggs of broadcast spawners can moderate the recognition and fusion of sperm at the gamete surface (Palumbi 1999; Levitan and Ferrell 2006), or select specific sperm nuclei when multiple sperm penetrate the egg (Carré and Sardet 1984). However, eggs can also influence sperm remotely (i.e., prior to the meeting of gametes) through the release of chemical attractants. This process, which is known as sperm chemotaxis, is often crucial in broadcast spawners for ensuring eggs are found and fertilized by conspecific sperm (Miller et al. 1994; Riffell et al. 2004). Moreover, it has been argued that when ejaculates from multiple conspecific males are present, such remote signal-

ing between eggs and sperm could be an important mediator of competitive fertilization success (Evans et al. 2012).

Although sperm chemotaxis is taxonomically widespread in both external and internal fertilizers (Miller 1985; Eisenbach 1999; Eisenbach and Giojalas 2006), its putative role in gamete-level sexual selection has only recently come to light. For example, recent studies on the broadcast spawning mussel *Mytilus galloprovincialis* have revealed that chemoattractants have differential effects on the swimming behavior (chemotactic responses, swimming trajectory, and speed; Evans et al. 2012; Oliver and Evans 2014) and physiology (acrosome reaction; Kekäläinen and Evans 2016) of sperm from different conspecific males. The strength of these effects correlate with differences in offspring survival among male–female crosses (Oliver and Evans 2014). These findings suggest that chemoattractants could promote fertilizations by genetically compatible sperm, but this has yet to be investigated under conditions of sperm competition. Moreover, the molecular processes underlying potential genetic compatibility effects are unknown. For example, differential sperm chemotaxis may be driven by gamete-level mechanisms that promote optimal levels of general offspring heterozygosity, which is often cited as an explanation of compatibility-based gamete choice (Firman et al. 2017). Alternatively, more specific patterns of genetic compatibility may apply in *M. galloprovincialis* populations, which typically contain multiple mitochondrial DNA lineages as a result of historical migration patterns (Westfall and Gardner 2010; Dias et al. 2014). What is clear, however, is that the intraspecific effects of chemoattractants on fertilization have important fitness implications for both males and females in this system.

In this study, we test whether differential sperm chemotaxis moderates gamete-level mate choice in *M. galloprovincialis*, and whether fertilization biases attributable to differential chemotactic responses reflect underlying patterns of genetic complementarity. Our experimental design allows us to measure competitive fertilization success directly, rather than the more usual method of estimating fertilization success indirectly from a male's paternity share. The latter method (paternity share) can be confounded by postfertilization effects on offspring viability that may not be related to sperm competitiveness (García-González 2008a; García-González and Evans 2011). Here, we overcome this problem using a fluorescent dye to label the mitochondria of sperm of competing males (Lymbery et al. 2016). In *M. galloprovincialis* and many other bivalves, embryos inherit both paternal and maternal mitochondria through a process termed doubly uniparental inheritance (DUI) (Zouros et al. 1994; Obata et al. 2006; Breton et al. 2007). In DUI, maternal mitochondria are inherited in the somatic tissue of all offspring, while the paternal mitochondria are ultimately transmitted to the germ line of male offspring (Breton et al. 2007). Initially, however, sperm mitochondria are transferred into all fertilized eggs (Obata et al. 2006). This feature of bivalve reproductive biology enables us to label sperm with a fluorescent mitochondrial vital dye and track their success during fertilization when labeled sperm from focal males compete with unlabeled rival ejaculates (Lymbery et al. 2016).

The primary aim of our study was to determine whether chemoattractants moderate competitive fertilization success in *M. galloprovincialis*. To test this we used a novel multistep experimental protocol involving multiple 2 × 2 factorial crosses to determine whether egg chemoattractants moderate the success of ejaculates when they compete to fertilize eggs (see Methods). We also tested whether fertilization biases induced by egg chemoattractants (ECs) reflect patterns of genetic complementarity between focal sperm competitors and female EC donors. Our highly controlled design enabled us to: (1) directly examine variation in competitive fertilization success using sperm dyes, therefore controlling for postfertilization effects on embryo viability; (2) separate the effects of males, females, and their interactions on competitive fertilization success; and (3) isolate the effect of differential chemical attraction as the female-moderated mechanism for biasing competitive fertilizations. Importantly, our design controls for stochastic variation in fertilization that could be caused by random sampling of rival males, by using sperm from a standard rival to compete with the dyed sperm of focal males within each factorial (García-González 2008b; García-González and Evans 2011). Our ensuing results provide the first direct evidence in any system that differential attraction of sperm up an egg chemoattractant gradient moderates intraspecific competitive fertilization success. Furthermore, we find that fertilization biases induced by egg chemoattractants reflect both preferences for unrelated males at nuclear loci and the selection of the same mitochondrial DNA lineage, thus revealing the putative genetic benefits of gamete-level mate choice in this system.

Methods

STUDY SPECIES AND SPAWNING

Mytilus galloprovincialis is a sessile, gonochoristic bivalve mollusc that forms large aggregations on intertidal substrates in temperate regions of both Hemispheres. *Mytilus galloprovincialis* is distributed across the southern coast of Australia (Westfall and Gardner 2010), with phylogenetic studies indicating that populations contain signatures of both a native Southern Hemisphere lineage and a more recent introduction of Northern Hemisphere individuals (Westfall and Gardner 2010; Colgan and Middelfart 2011; Dias et al. 2014). Nevertheless, there appears to have been extensive reproductive mixture of individuals from these different lineages in Australian populations (Westfall and Gardner 2013). We collected mussels from Woodman Point, Cockburn, Western Australia (32°14′ 03.6″S, 115°76′ 25″E) during the 2015 spawning season (June–September), and maintained them in aquaria of

recirculating seawater at the University of Western Australia until required (within one week of collection). Spawning was induced using a temperature increase from ambient to 28°C (Lymbery et al. 2016). Once an individual began spawning and its sex was determined, we immediately removed it from the spawning tank, washed it in filtered seawater (FSW) to remove possible contaminating gametes, placed it in an individual 250 mL cup and covered it in FSW. Once gametes were suitably dense, we removed the spawning individuals, estimated egg concentration by counting the number of cells in a homogenized 5 μL sample under a dissecting microscope, and estimated sperm concentration from subsamples (fixed in 1% formalin) using an improved *Neubauer haemocytometer*. We used these estimates to dilute gametes to their required concentrations for ensuing trials (see below).

EXPERIMENTAL OVERVIEW

We used a multistep cross-classified design with blocks of two focal males (M1 and M2) and two focal females (F1 and F2) (Fig. 1A; the steps involved in a trial from a single cell of the block are shown in Fig. 1B). The initial steps involved differential sperm chemotaxis assays, where sperm from each focal male (dyed sperm, see below) competed with undyed sperm from a standard rival (SR) male in the presence of a chemoattractant gradient from each of the two focal females (EC1 and EC2). Therefore, four competitions were performed per block; M1 versus SR in EC1, M1 versus SR in EC2, M2 versus SR in EC1, and M2 versus SR in EC2. The final step involved competitive fertilization assays, where eggs from a single standard female (different to the focal females used for chemoattractant gradients) were used to assess the competitive fertilization success of the focal male (in competition with the standard rival) in each cross. This latter step enabled us to attribute differences in competitive fertilization success between competing ejaculates exclusively to the action of chemoattractant (i.e., it allows us to directly link differential chemotactic movement with the fitness outcome of sperm competition). Using eggs from a separate standard female for the fertilizations enables us to make this link by ensuring that within each block, the only source of male × female variation in competitive fertilization rates is through differential chemoattraction. The standard female eggs, which were the same throughout all cells of the block, would have had no confounding effect on male × female variation. We performed each competition in replicate, that is eight competitions per block (Fig. 1A), and conducted a total of 11 blocks (i.e., $n = 22$ focal males, 22 focal females, 44 male–female combinations, 88 competitions).

COMPETITIVE CHEMOTAXIS AND FERTILIZATION TRIALS

In the first step of our experimental procedure, we established a chemoattractant gradient in an experimental chemotaxis chamber,

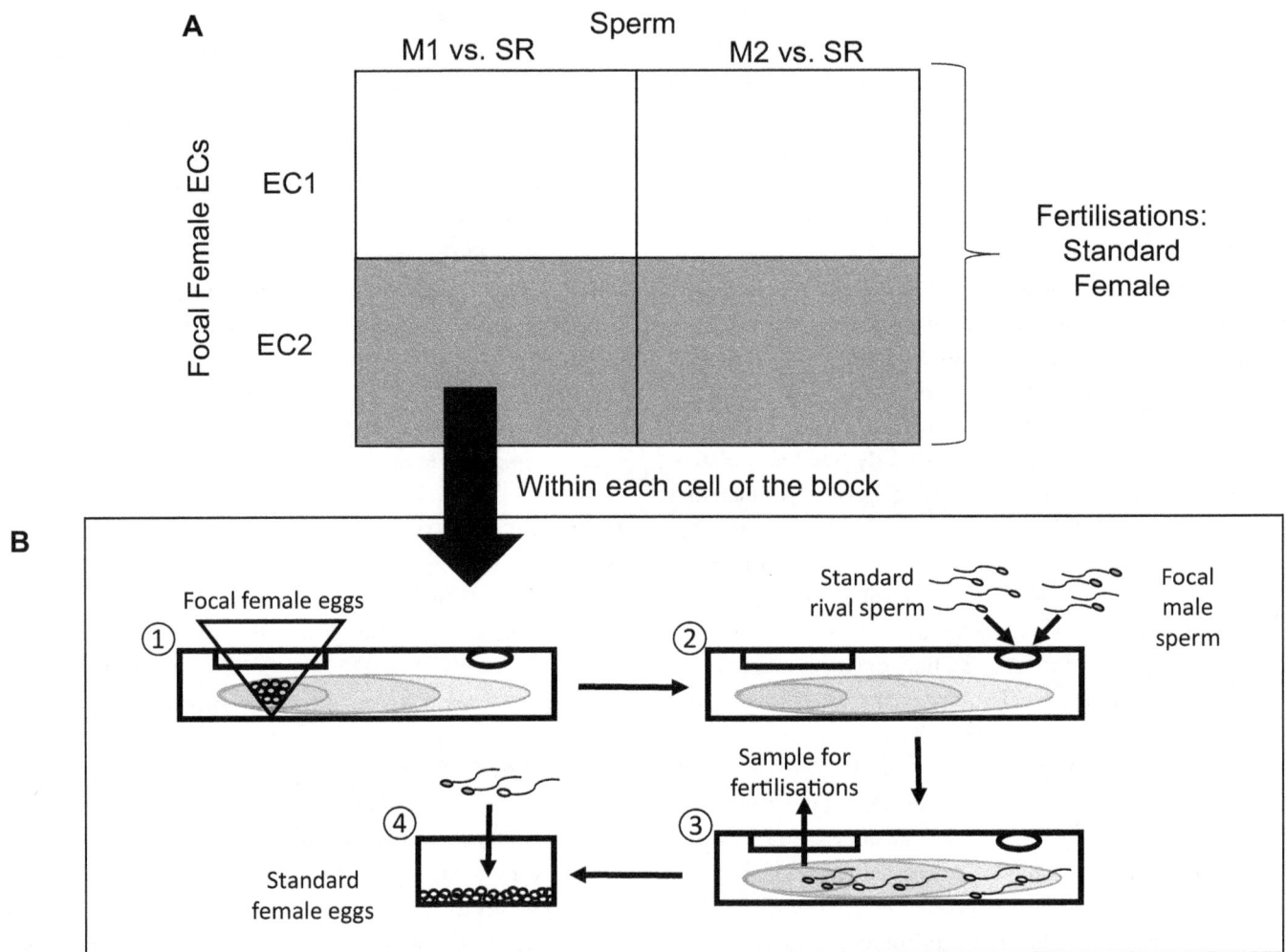

Figure 1. The overall design of an experimental block (A), and the steps performed within each cell of the block (B). (A) An example of one cross classified block, in which sperm from each of two focal males (M1 and M2) compete against sperm from a single standard rival (SR) in chemoattractant gradients from each of two focal females (F1 and F2). This generated four combinations per block, which were each replicated ($n = 11$ blocks, 44 combinations, 88 competitions total). Eggs from a single standard female per block were used to estimate competitive fertilization success. (B) The multistep competition assay illustrated using a single combination from within a block. (1) Eggs from the focal female were suspended in filter mesh to generate a chemoattractant gradient within the chamber. (2) The mesh and eggs were removed after 1 h, and dyed sperm from the focal male and undyed sperm from the standard rival added to the other end of the chamber. (3) After 10 minutes, a subsample was taken from the center of the chemoattractant gradient. (4) The subsample was added directly to eggs from the standard female, and competitive fertilization success of the focal male was measured.

then allowed dyed focal (M1 or M2) sperm and undyed rival (SR) sperm to swim in the chamber (Fig. 1B; these steps were performed for each cell of Fig. 1A). The chambers were made from sterile syringes (Terumo), with the ends of each syringe sawn off and sealed with parafilm (Bemis) to form a 10 mL tube. A ~2 cm^2 section was removed at one end of the chamber, and a small hole drilled in the other end. The chambers were fixed to a flat surface and a filter sack made of 30 µm filter mesh was inserted through the square opening. We added 5 mL of FSW to the chamber and 2 mL of egg solution (at 5×10^4 cells mL^{-1}) to the filter sack, which retained eggs but allowed chemoattractants to disperse into the chamber. We left the chambers for 1 h to establish

a chemoattractant gradient (this time frame has previously been used to establish a chemoattractant gradient in larger chambers and we confirmed in preliminary trials that it was sufficient for our chambers; Evans et al. 2012).

Aliquots of sperm from the focal males and the standard rival were standardized to the same concentration (see below) and prepared for each competitive chemotaxis trial. The focal male's sperm was labeled using MitoTracker Green FM (Molecular Probes), prepared as described in Lymbery et al. (2016). In our previous study, we showed that dyeing sperm has no effect on sperm behavior or competitive ability (Lymbery et al. 2016). Apart from the addition of dye, focal male and standard rival sperm were treated to the same procedure. Briefly, 950 μL aliquots of sperm at 1×10^6 cells mL^{-1} were prepared from each male, 50 μL of 500 nm dye solution added to focal male aliquots, and 50 μL of FSW added to rival aliquots. All samples (including undyed) were left in the dark (to prevent degradation of dye) for 10 minutes. The filter mesh containing focal female eggs was then removed from each chamber, and 500 μL each of focal male and standard rival sperm solution added to the drilled hole at the opposite end of the chamber (Fig. 1B). Sperm were allowed to swim in the gradient for 10 minutes. Preliminary trials confirmed that this assay did not result in any contamination of nonfocal sperm by excess dye from focal sperm (see Supplementary Methods).

After focal and rival sperm had been in the chemotaxis chamber for 10 minutes, 1 mL samples were taken from the center of the chemoattractant gradient (see Fig. 1B) and added to a separate petri dish containing 1 mL of FSW with eggs from the standard female (diluted to 1×10^4 cells mL^{-1}). Prior to the addition of sperm, we rinsed the standard eggs with FSW through 30 μm filter mesh to remove egg chemoattractants. However, even if these standard female eggs subsequently released chemoattractants, their impact (if any) would be to lessen our chance of detecting significant male-by-female effects (by obscuring patterns driven by the chemoattractants of focal females). Therefore, a significant male-by-female interaction in our analysis could only be attributable to the focal chemoattractants, which varied across the focal male samples. Moreover, fertilization occurs almost instantaneously upon the addition of sperm to the standard eggs (Lymbery et al. 2016), therefore decreasing the possibility that standard egg chemoattractants could reduce our power to detect effects. Although fertilization itself was instantaneous, we waited 10 minutes after the addition of sperm to allow dyed mitochondria to become visible inside fertilized eggs (Lymbery et al. 2016). We then estimated the fertilization success of the focal male under a fluorescent microscope by observing haphazard samples of 100 eggs, recording the numbers with and without dyed mitochondria.

Fertilizations from the rival (undyed) male were not scored, as estimating fertilizations from undyed sperm requires eggs to be left until they develop polar bodies, undergo cell division or until they can be assayed for survival. Therefore, the total numbers of fertilized eggs (dyed plus undyed) were not scored in this procedure. However, this is not required for the interpretation of the effects in our design, as we are not directly comparing the competitive success of focal males to rival males, but rather comparing the competitiveness of different focal males when they compete with a standard rival for standard eggs across different focal chemoattractants. Variation in the number of standard female eggs available for fertilization overall would only contribute to block-level variation (as all trials within a block used eggs from the same standard female) and therefore would not systematically change the relative share of paternity among focal males within

a block. Therefore, the male, female, and male x female effects (all nested within block) on competitive fertilization were not confounded by variation in proportion of standard female eggs available for fertilization.

NUCLEAR GENETIC RELATEDNESS

Foot tissue samples from all focal males and focal females (i.e., egg chemoattractant donors) were preserved in 100% ethanol. DNA was extracted using a salt-extraction method as described in Simmons et al. (2006) with the following alterations: tissue samples were incubated at 56°C overnight in the extraction buffer, and extracted DNA was resuspended in 100 μL of sterile water. DNA concentrations were estimated using a Nanodrop ND-1000 spectrophotometer (Thermo Fisher Scientific) and DNA samples were stored at −20°C until required for PCR amplification. Each individual was genotyped at 13 polymorphic microsatellite loci; MGE002, MGE005, MGE008 (Yu and Li 2007), Mgu3 (Presa et al. 2002), Med744 (Lallias et al. 2009), MT282 (Gardeström et al. 2008), MGES11 (Li et al. 2011), Mg-USC20, Mg-USC22, Mg-USC25, Mg-USC28, Mg-USC42, and Mg-USC43 (Pardo et al. 2011) (primer sequences provided in Table S1). Single-plex PCR reactions were run for each sample at each locus with a reaction volume of 5 μL, containing 1 μL MyTaq reaction buffer (Bioline), 0.2 μL primer mix (solution containing 10 nM each of forward and reverse primer, forward primer fluorescently labeled), 0.5 μL bovine serum albumin (Fisher Biotec), 0.1 μL MyTaq DNA Polymerase (Bioline), 2.2 μL sterile water, and 1 μL DNA sample (approximately 10 ng). PCRs were performed using an Eppendorf Mastercycler epGradient S, with an initial denaturation step at 95°C for 3 min, followed by 35 cycles of 95°C for 1 min, 54°C (MGE005 and MGE008) or 60°C (all other loci) for 1 min and 72°C for 1 min, with a final extension step of 72°C for 5 min. The PCR products were analyzed on an ABI 3730 96 capillary machine using a Genescan-500 LIZ internal size standard, and genotypes for each locus were scored using GEN-EMARKER software (SoftGenetics). Peaks identified by GENE-MARKER were checked manually and adjusted as necessary to minimize scoring errors.

One locus (MGES11) was monomorphic for our samples, with the number of alleles for the other 12 loci ranging from 3–20. We examined patterns of subpopulation variation and clustering of nuclear genotypes using the software program STRUCTURE (Pritchard et al. 2000, 2007; Falush et al. 2003; Supplementary Methods). Pairs of loci were tested for genetic linkage using likelihood ratio tests in GENEPOP (Raymond and Rousset 1995; Rousset 2008), with one pair of loci in significant linkage disequilibrium (Med744 and Mg-USC22, $P < 0.001$). We therefore removed one of these loci from the analysis, specifically Med744 as there was also evidence of null alleles at this locus (Table S2; null alleles estimated using MICROCHECKER software; Van Oosterhout et al. 2004). There were excess homozygotes and evidence for null alleles at seven other loci (Table S2). However, removing all loci with null alleles can considerably reduce the power to detect variation in genetic relatedness and result in less accurate relatedness estimates than when all loci are included (Supplementary Methods; see also Robinson et al. 2013). We therefore used a maximum likelihood estimator that can account for null alleles (Kalinowski et al. 2006) to calculate genetic relatedness from the remaining 11 loci between each focal male–female pair in each block. These estimates were calculated using the ML-RELATE software package (Kalinowski et al. 2006). We compared these estimates to a range of other relatedness estimators and found consistent patterns of variation in relatedness across different methods, increasing our confidence in the reported measures of nuclear genetic relatedness (see Supplementary Methods). Moreover, to determine whether any markers had a disproportionate effect on measures of relatedness, we examined whether relatedness changed when each marker was removed in turn, and found little variation across different combinations (Table S3).

MITOCHONDRIAL HAPLOTYPES

We sequenced female-type (F-type) CO1 mtDNA, which is generally considered to have a more reliable phylogenetic signal than male-type mtDNA and has multiple phylogenetic lineages in Australian *M. galloprovincialis* populations (Gérard et al. 2008; Colgan and Middelfart 2011; Dias et al. 2014). Using the DNA extracted as previously described, we amplified F-type CO1 haplotypes using PCR reagents and conditions as described in Dias et al. (2014). Samples were sequenced in both directions by the Australian Genome Research Facility, Perth. Consensus sequences were aligned, analyzed and trimmed in Geneious v 6.1.8 (Kearse et al. 2012) using the Geneious alignment feature with default parameters. A preliminary Neighbor-Joining tree was constructed from the 44 individuals to identify the number of unique sequences present ($n = 14$; Table S4). We added 105 northern and southern *Mytilus* haplotypes of the COI gene to our unique sequence set, as compiled in Dias et al. (2014). We inferred phylogenetic relationships using MRBAYES V3.1.2 (Huelsenbeck and Ronquist 2001) in Geneious v 6.1.8. We set the parameters and performed the Bayesian analyses as described in Dias et al. (2014), with the modification that we used a GTR+G substitution model. We determined phylogenetic relationships from 75% majority-rule consensus of postburn-in trees.

STATISTICAL ANALYSES

Analyses were performed using R version 3.3.2 (R Core Team 2016). We first analyzed competitive fertilization success of focal sperm as a binomial response variable (proportion of eggs successfully fertilized by dyed sperm in competition). We fit a GLMM with logit link function in the "lme4" package (Bates et al.

2014), using the Laplace approximation of the log-likelihood to estimate model parameters (Raudenbush et al. 2000). Our model included a fixed intercept term and random effects of male (overall variation among sperm of focal males), female (overall variation among focal female chemoattractants), male-by-female interaction (variation among sperm-chemoattractant combinations), and experimental block. There was no overdispersion in our model (residual deviance = 77.15 on 83 degrees of freedom, dispersion parameter = 0.93), and the scaled residuals (calculated using the "DHARMa" package; Hartig 2017) were uniformly distributed (Kolmogorov–Smirnov test; D = 0.053, $P = 0.967$). Focal male competitive fertilization success ranged from 0% to 44%, that is significantly lower than 50% (fixed intercept term of GLMM = -1.79 [95% CIs = –2.11, –1.47], Wald Z = -1.78, $P < 0.001$). This was expected given only the subset of sperm that successfully traveled to the center of the chemoattractant gradient was used for fertilizations. We assessed the significance of random effect terms by removing each from the model in turn and compared the fit of the reduced models against the full model with likelihood ratio tests ($-2 \times$ difference in log likelihoods compared against χ^2 distribution with 1 degree of freedom).

Next, we examined whether nuclear genetic relatedness and mitochondrial lineages of focal male and focal (i.e., chemoattractant-producing) female pairs were predictive of competitive fertilization success. The replicate measures of competitive fertilization success for each combination of focal sperm and focal chemoattractant were significantly repeatable ($R = 0.044$ [95% CIs 0.023, 0.069], $P < 0.001$; estimated using GLMM method in the "rptR" package; Nakagawa and Schielzeth 2010). Therefore, the replicate measures were combined into weighted means (i.e., total fertilized out of total number of eggs across the two replicates). We fit a GLMM with logit link function to competitive fertilization success, with a continuous fixed effect of nuclear relatedness and a fixed categorical factor specifying whether the focal male and focal female pair had the same mitochondrial lineage or a different lineage. We also fit random effects of male, female, and block. There was no evidence of overdispersion in our model (residual deviance = 11.91 on 37 degrees of freedom, dispersion parameter = 0.32), nor heteroscedasticity of scaled residuals (Kolmogorov–Smirnov test; D = 0.079, $P = 0.944$). We used Wald Chi-square tests to assess the significance of the fixed effects.

Results

COMPETITIVE FERTILIZATION SUCCESS

There were two sources of significant variation in focal male competitive fertilization success: (a) the male effect, and (b) the male-by-female interaction (Table 1). Although significant interactions often dictate that other effects must be interpreted cautiously,

Table 1. Results of log-likelihood ratio tests for random effects on focal male competitive fertilization success.

Model	Log likelihood	AICc	G^2	P
Full	−282.94	576.60		
(-Male)	−285.89	580.26	5.90	0.015*
(-Female)	−283.27	575.01	0.66	0.417
(-Male × Female)	−285.41	579.30	4.95	0.026*
(-Block)	−283.84	576.17	1.81	0.178

Full generalized linear-mixed effects model included the proportion of eggs successfully fertilized by the focal male as the response variable (with logit link function), with random effects of focal male ID, focal female ID, male-by-female interaction and experimental block. The fixed intercept of the full model was significantly negative (intercept = −1.79 [95% CIs = −2.11, −1.47], Wald Z = −1.78, $P < 0.001$). Estimated variance components associated with random effects are provided in Table S5. Reduced models were fit by excluding each random effect in turn. Aikaike information criteria with correction for finite sample sizes (AICc) are provided for full and reduced models. The likelihood ratio statistic (G^2) for each random effect was calculated as −2 × difference in log-likelihoods between the relevant reduced model and the full model. Probability (P) statistics were estimated by comparing G^2 to a χ^2 distribution with one degree of freedom.

Table 2. Effects of nuclear genetic relatedness and phylogenetic mitochondrial lineage on competitive fertilization success.

Fixed effect	Estimate	χ^2	P
Nuclear relatedness	−0.35 [−1.32, −0.02]	3.92	0.047
Mitochondrial lineage	0.35 [0.22, 0.65]	15.52	<0.001

Effects estimated from generalized linear-mixed effects models of the proportion of eggs successfully fertilized by the focal male (with logit link function), with fixed effects of nuclear relatedness and mitochondrial lineage and random effects of focal male ID, focal female ID, and experimental block. The final model did not include the interaction term of the fixed effects, as the interaction was nonsignificant in the full model (Wald $\chi^2 = 0.93$, $P = 0.335$) and its inclusion reduced model fit (see Table S6; although significance of the main effects did not change with inclusion of the interaction). The fixed intercept of the model was significantly negative (intercept = −1.58 [95% CIs = −1.95, −1.22], Wald Z = −9.08, $P < 0.001$). Nuclear relatedness of focal male and focal female pairs was estimated from microsatellite loci using maximum likelihood (higher values = more closely related). Mitochondrial lineage (Northern or Southern Hemisphere) was assigned based on female-type CO1 sequences, with focal male and focal female pairs scored as belonging to different or same lineage (estimate represents the mean change in fertilization success on the latent scale from different to same lineage). Hypothesis tests of main effects were conducted using Wald χ^2 tests (d.f. = 1 for each effect).

in this case the removal of both the male effect and the male-by-female interaction resulted in a significantly worse fit than removal of the male-by-female interaction alone (likelihood ratio statistic $G^2 = 68.80$, $P < 0.001$). Therefore, the significant male effect suggests that there was variation among males in their average competitive success (i.e., some males were intrinsically "better" sperm competitors than others). The male-by-female interaction, on the other hand, indicates that there was significant variation in the way chemoattractants of focal females affected the competitive success of different focal males. In other words, the success of each focal male within a block depended on the specific identity of the focal female chemoattractant.

GENETIC RELATIONSHIPS

The nuclear data indicated a well-mixed population (Fig. S1), despite F-type CO1 mtDNA haplotypes revealing signatures of two historical phylogenetic lineages (consistent with previously identified Northern and Southern Hemisphere lineages; Fig. S2; see also Dias et al. 2014). Nuclear genetic relatedness did not differ between focal male–female pairs that had the same mitochondrial lineage and those that had different mitochondrial lineages (two-sample t-test, $t_{42} = 0.31$, $P = 0.759$). We tested whether overall nuclear genetic relatedness or phylogenetic mtDNA lineages of focal male and focal (i.e., chemoattractant-producing) female pairs predicted patterns of gamete-level sexual selection (i.e., competitive fertilization success). We found significant main effects of both nuclear relatedness and mitochondrial lineage (Table 2). Specifically, competitive fertilization success was

higher when focal male and focal female nuclear genotypes were less related, but also when focal males and focal females had the same mitochondrial lineage.

Discussion

Our results reveal that differential attraction of sperm up a chemical gradient can act as a mechanism of gamete-level mate choice. To our knowledge, this is the first direct evidence that egg chemoattractants influence intraspecific sperm competition, supporting the previously documented differential effects of egg chemoattractants on sperm swimming direction (Evans et al. 2012), sperm motility (Oliver and Evans 2014), and sperm physiology (Kekäläinen and Evans 2016). We show that the effect of chemoattractants on competitive fertilization success depends upon the particular combination of focal male and focal female, specifically favoring certain genetic combinations over others. Previous work on this system has shown that the strength of sperm chemotactic responses for any given male–female pairing is positively correlated with offspring survival (Oliver and Evans 2014). These previous findings, together with the present results, suggest that egg chemoattractants allow females to promote fertilization by more compatible males when multiple ejaculates compete. This provides rare insight into the mechanisms used by females to gain control over the outcome of sperm competition.

Our results complement and extend recent evidence that female reproductive fluids more broadly can have important roles in gamete-level sexual selection. In particular, there has been considerable interest in the ovarian fluid (OF) produced by various female fishes. In externally fertilizing salmonids, for example, OF released with eggs can differentially mediate the swimming speed of conspecific sperm depending on the particular male–female pairing (Urbach et al. 2005; Rosengrave et al. 2008; Butts et al. 2012). Although OF has yet to be implicated in intraspecific gamete-level mate choice in salmonids (Evans et al. 2013), it has been shown to promote fertilization by conspecific sperm when in competition with those of sister species (Yeates et al. 2013). Intriguingly, however, there is evidence from an internally fertilizing poeciliid fish that OF within the female's reproductive tract can selectively bias fertilization in favor of sperm from unrelated males over related males (Gasparini and Pilastro 2011). Recent work on an externally fertilizing wrasse has also shown that OF can bias competitive fertilization success toward dominant "nest" males (i.e., directional cryptic female choice; Alonzo et al. 2016). Our findings for mussels complement these prior studies by showing that egg chemoattractants similarly play an important role in mediating intraspecific sperm competition, thus exposing a previously unforeseen mechanism of sexual selection that may occur more broadly in other taxa. We suggest that further investigation into the effects of female reproductive fluids, including egg chemoattractants, across a broader range of taxa will provide fruitful mechanistic insights into gamete-level mate choice.

We also found that the competitive fertilization biases induced by egg chemoattractants reflect complex genetic relationships between the focal males and focal (i.e., chemoattractant producing) females. These results may shed some light on patterns of genetic compatibility that underlie competitive fertilization biases, given previous findings that differential chemotaxis is correlated with offspring fitness of male–female pairs (Oliver and Evans 2014). Competitive fertilization success was higher for focal males that had a lower overall genetic relatedness to focal females (based on neutral nuclear markers), which complements recent evidence in other taxa that preferences for genetically dissimilar males may drive compatibility-based cryptic female choice (Gasparini and Pilastro 2011; Firman and Simmons 2015). Although we did not directly examine the extent of inbreeding in our population, homozygote excesses consistent with inbreeding are not uncommon in populations of broadcast spawners (Huang et al. 2000; Addison and Hart 2005; Kenchington et al. 2006), possibly due to the unpredictable patterns of spawning and recruitment in these systems (Hedgecock and Pudovkin 2011). Therefore, gamete-level mechanisms of maximizing offspring heterozygosity may be important for individual reproductive fitness.

In contrast to the patterns of overall genetic relatedness, we also found a competitive fertilization bias toward males that had the same phylogenetic mitochondrial lineage as the female. Preferences based on phylogenetic lineage are not unexpected in Australian *M. galloprovincialis* populations, as Northern and Southern Hemisphere lineages had diverged in allopatry from the Pleistocene before the more recent introduction of Northern individuals (Hilbish et al. 2000; Gérard et al. 2008). Nevertheless, it appears that such preferences have not maintained reproductive isolation between lineages, with the admixture of nuclear genotypes in our population supporting previous findings for Australian populations (Westfall and Gardner 2013). Possibly, this could be due to lineage-based patterns being offset by the preferences for less related nuclear genotypes. However, the precise fitness benefits of the mitochondrial lineage-based biases deserve further investigation. For example, one possibility is that fertilization biases reflect cyto-nuclear compatibilities brought about by the presence of divergent mitochondrial lineages; it would therefore be interesting to examine how preferences relate to nuclear genes involved in mitochondrial function. Moreover, we sequenced the female-type mtDNA common to somatic tissues of both males and females, but the occurrence and transmission of male-type mitochondria in sperm may further complicate patterns. Therefore, the precise genetic interactions between males and females that underlie chemoattractant-driven fertilization biases in these systems remain to be fully resolved.

To provide further mechanistic insights into gamete-level mate choice in this system we need to identify the chemical profiles of egg chemoattractants and determine how variation in these profiles correspond to patterns of differential sperm attraction. Chemoattractant molecules have not yet been identified in *M. galloprovincialis*, but several types of egg-derived chemicals have been described in other broadcast spawners (reviewed in Evans and Sherman 2013). For example, in echinoderms, peptides released from eggs bind to guanylyl cyclase receptors on the sperm surface, triggering a signaling pathway that results in influxes of extracellular calcium ions and a corresponding flagellar beat pattern (Kaupp et al. 2006; Alvarez et al. 2014). However, to our knowledge there has been no examination of intraspecific variation in such signaling pathways in any species. Recent evidence suggests that sperm-activating peptides are evolutionarily conserved and vary little within genera (Jagadeeshan et al. 2015). Therefore, it may be unlikely that a single molecule type (such as a particular peptide) is responsible for intraspecific variation in sperm chemoattraction. Instead, it is possible that eggs release a variety of molecules that affect such signaling pathways. Our finding that the interacting effects of parental genotypes drive chemoattractant preferences suggests that these chemical signals are likely to be complex. Clearly there is a need to characterize intraspecific variation in egg chemoattractant chemical profiles to address these questions.

In conclusion, we provide the first direct evidence that egg chemoattractants moderate sperm competition and complement these findings with genetic data that may explain the previously documented offspring fitness benefits associated with differential sperm chemotaxis (Oliver and Evans 2014). Given our focus on a species exhibiting the ancestral mating strategy of broadcast spawning, and the fact that egg chemoattractants are found throughout a diverse range of taxa (Miller 1985; Eisenbach 1999; Teves et al. 2009), we anticipate that such mechanisms of gamete-level mate choice may be prevalent in other species. However, until now the putative role of sperm chemotaxis in mediating intraspecific sperm competition has been largely untested. This is likely due in part to the empirical difficulty of linking the effect of putative mechanisms of gamete-level mate choice directly to variation in competitive fertilization success. We demonstrate that powerful and tightly controlled experimental designs can provide detailed insights into the intricacies of gamete-level sexual selection.

AUTHOR CONTRIBUTIONS

R.A.L., W.J.K., and J.P.E. designed the study. R.A.L. conducted the experiments and data collection. All authors discussed analysis of the data, and R.A.L. performed the formal statistical analyses. R.A.L. wrote the first draft of the manuscript, and all authors contributed to the final version.

ACKNOWLEDGMENTS

We thank Jukka Kekäläinen for assistance with mussel collections, Stephen Robinson and Catherine Seed for helping design and create the chemotaxis chambers, Yvette Hitchen for assistance with microsatellite and mtDNA analyses, and Alan Lymbery and two anonymous reviewers for helpful comments on the manuscript. Funding was provided by the UWA School of Animal Biology (R.A.L.) and the Australian Research Council (J.P.E.; Grant Number DP150103266). R.A.L. was supported by the Hackett Postgraduate Research Scholarship and the Bruce and Betty Green Postgraduate Research Top-Up Scholarship.

LITERATURE CITED

Addison, J., and M. Hart. 2005. Spawning, copulation and inbreeding coefficients in marine invertebrates. Biol. Lett. 1:450–453.

Alonzo, S. H., K. A. Stiver, and S. E. Marsh-Rollo. 2016. Ovarian fluid allows directional cryptic female choice despite external fertilization. Nat. Commun. 7:12452.

Alvarez, L., B. M. Friedrich, G. Gompper, and U. B. Kaupp. 2014. The computational sperm cell. Trends Cell Biol. 24:198–207.

Bates, D., M. Macechler, B. Bolker, and S. Walker. 2014. lme4: linear mixed-effects models using Eigen and S4. R package version 1:1–7.

Breton, S., H. D. Beaupre, D. T. Stewart, W. R. Hoeh, and P. U. Blier. 2007. The unusual system of doubly uniparental inheritance of mtDNA: isn't one enough? Trends Genet. 23:465–474.

Butts, I. A. E., K. Johnson, C. C. Wilson, and T. E. Pitcher. 2012. Ovarian fluid enhances sperm velocity based on relatedness in lake trout, *Salvelinus namaycush*. Theriogenology 78:2105–2109.

Carré, D., and C. Sardet. 1984. Fertilization and early development in *Beroë ovata*. Dev. Biol. 105:188–195.

Clark, A. G., D. J. Begun, and T. Prout. 1999. Female x male interactions in *Drosophila* sperm competition. Science 283:217–220.

Colgan, D. J., and P. Middelfart. 2011. *Mytilus* mitochondrial DNA haplotypes in southeastern Australia. Aquat. Biol. 12:47–53.

Dias, P. J., S. Fotedar, and M. Snow. 2014. Characterisation of mussel (*Mytilus* sp.) populations in Western Australia and evaluation of potential genetic impacts of mussel spat translocation from interstate. Mar. Freshw. Res. 65:486–496.

Eberhard, W. G. 1996. Female control: Sexual selection by cryptic female choice. Princeton Univ. Press, Princeton, NJ.

Eisenbach, M. 1999. Sperm chemotaxis. Rev. Reprod. 4:56–66.

Eisenbach, M., and L. C. Giojalas. 2006. Sperm guidance in mammals—an unpaved road to the egg. Nat. Rev. Mol. Cell Biol. 7:276–285.

Evans, J. P., F. García-González, M. Almbro, O. Robinson, and J. L. Fitzpatrick. 2012. Assessing the potential for egg chemoattractants to mediate sexual selection in a broadcast spawning marine invertebrate. Proc. R Soc. B 279:20120181.

Evans, J. P., P. Rosengrave, C. Gasparini, and N. J. Gemmell. 2013. Delineating the roles of males and females in sperm competition. Proc. R Soc. B 280:20132047.

Evans, J. P., and C. D. H. Sherman. 2013. Sexual selection and the evolution of egg-sperm interactions in broadcast-spawning invertebrates. Biol. Bull. 224:166–183.

Falush, D., M. Stephens, and J. K. Pritchard. 2003. Inference of population structure using multilocus genotype data: linked loci and correlated allele frequencies. Genetics 164:1567–1587.

Firman, R. C., C. Gasparini, M. K. Manier, and T. Pizzari. 2017. Postmating female control: 20 years of cryptic female choice. Trends Ecol. Evol. 32:368–382.

Firman, R. C., and L. W. Simmons. 2015. Gametic interactions promote inbreeding avoidance in house mice. Ecol. Lett. 18:937–943.

García-González, F. 2008a. Male genetic quality and the inequality between paternity success and fertilization success: consequences for studies of sperm competition and the evolution of polyandry. Evolution 62:1653–1665.

———. 2008b. The relative nature of fertilization success: implications for the study of post-copulatory sexual selection. BMC Evol. Biol. 8:9.

García-González, F., and J. P. Evans. 2011. Fertilization success and the estimation of genetic variance in sperm competitiveness. Evolution 65:746–756.

Gardeström, J., R. T. Pereyra, and C. André. 2008. Characterization of six microsatellite loci in the Baltic blue mussel *Mytilus trossulus* and cross-species amplification in North Sea *Mytilus edulis*. Conserv. Genet. 9:1003–1005.

Gasparini, C., and A. Pilastro. 2011. Cryptic female preference for genetically unrelated males is mediated by ovarian fluid in the guppy. Proc. R Soc. B 278:2495–2501.

Gérard, K., N. Bierne, P. Borsa, A. Chenuil, and J. P. Féral. 2008. Pleistocene separation of mitochondrial lineages of *Mytilus* spp. mussels from Northern and Southern Hemispheres and strong genetic differentiation among southern populations. Mol. Phylogenet. Evol. 49:84–91.

Hartig, F. 2017. DHARMa: residual diagnostics for hierarchical (multilevel/mixed) regression models. R package version 0.1.5.

Hedgecock, D., and A. I. Pudovkin. 2011. Sweepstakes reproductive success in highly fecund marine fish and shellfish: a review and commentary. Bull. Mar. Sci. 87:971–1002.

Hilbish, T. J., A. Mullinax, S. I. Dolven, A. Meyer, R. K. Koehn, and P. D. Rawson. 2000. Origin of the antitropical distribution pattern in marine mussels (*Mytilus* spp.): routes and timing of transequatorial migration. Mar. Biol. 136:69–77.

Huang, B. X., R. Peakall, and P. J. Hanna. 2000. Analysis of genetic structure of blacklip abalone (*Haliotis rubra*) populations using RAPD, minisatellite and microsatellite markers. Mar. Biol. 136:207–216.

Huelsenbeck, J. P., and F. Ronquist. 2001. MRBAYES: Bayesian inference of phylogenetic trees. Bioinformatics 17:754–755.

Jagadeeshan, S., S. E. Coppard, and H. A. Lessios. 2015. Evolution of gamete attraction molecules: evidence for purifying selection in speract and its receptor, in the pantropical sea urchin *Diadema*. Evol. Dev. 17:92–108.

Kalinowski, S. T., A. P. Wagner, and M. L. Taper. 2006. ML-RELATE: a computer program for maximum likelihood estimation of relatedness and relationship. Mol. Ecol. Notes 6:576–579.

Kaupp, U. B., E. Hildebrand, and I. Weyand. 2006. Sperm chemotaxis in marine invertebrates—molecules and mechanisms. J. Cell. Physiol. 208:487–494.

Kearse, M., R. Moir, A. Wilson, S. Stones-Havas, M. Cheung, S. Sturrock, S. Buxton, A. Cooper, S. Markowitz, C. Duran, et al. 2012. Geneious basic: an integrated and extendable desktop software platform for the organization and analysis of sequence data. Bioinformatics 28:1647–1649.

Kekäläinen, J., and J. P. Evans. 2016. Female-induced remote regulation of sperm physiology may provide opportunities for gamete-level mate choice. Evolution 71:238–248.

Kenchington, E. L., M. U. Patwary, E. Zouros, and C. J. Bird. 2006. Genetic differentiation in relation to marine landscape in a broadcast-spawning bivalve mollusc (*Placopecten magellanicus*). Mol. Ecol. 15:1781–1796.

Lallias, D., R. Stockdale, P. Boudry, S. Lapegue, and A. R. Beaumont. 2009. Characterization of 10 microsatellite loci in the blue mussel *Mytilus edulis*. J. Shellfish Res. 28:547–551.

Levitan, D. R. 2010. Sexual selection in external fertilizers. Pp. 365–378 *in* D. F. Westneat and C. W. Fox, eds. Evolutionary behavioral ecology. Oxford Univ. Press, Oxford.

Levitan, D. R., and D. L. Ferrell. 2006. Selection on gamete recognition proteins depends on sex, density, and genotype frequency. Science 312:267–269.

Li, H. J., Y. Liang, L. J. Sui, X. G. Gao, and C. B. He. 2011. Characterization of 10 polymorphic microsatellite markers for Mediterranean blue mussel *Mytilus galloprovincialis* by EST database mining and cross-species amplification. J. Genet. 90:E30–E33.

Lovlie, H., M. A. F. Gillingham, K. Worley, T. Pizzari, and D. S. Richardson. 2013. Cryptic female choice favours sperm from major histocompatibility complex-dissimilar males. Proc. R Soc. B 280:20131296.

Lymbery, R. A., W. J. Kennington, and J. P. Evans. 2016. Fluorescent sperm offer a method for tracking the real-time success of ejaculates when they compete to fertilise eggs. Sci. Rep. 6:22689.

Miller, R. L. 1985. Sperm chemo-orientation in the metazoa. Pp. 274–337 *in* C. B. Metz and A. Monroy, eds. Biology of fertilization V2: Biology of sperm. Academic Press, New York.

Miller, R. L., J. J. Mojares, and J. L. Ram. 1994. Species-specific sperm attraction in the zebra mussel, *Dreissena polymorpha*, and the quagga mussel, *Dreissena bugensis*. Can. J. Zool. 72:1764–1770.

Nakagawa, S., and H. Schielzeth. 2010. Repeatability for Gaussian and non-Gaussian data: a practical guide for biologists. Biol. Rev. 85:935–956.

Nilsson, T., C. Fricke, and G. Arnqvist. 2003. The effects of male and female genotype on variance in male fertilization success in the red flour beetle (*Tribolium castaneum*). Behav. Ecol. Sociobiol. 53:227–233.

Obata, M., C. Kamiya, K. Kawamura, and A. Komaru. 2006. Sperm mitochondrial DNA transmission to both male and female offspring in the blue mussel *Mytilus galloprovincialis*. Dev. Growth Differ. 48:253–261.

Oliver, M., and J. P. Evans. 2014. Chemically moderated gamete preferences predict offspring fitness in a broadcast spawning invertebrate. Proc. R Soc. B 281:20140148.

Palumbi, S. R. 1999. All males are not created equal: fertility differences depend on gamete recognition polymorphisms in sea urchins. Proc. Natl. Acad. Sci. USA 96:12632–12637.

Pardo, B. G., M. Vera, A. Pino-Querido, J. A. Álvarez-Dios, and P. Martinez. 2011. Development of microsatellite loci in the Mediterranean mussel *Mytilus galloprovincialis*. Mol. Ecol. Resour. 11:586–589.

Parker, G. A. 1970. Sperm competition and its evolutionary consequences in the insects. Biol. Rev. 45:525–567.

———. 2014. The sexual cascade and the rise of pre-ejaculatory (Darwinian) sexual selection, sex roles, and sexual conflict. Cold Spring Harb. Perspect. Biol. 6:a017509.

Pilastro, A., M. Simonato, A. Bisazza, and J. P. Evans. 2004. Cryptic female preference for colorful males in guppies. Evolution 58:665–669.

Pitnick, S., and D. J. Hosken. 2010. Postcopulatory sexual selection. Pp. 379–399 *in* D. F. Westneat and C. W. Fox, eds. Evolutionary behavioral ecology. Oxford Univ. Press, Oxford.

Pitnick, S., M. F. Wolfner, and S. S. Suarez. 2009. Ejaculate-female and sperm-female interactions. Pp. 247–304 *in* T. R. Birkhead, D. J. Hosken, and S. Pitnick, eds. Sperm biology: An evolutionary perspective. Elsevier, Burlington, MA.

Presa, P., M. Perez, and A. P. Diz. 2002. Polymorphic microsatellite markers for blue mussels (*Mytilus* spp.). Conserv. Genet. 3:441–443.

Pritchard, J. K., M. Stephens, and P. Donnelly. 2000. Inference of population structure using multilocus genotype data. Genetics 155:945–959.

Pritchard, J. K., X. Wen, and D. Falush. 2007. Documentation for STRUCTURE software: version 2.2.

R Core Team. 2016. R: a language and environment for statistical computing. R Foundation for Statistical Computing, Vienna, Austria. URL http://www.R-project.org.

Raudenbush, S. W., M. L. Yang, and M. Yosef. 2000. Maximum likelihood for generalized linear models with nested random effects via high-order, multivariate Laplace approximation. J. Comput. Graph. Stat. 9:141–157.

Raymond, M., and F. Rousset. 1995. GENEPOP (version 1.2): population genetics software for exact tests and ecumenicism. J. Hered. 66:248–249.

Riffell, J. A., P. J. Krug, and R. K. Zimmer. 2004. The ecological and evolutionary consequences of sperm chemoattraction. Proc. Natl. Acad. Sci. USA 101:4501–4506.

Robinson, S. P., L. W. Simmons, and W. J. Kennington. 2013. Estimating relatedness and inbreeding using molecular markers and pedigrees: the effect of demographic history. Mol. Ecol. 22:5779–5792.

Rosengrave, P., N. J. Gemmell, V. Metcalf, K. McBride, and R. Montgomerie. 2008. A mechanism for cryptic female choice in chinook salmon. Behav. Ecol. 19:1179–1185.

Rousset, F. 2008. Genepop'007: a complete reimplementation of the Genepop software for Windows and Linux. Mol. Ecol. Resour. 8:103–106.

Simmons, L. W., M. Beveridge, N. Wedell, and T. Tregenza. 2006. Postcopulatory inbreeding avoidance by female crickets only revealed by molecular markers. Mol. Ecol. 15:3817–3824.

Teves, M. E., H. A. Guidobaldi, D. R. Unates, R. Sanchez, W. Miska, S. J. Publicover, A. A. M. Garcia, and L. C. Giojalas. 2009. Molecular mechanism for human sperm chemotaxis mediated by progesterone. PLoS One 4:11.

Thornhill, R. 1983. Cryptic female choice and its implications in the scorpionfly *Harpobittacus nigriceps*. Am. Nat. 122:765–788.

Urbach, D., I. Folstad, and G. Rudolfsen. 2005. Effects of ovarian fluid on sperm velocity in Arctic charr (*Salvelinus alpinus*). Behav. Ecol. Sociobiol. 57:438–444.

Van Oosterhout, C., W. F. Hutchinson, D. P. M. Wills, and P. Shipley. 2004. MICRO-CHECKER: software for identifying and correcting genotyping errors in microsatellite data. Mol. Ecol. Notes 4:535–538.

Westfall, K. M., and J. P. A. Gardner. 2010. Genetic diversity of Southern hemisphere blue mussels (Bivalvia: Mytilidae) and the identification of non-indigenous taxa. Biol. J. Linn. Soc. 101:898–909.

Westfall, K. M., and J. P. A. Gardner. 2013. Interlineage *Mytilus galloprovincialis* Lmk. 1819 hybridization yields inconsistent genetic outcomes in the Southern hemisphere. Biol. Invasions 15:1493–1506.

Yeates, S. E., S. E. Diamond, S. Einum, B. C. Emerson, W. V Holt, and M. J. G. Gage. 2013. Cryptic choice of conspecific sperm controlled by the impact of ovarian fluid on sperm swimming behavior. Evolution 67:3523–3536.

Young, B., D. V Conti, and M. D. Dean. 2013. Sneaker "jack" males outcompete dominant "hooknose" males under sperm competition in Chinook salmon (*Oncorhynchus tshawytscha*). Ecol. Evol. 3:4987–4997.

Yu, H., and Q. Li. 2007. Development of EST-SSRs in the Mediterranean blue mussel, *Mytilus galloproviancialis*. Mol. Ecol. Notes 7:1308–1310.

Zouros, E., A. O. Ball, C. Saavedra, and K. R. Freeman. 1994. An unusual type of mitichondrial DNA inheritence in the blue mussel *Mytilus*. Proc. Natl. Acad. Sci. USA 91:7463–7467.

PERMISSIONS

All chapters in this book were first published in EL, by John Wiley & Sons Ltd.; hereby published with permission under the Creative Commons Attribution License or equivalent. Every chapter published in this book has been scrutinized by our experts. Their significance has been extensively debated. The topics covered herein carry significant findings which will fuel the growth of the discipline. They may even be implemented as practical applications or may be referred to as a beginning point for another development.

The contributors of this book come from diverse backgrounds, making this book a truly international effort. This book will bring forth new frontiers with its revolutionizing research information and detailed analysis of the nascent developments around the world.

We would like to thank all the contributing authors for lending their expertise to make the book truly unique. They have played a crucial role in the development of this book. Without their invaluable contributions this book wouldn't have been possible. They have made vital efforts to compile up to date information on the varied aspects of this subject to make this book a valuable addition to the collection of many professionals and students.

This book was conceptualized with the vision of imparting up-to-date information and advanced data in this field. To ensure the same, a matchless editorial board was set up. Every individual on the board went through rigorous rounds of assessment to prove their worth. After which they invested a large part of their time researching and compiling the most relevant data for our readers.

The editorial board has been involved in producing this book since its inception. They have spent rigorous hours researching and exploring the diverse topics which have resulted in the successful publishing of this book. They have passed on their knowledge of decades through this book. To expedite this challenging task, the publisher supported the team at every step. A small team of assistant editors was also appointed to further simplify the editing procedure and attain best results for the readers.

Apart from the editorial board, the designing team has also invested a significant amount of their time in understanding the subject and creating the most relevant covers. They scrutinized every image to scout for the most suitable representation of the subject and create an appropriate cover for the book.

The publishing team has been an ardent support to the editorial, designing and production team. Their endless efforts to recruit the best for this project, has resulted in the accomplishment of this book. They are a veteran in the field of academics and their pool of knowledge is as vast as their experience in printing. Their expertise and guidance has proved useful at every step. Their uncompromising quality standards have made this book an exceptional effort. Their encouragement from time to time has been an inspiration for everyone.

The publisher and the editorial board hope that this book will prove to be a valuable piece of knowledge for researchers, students, practitioners and scholars across the globe.

LIST OF CONTRIBUTORS

Lynda F. Delph and Curtis M. Lively
Department of Biology, Indiana University, Bloomington, Indiana 47405

Amanda K. Gibson
Department of Biology, Indiana University, Bloomington, Indiana 47405
Department of Biology, Emory University, Atlanta, Georgia 30322

Susana M. Wadgymar and Jill T. Anderson
Department of Genetics and Odum School of Ecology, University of Georgia, Athens, Georgia 30602

Caroline Daws
Department of Ecology, Evolution and Behavior, University of Minnesota, St. Paul, Minnesota 55108

Dan A. Greenberg
Department of Biological Sciences, Simon Fraser University, Burnaby, British Columbia V5A 1S6, Canada
Earth-to-Ocean Research Group, Simon Fraser University, Burnaby, British Columbia V5A 1S6, Canada
Crawford Lab for Evolutionary Studies, Simon Fraser University, Burnaby, British Columbia V5A 1S6, Canada

Arne Ø. Mooers
Department of Biological Sciences, Simon Fraser University, Burnaby, British Columbia V5A 1S6, Canada
Crawford Lab for Evolutionary Studies, Simon Fraser University, Burnaby, British Columbia V5A 1S6, Canada

Barker Pasi M. Rastas Simon H. Martin
Department of Zoology, University of Cambridge, Downing Street, Cambridge CB2 3EJ, United Kingdom

John W. Davey, Ana Pinharanda, Richard M. Merrill and Chris D. Jiggins
Department of Zoology, University of Cambridge, Downing Street, Cambridge CB2 3EJ, United Kingdom
Smithsonian Tropical Research Institute, Gamboa, Panama

Owen McMillan
Smithsonian Tropical Research Institute, Gamboa, Panama

Richard Durbin
Wellcome Trust Sanger Institute, Cambridge CB10 1SA, United Kingdom

Joanne L. Godwin, Ramakrishnan Vasudeva, Alyson J. Lumley, Tracey Chapman and Matthew J. G. Gage
School of Biological Sciences, University of East Anglia, Norwich Research Park, Norwich NR4 7TJ, United Kingdom

Łukasz Michalczyk
Institute of Zoology, Jagiellonian University, Kraków, Poland

Oliver Y. Martin
ETH Zürich, Institute of Integrative Biology, Zürich, Switzerland

Reto Burri
Department of Population Ecology, Friedrich Schiller University Jena, Dornburger Strasse 159, D-07743 Jena, Germany

Pascal Milesi, Mylène Weill and Pierrick Labbé
Institut des Sciences de l'Evolution de Montpellier (UMR 5554, CNRS-Universite´ de Montpellier-IRD-EPHE), Campus

Thomas Lenormand
Centre d'Ecologie Fonctionnelle et Evolutive (UMR 5175, CNRS-Université de Montpellier-Université Paul-Valéry
Montpellier-EPHE) 1919 route de Mende, F-34293 Montpellier, CEDEX 05, France

Kat Bebbington
School of Biological Sciences, University of East Anglia, Norwich Research Park, Norwich NR4 7TJ, UK
Behavioural and Physiological Ecology, GELIFES, University of Groningen, 9700CC Groningen, The Netherlands

Sjouke A. Kingma
Behavioural and Physiological Ecology, GELIFES, University of Groningen, 9700CC Groningen, The Netherlands

Ben Ashby
Department of Mathematical Sciences, University of Bath, Bath BA2 7AY, United Kingdom
Department of Integrative Biology, University of California Berkeley, Berkeley 94720, California

Kayla C. King
Department of Zoology, University of Oxford, Oxford
OX1 3PS, United Kingdom

Rodrigo Pracana, Ilya Levantis, Carlos Martínez-Ruiz, Eckart Stolle, Anurag Priyam and Yannick Wurm
School of Biological and Chemical Sciences, Queen
Mary University of London, E1 4NS London, United
Kingdom

Stuart K. J. R. Auld and June Brand
Biological and Environmental Sciences, University of
Stirling, Stirling, United Kingdom

Daisuke Kageyama, Akiya Jouraku and Seigo Kuwazaki
Institute of Agrobiological Sciences, National
Agriculture and Food Research Organization, Tsukuba,
Ibaraki 305–0854, Japan

Narita
Institute of Agrobiological Sciences, National Agriculture
and Food Research Organization, Tsukuba, Ibaraki 305–
0854, Japan
Tsukuba Primate Research Center, National Institute
of Biomedical Innovation, Hachimandai, Tsukuba,
Ibaraki 305–0843, Japan

Mizuki Ohno, Tatsushi Sasaki, Atsuo Yoshido and Ken Sahara
Laboratory of Applied Entomology, Faculty of
Agriculture, Iwate University, Morioka 020–8550,
Japan

Konagaya
Graduate School of Science, Kyoto University, Kyoto
606–8502, Japan

Hiroyuki Kanamori and Yuichi Katayose
Institute of Crop Science, National Agriculture and
Food Research Organization, Tsukuba, Ibaraki 305–
0854, Japan

Mai Miyata
Graduate School of Horticulture, Chiba University,
Matsudo, Chiba 271–8510, Japan

Markus Riegler
Hawkesbury Institute for the Environment, Western
Sydney University, Penrith, New South Wales 2751,
Australia

Elizabeth G. Mandeville and C. Alex Buerkle
Department of Botany and Program in Ecology,
University of Wyoming, Laramie Wyoming 82071

Thomas L. Parchman
Department of Biology, University of Nevada, Reno
Nevada 89557

Kevin G. Thompson
Colorado Parks and Wildlife, Montrose Colorado
81401

Robert I. Compton and Kevin R. Gelwicks
Wyoming Game and Fish Department, Laramie
Wyoming 82070

Se Jin Song
Department of Ecology and Evolutionary Biology,
University of Colorado, Boulder Colorado 80309

Casper J. van der Kooi, Cyril Matthey-Doret and Tanja Schwander
Department of Ecology and Evolution, University of
Lausanne, Lausanne, Switzerland
980–8578, Japan

Paul M. Richards and Angus Davison
School of Life Sciences, University of Nottingham,
Nottingham NG7 2RD, United Kingdom

Yuta Morii, Kazuki Kimura, Takahiro Hirano and Satoshi Chiba
Division of Ecology and Evolutionary Biology,
Graduate School of Life Sciences, Tohoku University,
Aobayama, Sendai

David Díez-del-Molino and Love Dalén
Department of Bioinformatics and Genetics, Swedish
Museum of Natural History, Stockholm, Sweden

Patrícia Pečnerová
Department of Bioinformatics and Genetics, Swedish
Museum of Natural History, Stockholm, Sweden
Department of Zoology, Stockholm University,
Stockholm, Sweden

Eleftheria Palkopoulou
Department of Bioinformatics and Genetics, Swedish
Museum of Natural History, Stockholm, Sweden
Department of Zoology, Stockholm University,
Stockholm, Sweden
Department of Genetics, Harvard Medical School,
Boston, Massachusetts 02115

Christopher W. Wheat
Department of Zoology, Stockholm University,
Stockholm, Sweden

Pontus Skoglund
Department of Genetics, Harvard Medical School,
Boston, Massachusetts 02115

Broad Institute of Harvard and MIT, Cambridge, Massachusetts 02142

Sergey Vartanyan
North-East Interdisciplinary Scientific Research Institute N.A.N.A. Shilo, Far East Branch, Russian Academy of Sciences
(NEISRI FEB RAS), Magadan, Russia

Tikhonov
Zoological Institute of Russian Academy of Sciences, Saint-Petersburg, Russia
Institute of Applied Ecology of the North, North-Eastern Federal University, Yakutsk, Russia

Nikolskiy
Geological Institute of the Russian Academy of Sciences, Moscow, Russia

Johannes van der Plicht
Centre for Isotope Research, Groningen University, Groningen, The Netherlands
Faculty of Archaeology, Leiden University, Leiden, The Netherlands

Rowan A. Lymbery, W. Jason Kennington and Jonathan P. Evans
School of Biological Sciences, The University of Western Australia, Crawley, WA 6009, Australia

Index

www.ingramcontent.com/pod-product-compliance
Lightning Source LLC
Chambersburg PA
CBHW050444200326
41458CB00014B/5056